The Invention of the Countryside

Also by Donna Landry

THE COUNTRY AND THE CITY REVISITED: England and the Politics of Culture, 1550–1850 (*co-editor with Gerald MacLean and Joseph P. Ward*)

MATERIALIST FEMINISMS (*co-author with Gerald MacLean*)

THE MUSES OF RESISTANCE: Labouring-Class Women's Poetry in Britain, 1739–1796

THE SPIVAK READER (*co-editor with Gerald MacLean*)

The Invention of the Countryside

Hunting, Walking and Ecology in English Literature, 1671–1831

Donna Landry

© Donna Landry 2001

All rights reserved. No reproduction, copy or transmission of this publication may be made without written permission.

No paragraph of this publication may be reproduced, copied or transmitted save with written permission or in accordance with the provisions of the Copyright, Designs and Patents Act 1988, or under the terms of any licence permitting limited copying issued by the Copyright Licensing Agency, 90 Tottenham Court Road, London W1T 4LP.

Any person who does any unauthorised act in relation to this publication may be liable to criminal prosecution and civil claims for damages.

The author has asserted her right to be identified as the author of this work in accordance with the Copyright, Designs and Patents Act 1988.

First published 2001 by
PALGRAVE
Houndmills, Basingstoke, Hampshire RG21 6XS and
175 Fifth Avenue, New York, N. Y. 10010
Companies and representatives throughout the world

PALGRAVE is the new global academic imprint of
St. Martin's Press LLC Scholarly and Reference Division and
Palgrave Publishers Ltd (formerly Macmillan Press Ltd).

ISBN 0-333-96154-4

This book is printed on paper suitable for recycling and made from fully managed and sustained forest sources.

A catalogue record for this book is available from the British Library.

Library of Congress Cataloging-in-Publication Data
Landry, Donna.
 The invention of the countryside : hunting, walking, and
  ecology in English literature, 1671–1831 / Donna Landry.
   p. cm.
 Includes bibliographical references (p.) and index.
 ISBN 0-333-96154-4 (cloth)
  1. English literature—History and criticism. 2. Country life in literature. 3. Pastoral literature, English—History and criticism. 4. Landscape in literature. 5. Hunting in literature. 6. Walking in literature. 7. Ecology in literature. 8. Nature in literature. I. Title.
  PR408.C66 L36 2001
  820.9′321734—dc21
                                    2001024552

10 9 8 7 6 5 4 3 2
10 09 08 07 06 05 04 03 02

Printed and bound in Great Britain by
Antony Rowe Ltd, Chippenham, Wiltshire

*For Gerald MacLean, Rosemary Hooley
and Basil*

Contents

List of Color Plates	x
List of Plates	xi
List of Figures	xii
Preface and Acknowledgements	xiii

Inventing the Countryside: An Introduction	1
Inventing Modern Fox-Hunting, Inventing the Countryside	12
'Green Pastoral Landscapes' and the Picturesque Countryside	15
This Peculiar Pastoral, the 'Idea of England'	22

PART I FROM COUNTRY TO COUNTRYSIDE	27
1 The Greenness of Hunting	29
Some Green Paradoxes in Hunting Writing	35
Hunting and Social Ecology	44
A Modern Coda	48
2 Land, and Writing about Land	54
The Poetry of Agrarian Improvement	55
The Science of Agriculture and the Science of Scent	62
The Closed Yet Open Landscape and the Ideology of the Landscaped Park	65
Ecology and the Landlords	68
3 Game and the Poacher	73
Game Legislation Described	74
The Politics of Game Legislation	75
Game, Politics and Ecology	76
The Mutual Entanglements of Country and City	78
Rights of Common: A Long History of Struggle	80
The Poacher in Art and Literature	86
John Clare, Robert Bloomfield and the Science of the Greenwood	

4 The Sporting Life   92

Rural Sports   94
The Language of Sport   95
Changes in Field Sports, 1671–1831: Fishing, Shooting, Coursing   97
Political Governance and the Sporting Ideal   102
Coopers Hill as Touchstone: Hunting   104
Men and Dogs   106

5 The Origins of the Anti-Hunting Campaign   113

The Vegetarian Ideal   115
Metropolitan Intellectuals   117
Poetry, Pets and the Beginnings of Animal Rights   119
Anti-Hunting Sentiments and the Rise of Pedestrianism   125
Coopers Hill as Touchstone: Walking   128
Sporting Culture and the Rise of Pedestrianism   132
Romantic Walking as Anti-Hunting   133

PART II HUNTING A COUNTRY   143

6 Sportswomen   145

Science and Sporting Culture: Women in the Field   146
Women, Hunting and Poetry in the Seventeenth Century   147
Hunting Women in the Eighteenth Century   153
What the Women Said   157
The Side-Saddle as a Gender Machine   163

7 The Pleasures of the Chase *circa* 1735 to *circa* 1831   168

Pleasures of the Grass   172
Equestrian Pleasures   173
Talking of Hounds   175
Sartorial and Linguistic Pleasures   176
The Pleasures of Somervile   179
Our Hunting Fathers   182
The Sadness of the Creatures   184
Somervile via Beckford   189
The Complexity of Somervile's Achievement   191

8 **The Pleasures of Surtees**	195
'Smoke-Dried Cits'	198
Hunting and the Spa Town	199

PART III   WALKING IN THE COUNTRYSIDE   203

9 **The Pleasures of Perambulation**	205
Gear	205
Botanizing, etc.	206
'Hunting the borough'	209
Walkers' Palimpsests	211
Walking as Trespassing, but also Writing: Wordsworth	213
Walking as Hunting: Clare	216

10 **'This Lime-Tree Bower' as Walking Poem**	220
Walkers for Pantisocracy	222

11 **Dartmoor Visible**	230

Notes	245
Index	293

List of Color Plates

1. George Morland (1763–1804), *The Benevolent Sportsman* (1792). By courtesy of the Syndics of the Fitzwilliam Museum, University of Cambridge.
2. Thomas Gainsborough (1727–88), *Mr and Mrs Robert Andrews* (c. 1748–50). By courtesy of the National Gallery, London.
3. George Stubbs (1724–1806), *Laetitia, Lady Lade* (1793). By permission of The Royal Collection © 2000, Her Majesty the Queen Elizabeth II.
4. John Wootton (1682/3–1764), *The Bloody Shoulder'd Arabian* (1724). By courtesy of private collection.
5. Jacques-Laurent Agasse (1767–1849), *Sleeping Fox* (1794). By courtesy of the Oskar Reinhart Collection, Winterthur, Switzerland.

List of Plates

1. John Wootton, *Hare Hunting on Salisbury Plain* (*c.* 1700). Reproduced by kind permission of the Duke of Beaufort. Photograph: Photographic Survey, Courtauld Institute of Art, London.
2. George Morland, *Tavern Interior with a Sportsman Refreshing* (*c.* 1790). By courtesy of John Barrell. Photograph: Cambridge University Press.
3. James Seymour (1702–52), *A Coursing Party* (1738). By courtesy of the Richard Green Gallery, London.
4. William Henry Davis (*c.* 1786–1865), *Colonel Newport Charlett's Favourite Greyhounds at Exercise* (1831). By courtesy of the Richard Green Gallery, London.
5. John Wootton, *Lady Henrietta Cavendish Holles, Countess of Oxford hunting at Wimpole Park* (1716). By courtesy of private collection. Photograph: The Paul Mellon Centre for Studies in British Art, London.
6. John Ferneley (1782–1860), *Sir John Palmer on his favourite mare with his shepherd John Green and prize Leicester long wool sheep* (1823). By courtesy of Leicestershire Museums, Arts and Records Service.
7. Thomas Barker of Bath (1769–1847), *The Woodman and his Dog in a Storm* (*c.* 1790). © Tate, London 2001.

List of Figures

1.	Pro-Hunting Rally at Wincanton Race Course (March 1995). By courtesy of Rosemary Hooley.	xix
2.	John Wells, Joint Secretary of the Mid Devon Foxhounds, on Minnow (April 1995). By courtesy of Rosemary Hooley.	xix
3.	John Wells, Wincanton Race Course (March 1995). By courtesy of Rosemary Hooley.	xix
4.	Advertisement for the National Farmers' Union Mutual Insurance 'Countryside' Policy (August 1993). By courtesy of the NFU. Photograph: Kevin Shea.	18
5.	Anonymous illustration for 'The Otters oration', opposite page 202, from George Gascoigne, *The Noble Arte of Venerie or Hunting* (1575). Reference shelfmark, Douce T 247(1). By courtesy of the Bodleian Library, University of Oxford.	42
6.	'The Accomplished Sportswoman', opposite page 154 from *The Sporting Magazine*, Volume 4, June 1794. Reference shelfmark, Vet.A5 e.875. By courtesy of the Bodleian Library, University of Oxford.	162
7.	John Leech (1817–64), 'Lucy Glitters showing the way', opposite page 379 from R. S. Surtees, *Mr Sponge's Sporting Tour* (London: Bradbury & Evans, 1853). Reference shelfmark, C. 70 d. 8. By courtesy of the British Library, London.	164
8.	Thomas Rowlandson (1756–1827), *The Chase* (*c.* 1788). By courtesy of the Witt Library, Courtauld Institute of Art, London.	171

Preface and Acknowledgements

Across the political spectrum in Britain today the rhetoric of a battle for the future of the countryside can be heard. The New Labour government's pledge to ban 'hunting with dogs' has met with resistance, though in January 2001 the House of Commons voted overwhelmingly in favor of a ban.[1] What was once regarded as a national sport has become the object of public scrutiny, if not outrage. The yoking in public debate of the hunting ban with the statutory Right to Roam, instituted by Tony Blair in March 1999, guaranteeing access to open mountain, moor, heath, down and registered common land, amounting to up to 4.5 million acres and covering about 10 per cent of England and Wales,[2] is no coincidence, as this book will show.

How have hunting and walking come to be viewed as antithetical? In the last two decades of the eighteenth century, long-distance pedestrianism was first popularized as an outgrowth of, and then as an alternative to, hunting and shooting. Famous walkers first achieved celebrity status in the journal of hunting, shooting and racing culture, *The Sporting Magazine*.[3] Seeking out botanical specimens and capturing in pencil picturesque views were first promoted as pastimes akin to hunting. But some pioneering pedestrians took a more direct approach, announcing that it was no longer quite the thing to kill something in order to make the most of being in the country. Consuming the beauties of nature visually and non-violently was sport enough. Walking, once the last resort of the poor in need of transport, or a pastime of the benighted rustic elite, became a popular form of leisure, enabling the walker to apprehend nature at ground level while rejecting the gentlemanly privileges associated with riding. In that social and semiotic shift, brought about by changes in demography and social relations, agricultural practice and urbanization, but also in sensibility and artistic representation, the countryside as we know it came into being.

Indebted to aesthetic theories of the picturesque and to what we now call Romanticism, this was initially a countercultural movement keenly embraced by disaffected middle-class people, especially radically inclined undergraduates, poets and members of the lower clergy. Its followers set themselves apart from the game-pursuing, hunting-mad aristocracy and gentry, their middle-class imitators and the poaching masses. The movement had a certain affinity with democratic politics, anti-slavery activism and anti-cruelty sentiments. By the early nineteenth century, the culture of hunting and field sports began to give way, as the dominant mode of experiencing being in the country, to landscape tourism.

But the story does not end there. Field sports themselves also changed during this time. Fox-hunting, once merely a form of vermin control, 'a sort of glorified rat-catching',[4] displaced stag- and hare-hunting and coursing with greyhounds as the most fashionable recreation of aristocrats and country gentlemen alike. Rendered increasingly fast-paced and physically challenging as the technology of hounds and horses improved, in tandem with the improvement of the privatized, enclosed landscape of the agricultural revolution, modern fox-hunting came to signify Englishness on an imperial playing field.

The splitting of views that began during the late eighteenth century is with us still, embodied in today's arguments among animal welfarists or rights activists and farmers, with ramblers usually on the welfarists' side and fox-hunters on the farmers'.[5] Now, at the beginning of the twenty-first century, what appeared eccentric in the 1790s has become a prevailing, if not *the* prevailing, view.

The intertwined history of hunting and walking is therefore more complicated than it may at first seem, and this book attempts to identify and untangle the most important strands. Once upon a time, hunting and field sports were the pastimes complementary to a life of traditional husbandry, the antidote to constant vigilance over livestock and worry about the weather, constant anxiety concerning crops, lambing, calving, rent-paying and the fatness of pigs. So long as the population remained rural, 'Rideing, Hunting, Courseing, Setting and Shooteing' provided the language and symbolic field for most English people's relations with the natural world.[6]

As recently as the seventeenth and eighteenth centuries, hunting and natural history went hand in hand. The greatest field naturalists were often sportsmen or women if they were gentlefolk, and poachers if they were not. As recently as the seventeenth and eighteenth centuries, coursing with greyhounds and hunting with hounds were forms of naturalistic fieldwork as well as rationales for ecological conservation. Hawking, netting and using a crossbow or a longbow required great technical skill as well as knowledge of woodcraft and animal behavior, and even then bags were far from big. Given the unreliability of firearms, shooting too was more about stalking than about killing. Even the later eighteenth-century natural history writing of Gilbert White owes everything to his experiences as a hunting and shooting man.

Whether Royalists or Levellers, qualified sportsmen or poachers, people of rank or the laboring poor, all were likely to employ hunting metaphors as a matter of course and to be familiar with hunting practice and the taking of game, whether legally or illegally. In 1641 a king might be rendered sympathetically in verse, while hunting: 'Here have I seene our Charles (when greate affaires / Give leave to slacken & unbend his Cares) / Chasing the Royall Stagge, the gallant beast',[7] but in 1649 a Leveller satirist would criticize Cromwell's government for intolerance by casting himself as a hunted beast: 'Now must the *Beagles* go a *hunting* again, and I must be the *Hare*.'[8] Animal fables and hunting metaphors resonated with moral and political allegory, but they were also grounded in everyday experience of the natural world.

The agricultural revolution and urbanization changed all that. By the early nineteenth century, modern, fast-paced fox-hunting had eclipsed other field sports in fashionability. If hunting had once been a matter of naturalistic knowledge and the ecological science of the greenwood, modern fox-hunting could be said almost not to be hunting at all. And the same could be said of organized formal shooting, with competitive marksmanship and size of bag having replaced the naturalistic informality of rough shooting. The fox-hunter, the modern English sportsman, was a suitable image for export to the empire, exemplifying gentlemanly virility. But it was an image that had little to do with hunting in its earlier non-elite, more ecologically alert forms. And it reduced to a squires' conspiracy what had formerly united many members of the rural upper and lower classes: their mutual obsession with hunting and poaching.

Accounting for the controversial status of fox-hunting in Britain today may seem too obvious to need repeating. Cruelty to animals has become anathema to many. For many people, hunting means a fox exhausted and torn apart by hounds. That hunting is only occasionally, and even then remotely, like that is difficult for most people unacquainted with it to believe. Their antagonism is hard to shift. It is fueled by something far more deeply interfused than the ostensible desire to save foxes.

To attribute anti-hunting sentiments entirely to the animal rights and animal welfare movements would be misleading, though there are historical connections. To interpret hostility to hunting as a form of class war is likewise too simplistic, though class antagonism plays a part in this hostility. What the anti-hunting campaign reveals most clearly is how contested a symbol of Britishness or Englishness fox-hunting has become. The history of hunting and other field sports suggests that these practices have most often been about something other than a view to a kill. Even today, when modern fox-hunting has lost much of its naturalistic content, and many members of the mounted field never even know if hounds are hunting or not, we could say that killing foxes is only a secondary activity, a way of inducing farmers to allow the hunt to ride over their land. Even so, a small proportion of farmers discourage, or even ban, hunting on their land.[9]

Rather, the continuing symbolic power of hunting for English people, and to some extent other Britons, has derived from its figurative excess and irresolution, its deep ambivalence, its capacity to mean more and different things to successive generations than they can clearly say. This book is not so much a defense of modern fox-hunting as an inquiry into its history.[10] But writing it has persuaded me that both sides of the hunting debate have foreclosed too much of this history to arrive at a satisfactory conclusion. We need to reopen the past to scrutiny.

Even today, not even the membership of the Labour Party is uniformly behind a ban on hunting. At the demonstration advertised as the 'Countryside Rally' in Hyde Park on 10 July 1997, the barrister Ann Mallalieu QC, who sits as a Labour peer (Baroness Studdridge), took the microphone wearing what she described as 'socialist pink, not hunting pink', and said:

> Hunting is often described as a sport. But to those of us who have heard the music of the hounds and love it, it is far more than that. Hunting is our music, it is our poetry, it is our art, it is our pleasure. It is where many of our best friendships are made, it is our community. It is our whole way of life. We will fight for these things with all the strength and dedication we possess because we love them.

Do not forget us, or what we have done today. We have made history. The countryside has come to London to speak out for freedom.[11]

In Mallalieu's speech, the voice of the countryside and the voice of riders to hounds become one and the same. Mallalieu historicized the rally: history had been made, because a practically unprecedented number of people had converged on London from country districts. Hunting as an expression of freedom, or freeborn liberty; hunting as music, poetry, art and pleasure; hunting as friendship, community and a whole way of life: this is a potent rhetoric, with compelling imagery. Mallalieu has appealed to a version of the ideology that Patrick Wright calls 'Deep England'.[12] And it is an ideology and a set of symbols no longer confined to the Conservative Party. In Wright's terms, despite the potential for right-wing manipulation of rural imagery, and the fact that the 'approved and dominant images of Deep England are pastoral and green', there is 'something "green" about everyday life, whatever the situation in which it is lived'.[13]

Environmental conservationists speak of 'solitude and personal discovery' and 'physical and spiritual refreshment' as among the greatest needs of people today, justifying public access to agricultural land and open moorland, especially in National Parks.[14] Mallalieu on other occasions has alluded to this need as part of her justification for hunting. She suggests that studying the fox with 'a naturalist's eye' will yield certain private pleasures, such as 'seeing the way a flock of sheep react to a fox running through them, or birds swooping down from the sky as a fox emerges from a wood'. Such incidents will not be witnessed by everybody, she admits; indeed '"you may be the only person to notice"' them. 'Solitude and personal discovery' are part of the pleasure: '"You feel very privileged."' Hunting today, as represented by Mallalieu, dramatizes the paradox of the tender-hearted, left-leaning, liberty-loving professional woman or man identifying with the fox while hunting him or her: '"I'm totally on the fox's side. My attitude is 'Come on, little fox'."' The paradox of willing and wishing the fox's escape while running with the hounds is justified by a conservationist argument – that hunting is necessary to control the fox population, to keep it within '"reasonable limits"', without causing undue suffering or extermination.[15] The concerns of farmers anxious to limit the number of foxes, hunting people eager for sport, animal lovers dedicated to preventing suffering, and ecologists and conservationists keen to preserve habitats and biodiversity in the face of agricultural pressures, are often at odds. Hunting as Mallalieu envisages it goes some way toward accommodating the interests of these parties. But the justification of modern mounted fox-hunting on the grounds of keeping the fox population within tolerable limits is not quite as straightforward as Mallalieu implies. Hunting practices and their rationales differ widely in different circumstances, with lowland and upland hunting representing one major difference, as we shall see.

Today we tend to associate naturalistic field study with bearded academics on foot rather than with hunt followers on horseback. But it was not always so, and Mallalieu draws upon the legacy of a long, if now largely forgotten, tradition linking hunting and conservation. The past can be a rich repository of undeveloped possibilities. Today, intensive agriculture and development inattentive to the wider ecological picture threaten not only British ecology but ecology on a global scale. Utility and beauty, or agricultural productivity and amenity use, combined in landscape aesthetics until the end of the eighteenth century; they could be recombined today in the interests of ecology, sustainable agriculture and recreation.

The appearance of a flourishing countryside, marked by a mixed, not monocultural, agriculture achieved through traditionally sustainable means, and a preponderance of wildlife, including game animals, was once highly valued. Although modern urbanization is hardly a reversible process, a return to an aesthetics as well as economics of agriculture is still possible. But it could only be ethically and socially realized if the concerns of small as well as big farmers, and of farm workers, were to count within agricultural and conservationist policy-making, if substantial changes in farming practice were to become economically viable, and if a healthy ecology were to determine this calculus. Studying the history of land use and human/animal relations since the seventeenth century may give us some important cultural levers to use to conserve a working countryside still comparatively rich in history and biodiversity while distributing its benefits more equitably.

Bridging the gap between walking and hunting cultures may not seem farfetched once we have begun to explore the historical record. There have been some recent attempts at rapprochement. At a pro-hunting demonstration in March 1995 at Wincanton racecourse in Somerset (Fig. 1), one of the joint secretaries of the Mid Devon foxhounds exchanged his usual hunting kit for a costume that would represent 'Walkers for Hunting' (Fig. 2 and Fig. 3). The Countryside March on 1 March 1998 followed the same strategy, drawing as it did at least 350,000 people to central London, more than 40,000 from the West Country. More pointedly than the Hyde Park rally, the march claimed to embrace a broader range of rural issues than merely protecting hunting, including the continuing financial woes of British farming, conflicts over the building of new houses in rural areas, and the then promised Right to Roam, now legislatively assured. Instead of claiming hunting as a prerogative of the privileged class, hunt supporters represented themselves as a minority, seeking equal protection under the law. Fox-hunting, once the very image of a national identity and so often equated with the dominance of the landed interest, could now be defended on the grounds of a minority identity politics.

There is, to be sure, something disingenuous about this claiming of minority status, given the social composition of many modern mounted hunting fields, especially in the fashionable Midland shires, but in another sense, hunting and other field sports are remnants of a past, and of a rural culture, with which only a minority remain in touch. Yet these practices have contributed to preserving the beauty and amenity value of the English landscape, so highly valued by the urban majority, by encouraging the maintenance of wildlife habitats and food supplies, bridleways and footpaths, and traditional landscape features such as hedgerows, copses, shelter belts and woodlands, that run contrary to agricultural profit logic. 'The behaviour of farmers with respect to habitats of wildlife value is crucial for biodiversity in the agricultural landscape,' the zoologists David W. Macdonald and Paul J. Johnson remark in a recent study.[16] They report that hedgerows are valuable for birds, small mammals, vascular plant species and their associated communities of phytophageous insects, while other economically non-productive habitats, such as small spinneys in field corners, coverts, shelter belts, trees, ponds, woodland and scrubland, which are 'more sparsely distributed than are hedgerows', nevertheless 'may have a more significant impact on the landscape'.[17] Macdonald and Johnson conclude that 'The view that farmers who are interested in field sports are more concerned for the environment finds some support in our data; there was evidence that some unproductive habitats (coverts, corners, trees) may have been better treated by hunting farmers', while farmers 'whose primary interest was in hunting had removed the least hedgerow', particularly in the earlier period of the study when hedgerow removal was 'most prevalent'.[18]

If hunting were more widely understood by its supporters as well as by its critics, in its full historical complexity – social, animal and ecological – I strongly suspect that most people, even if they had no wish to take part,

Preface and Acknowledgements xix

Figure 1 Pro-Hunting Rally at Wincanton Race Course (March 1995). By courtesy of Rosemary Hooley.

Figure 2 John Wells, Joint Secretary of the Mid Devon Foxhounds, on Minnow (April 1995). By courtesy of Rosemary Hooley.

Figure 3 John Wells, Wincanton Race Course (March 1995). By courtesy of Rosemary Hooley.

xx *Preface and Acknowledgements*

might agree that a ban was unnecessary. In the present climate, in which 'there is an increasing desire by urban-dwellers to participate in rural pastimes, a populace that is ever more knowledgeable about wildlife and eager to be out-of-doors, and a farming economy under mounting pressure to diversify', as Macdonald and Johnson observe, 'there would seem to be a risk in banning foxhunting that the proverbial baby be thrown out with the bathwater'.[19]

Chagford, Devon, August 2000; and Detroit, Michigan, January 2001

The longer the making of a book, the more debts are incurred, and the greater the likelihood of defaulting on them. I must thank the John Simon Guggenheim Foundation for a fellowship in 1995, and Wayne State University for two semester sabbaticals, a summer research grant, a Josephine Nevins Keal award and a Distinguished Faculty Fellowship. The staff of the Bodleian Library, Oxford, of the Rare Books Room and the Manuscripts Room at the British Library, London, and of the Devon and Exeter Institute have been congenial fellow travelers. I am indebted to especially acute audiences at the University of York, Manchester Metropolitan University, Loughborough University, Rutgers University, the University of Michigan, the William Andrews Clark Library of UCLA, the University of Colorado at Boulder, MIT, Goldsmith's College, University of London and the University of Exeter. Friends and colleagues who have read all or parts of the book have earned my eternal gratitude for their suggestions: John Barrell, Harriet Guest, David W. Macdonald and Paul J. Johnson, Howard Newby, Felicity Nussbaum, Kevin Sharpe, Nigel Smith and James Grantham Turner. I thank Josie Dixon for sublime expertise, and for being so quick off the mark; the anonymous Palgrave readers for sound advice and inspiration; and Eleanor Birne and Ruth Willats for recognizing the book's value and working tirelessly to realize it. Without the friendship of John and Alice Wells, other members of the Mid Devon, and the regulars of the Northmore Arms, hunting would have been one blank day after another. Without Ian and Christine Shalders, I would never have retraced Coleridge's footsteps or enjoyed the company of lurchers and deerhounds. For exceeding all my hopes of readerly patience and companionship, I thank Gerald MacLean and Rosemary Hooley. Basil contributed pace and cadence. Had I not spent time living on Dartmoor, seeing foxes so frequently and hearing hounds singing in the night, this book might never have been written.

Inventing the Countryside: An Introduction

> How much English poetry depends upon English hunting this is not the place to enquire.
>
> Virginia Woolf, 'Jack Mytton' (1926)

This book investigates the invention of the 'countryside' in England between the Game Act of 1671 and its repeal in 1831. For it was during this period of most controversial game legislation that 'countryside' ceased to refer to a specific side – east or west, north or south – of a piece of country, or a river valley, or a range of hills, and became 'the countryside', an imaginary, generalized space. Henceforth 'countryside' would connote not specific, local but aesthetic, global 'natural unity' (*OED*). 'The countryside' became a favorite word of descriptive writers.

In a movement exactly parallel to what John Barrell and Anne Janowitz have observed of the terms 'landscape' and 'country', respectively, 'countryside' no longer denoted a piece of knowable terrain, viewable as a prospect or rideable or walkable as a tract of country.[1] It became an idea, and a way of giving an imaginary, yet material, form to a unified, homogeneous vision of the nation. The very term countryside became a symbolic repository of all that was and is most cherished about being English. To be English was to be at the very center of Britishness, an identity forged in rivalry with other nations, particularly with France.[2] While the British Empire extended itself abroad, the English countryside glowed greenly at its heart.

In earlier periods, other terms had served similar purposes. In the sixteenth and seventeenth centuries, the 'country', usually apprehended as a particular landscape – a country estate or the prospect offered by a hill – had been contrasted with the corruptions of city and court and invested with moral and patriotic associations. The goodness of the country guaranteed the health of the English nation. Between 1671 and 1831 the term 'countryside' began to replace the term 'country' in this sense, as English society became more urban. The country continued to designate rural places in the back-of-beyond, where rustic folk lived and agriculture happened, increasingly intensively. The countryside, by contrast, now began to evoke images of unchanging rural beauty arousing protective, patriotic sentiments in metropolitan minds, often from a considerable distance. From its beginnings as a linguistic signifier for the heart of the nation, the countryside needed protecting because it was 'out there', and perpetually endangered. The long agricultural revolution, spanning the late sixteenth to mid-nineteenth centuries, and devoted to extracting maximum productivity from the land and highest profits from the market, produced the timeless countryside as its imaginary Other.[3] Not production, but consumption and pleasure, recreation and retreat, were the goods associated with the countryside.

From the beginning there was a contest over meaning and the proper uses of the countryside, in which class differences played themselves out within and sometimes against the urban–rural divide. Until at least the late seventeenth century, the country had been indisputably the place where field sports – or 'Countrey Contentments' as Gervase Markham called them in 1615 – had been pursued by all social ranks, from aristocrats and members of the gentry to the poorest cottagers and laborers.[4] Historians have documented how heroic self-assertions by members of the metropolitan mercantile classes during the seventeenth and eighteenth centuries helped produce a code of manners for those outside the aristocracy and gentry, conveying 'upper-class gentility, enlightenment, and sociability to a wider élite whose only qualification was money'.[5] The English were recruited into being that 'polite and commercial people' described by William Blackstone.[6] Commerce, we should note, was 'not just about exchange but more fundamentally about consumption',[7] especially in the 'plutocracy' of eighteenth-century Britain.[8] As foreign visitors observed, nothing distinguished eighteenth-century English society more than 'the unceasing struggle of all ranks to emulate those above them'.[9] The new politeness promoted in the pages of periodicals such as the *Spectator* was urbane, with a strong influence of mercantile taste. Readers were meant to prefer the enlightened Whig, Sir Andrew Freeport, a self-made man, to the huntin'-mad Tory traditionalist Sir Roger de Coverley, who ruled 'servants, tenants, and neighbours with bluff good humour, an autocratic manner and a ceaseless concern for their welfare'.[10] As prosperity derived from the wealth of empire combined with these metropolitan influences and the rise of provincial towns as centers of leisure and culture, members of the aristocracy as well as many wealthier gentry and farmers were converted to a new 'polite' culture.[11]

In *The Farmer's Boy* of 1800, the farm worker and shoemaker-turned-poet, Robert Bloomfield, lamented the coming of 'Refinement', which had polarized rural culture along class lines, dividing the plebeian from the polite. Rising grain prices since the beginning of the French war in 1793 had made farmers especially prosperous and opened a widening gulf in the standard of living of the farmer and his workers.[12] Refinement, brought about by new wealth, was 'the peasant's curse, / That hourly makes his wretched station worse' ('Summer', ll. 338, 339–40).[13] The new polite culture had destroyed 'life's intercourse; the social plan / That rank to rank cements, as man to man' (ll. 341–2), rudely separating farmers and landowners from their laborers:

> 'The wid'ning distance which I daily see,
> 'Has Wealth done this? . . . then wealth's a foe to me;
> 'Foe to our rights . . .'
>
> (ll. 349–551)

But old habits die hard, and in some provincial districts a more egalitarian hunting and husbanding culture persisted, in which obvious inequalities of distribution were sometimes mitigated by shared pastimes.

It was a truth universally acknowledged by English gentlemen that they had the leisure to enjoy what everyone else wished to. Commoners relished hunting, coursing, and fishing as much as their social superiors, and resented any attempt to restrict these pursuits to the gentry.[14] This is why for centuries every popular uprising in England 'brought forth a Wat Tyler or a Gerrard Winstanley who proclaimed the principle of common access to hunting and fishing'.[15] The English were not alone, for 'hunting and fishing rights were among the first demands made by German peasants during the Peasants' War of 1524–5 and by French peasants in the early stages of the French Revolution.'[16] Wat Tyler demanded during the Great Revolt of 1381 that all warrens, parks and chases should be free '"so that throughout the realm, in . . . the woods and forests, poor as well as rich might take wild beasts and hunt the hare in the field."'[17] The 'mass invasions of deer parks in south-east England' between 1641 and 1643 during the Civil War produced some similar demands, like that of the mob at Farnham, who told a keeper that ' "they cared not what Parliament did or said" ', they would have back the right to resources unjustly appropriated.[18] ' "They came for venison and venison they would have", rioters in Waltham Forest announced, "for there was no law settled at this time." '[19] Tyler's demands, we notice, embraced both food and sport: the poor should be entitled to 'take' wild beasts for meat – and venison was the most highly prized – but to 'hunt the hare in the field' for sporting pleasure was an entitlement as well. In 1792, Thomas Paine testified to Tyler's undying fame for having made 'proposals' to Richard II 'on a more just and public ground, than those which had been made to John by the Barons': 'If the Barons merited a monument to be erected in Runnymeade, Tyler merits one in Smithfield.'[20] Paine's inclusion of hunting among his *Rights of Man* would have been particularly contentious during the poaching wars of the late eighteenth century.

This is why the Game Act of 1671 was so socially divisive, and its harsher enforcement after 1750 even more so. The Game Act legislated that sporting rights were restricted to those with an income of £100 a year from a freehold estate, or leases of 99 years worth £150, and to the sons and heirs of esquires, and the owners of parks, warrens, chases or free fisheries (22 & 23 Chas. II c. 25). Thus the Act evidenced symbolically that the gentry as a class were no longer willing to tolerate sharing game and 'Countrey Contentments' with their social inferiors. As the term 'countryside' came to replace 'country' in describing rural places during the years between the Game Act of 1671 and its repeal in 1831, those places become sites of greater social antagonism and unrest, with many magistrates devoting themselves to punishing as poaching and trespassing what had been common country practices for centuries. The Game Act of 1671 shows us the gentry looking up, elevating themselves by following the progress of royalty round the kingdom and its sporting preserves, while also looking down on lesser mortals. The *OED* points us toward a curious concatenation of words here, in that 'progress' and the notion of viewing a 'prospect' from a high point in the landscape are significantly connected in seventeenth-century usage. A region or distance traversed can be synonymous with dominion, and with a commanding view from the heights of that dominion. Take this example from 1601: 'His dominion . . . stretcheth from the promontorie Bayador to Tanger, and from the Atlantike Ocean to the riuer Muluia. In which progresse is conteined the best portion of all Afrike' (*OED*). Here 'progress' as a region traversed is conflated with an imagined 'prospect' view of that region as lordly dominion.

Much has been written about the prospect view as a key to the disinterestedness or public-spiritedness of landed gentlemen, as opposed to everybody else. The argument that landscape aesthetics reveals the ideology by which the landed classes justified their entitlement to govern has been made most effectively by John Barrell, who also notices the connection to hunting and shooting. Barrell quotes the opening lines from the first epistle of Alexander Pope's *Essay on Man* (1733):

> AWAKE, my ST. JOHN! leave all meaner things
> To low ambition, and the pride of Kings.
> Let us (since Life can little more supply
> Than just to look about us and to die)
> Expatiate free o'er all this scene of Man;
> A mighty maze! but not without a plan;
> . . .
> Together let us beat this ample field,

> Try what the open, what the covert yield;
> The latent tracts, the giddy heights explore
> Of all who blindly creep, or sightless soar;
> Eye Nature's walks, shoot Folly as it flies,
> And catch the Manners living as they rise.
>
> (1: 1–6, 9–14)[21]

Barrell observes that it is 'as a man of landed property', and hence, a sportsman, that Henry St. John, Lord Bolingbroke, and with him his friend the poet Pope, claim entitlement to 'expatiate free' over the human social scene and grasp the social 'plan' from a commanding height, which affords a comprehensive prospect view inaccessible to others with more partial viewpoints (whether 'latent tracts' or 'giddy heights'). Bolingbroke and Pope '"beat" the field, and "catch" the Manners, they thus put up,' Barrell concludes, in a 'thoroughly appropriate' way.[22] Entitlement to hunt as much as knowing how to read a landscape symbolized and enacted this imagined difference of view possessed by landed gentlemen.

Is this view so really different from all others? In what sense is the landed interest operating without self-interest? The commoners' view of traversing the land in order to possess or wield power over it took the form of the beating of bounds in a parish, manor, forest, or holding – a perambulation of the district that reiterated its boundaries. This practice of 'hunting the borough', as it was known in some places in Wiltshire,[23] put custom in action, reinforcing the collective memory of where customary rights lay. The perambulation of the bounds might follow an ancient watercourse, or the hedges of a close, and at each boundary point a cross or mark would be made in the ground. As E.P. Thompson noted, 'Small boys were sometimes ducked in the ditch or given a clout to imprint the spot upon their memories.'[24] Agricultural improvement and enclosure often jeopardized this practice. The historian David Underdown concludes, 'The new hedges and enclosures of the market economy obliterated both the old physical landmarks and the traditions of sociability which perambulation of the bounds had preserved.'[25] Another historian, Bob Bushaway, goes even further: 'The suppression of the parish bounds walk aimed also to suppress the memories of parishioners who could recall rights and customs associated with open fields and commons before enclosure.'[26] Sporting culture and walking culture both bear traces of this practice of perambulation which ran counter to enclosure and lordly dominion.

Through the Game Act of 1671, the gentry attempted to settle the law in their favor, elevate land over other forms of property and make themselves, rather than the Crown, stewards of the game. Agricultural improvement, deforestation and the Civil War had all taken their toll upon the wild animal population, and by 1671 there was a crisis of scarcity. The agrarian historian Joan Thirsk has argued, 'Hitherto the theory about gentlemen's exclusive rights to game, and the practice whereby all classes hunted it, had remained happily divergent; now they were converging uncomfortably.'[27] 'The game acts may thus be seen,' concludes Thirsk, 'as a mirror of the successful seventeenth-century campaign to improve wastes, forests, and chases.'[28] Maximal extraction of excitement and spoilage in the field could be made to match maximal extraction from the soil. Deer, the largest wild mammal, were the first to become scarce in the wild and require breeding and rearing in deer parks; deer were thus the first beast of chase to be classified as private property rather than game, which remained strictly defined as hare, partridge, pheasant and moor fowl.

After 1671, whenever the exclusivity of the eighteenth-century game laws was contested, the gentry defended their privileges with increasing vehemence. P.B. Munsche has discovered that 'The willingness of country gentlemen to "indulge" their tenants and neighbours with sporting rights slackened noticeably after 1750' and by the end of the century, ' "indulgences" were being granted far less frequently than they had been fifty years earlier.'[29] The ascendancy of a Whig oligarchy with a distinctive landscaping style and a detachment from local interests contributed to this change in attitude, as did troubling memories of the radical redistribution of land advocated by the Diggers in the previous century.[30] The game laws could always be deployed to crush political dissent, as Sir Robert Walpole demonstrated. His administration's notorious Black Act of 1723 (George I c. 22) made hunting the king's deer or poaching hares, conies or fish in royal forests potentially capital offenses, especially if the persons accused had blacked their faces or were otherwise disguised and armed.[31] Hunting for Jacobites was followed, after the failure of the Jacobite Rebellion of 1745, by prosecuting Jacobins and other perceived threats to social order. Beginning in the politically and socially 'nervous' decade of the 1790s, and intensifying during the bad harvests and economic hardships of the Napoleonic Wars, poaching gangs dealt heavily in the blackmarket in game, and landowners resorted to spring guns, man-traps and armed gamekeepers to preserve their sporting rights.[32] Not surprisingly, attacks on the game laws and on sporting culture acquired an increasingly anti-gentry class fix.

Though much resented, protested against, subverted and flouted, the Game Act of 1671 was not repealed until 1831, the year before the Reform Bill. And it would still take until the Ground Game Act of 1880 before tenant farmers were given the right to shoot rabbits and hares on the farms they occupied, a change that gave villagers legal access to rabbits with the farmers' permission,[33] but which was anathema to sportsmen, who claimed that the Act had made the hare extinct over thousands of acres.[34]

What changed sporting culture to a minority interest? Evidence of agrarian complaint against hunters and of sympathy for hunted animals exists for earlier periods than the late eighteenth century. In *As You Like It* Shakespeare gives us Jaques's grieving for a wounded stag (2.1.25–63) and Duke Senior announcing, 'Come, shall we go and kill us venison?' at the same time that it 'irks' him that 'the poor dappled fools, / Being native burghers of this desert city, / Should, in their own confines, with forked heads / Have their round haunches gored' (2.1.21–5). (On the ambivalences and empathy of early modern hunting discourse, see Chapter 1.) In *Joseph Andrews* (1742), Henry Fielding sketched in the episode of Mr Wilson, whose daughter's spaniel is shot by the son of the lord of the manor, a portrait of the egregious hunting-mad preserver of game – 'as absolute as any tyrant in the universe, and had killed all the dogs, and taken away all the guns in the neighbourhood, and not only that, but he trampled down hedges, and rode over corn and gardens, with no more regard than if they were in the highway.'[35] Violent and abusive, this young squire has even less to recommend him than the comically offensive Squire Western of *Tom Jones* (1749). By mid-century, polite culture was beginning to disapprove of singleminded sportsmen.

It was, however, during the last two decades of the eighteenth century that the greatest shift in sensibility regarding hunting and field sports occurred, when the urbanization of the English population, a final consequence of the landlords' victory in the long agricultural revolution, had removed many people from their former proximity to animals in economic production and traditional husbandry. The English became a nation of pet-owners.[36] As Keith Thomas, citing John Berger, has shown, the triumph of the new attitude toward cruelty to animals – and criticism of hunting was one of the consequences of this new attitude – 'was closely linked to the growth of towns and the emergence of an industrial order in which animals became increasingly marginal to the processes of production'.[37] Harriet Ritvo has argued that between the seventeenth and nineteenth centuries, animals ceased to be regarded as independent agents and became the property of humans, whether as pets or livestock. When in 1679 a London woman was hanged for bestiality, 'her canine partner in crime suffered the same punishment on the same grounds', Ritvo claims.[38] By the nineteenth century, however, British authorities had stopped sentencing animals to death, not merely on humanitarian grounds but because animals were no longer regarded as capable of certain kinds of agency. Ritvo concludes that this 'circumscription of the legal role of animals' reflected a fundamental shift in the relationship between humans and their fellow creatures, as humans 'systematically appropriated power they had previously attributed to animals, and animals became significant primarily as the object of human manipulation'.[39]

This increasingly proprietorial relation to animals exemplifies the desacralization of nature, but also the ideological power of private property as more than a legal concept – as the representation of an imaginary, yet lived, relation to the world.[40] Earlier periods had found animals morally and politically instructive, in the early Middle Ages by sticking close 'to the reality of animal behavior to give lessons to people',[41] then beginning in the twelfth century, by metaphorizing them as human exemplars.[42] In 1532 Machiavelli offered both the fox and the lion as effective political strategizers, the former capable of recognizing traps in order to avoid them, the latter able to dismay the wolves by showing greater strength.[43] Aesop's and other beast fables could be said to address unequal power relations by mediating between 'human consciousness and human survival', as Annabel Patterson puts it, the mind recognizing 'rock bottom, the irreducibly material, by rejoining the animals, one of whom is the human body'.[44] As recently as the seventeenth century, animals were regarded as both the antithesis of humans and as fellow beings perilously close to humans, who could instruct humanity or set bad examples. Significantly, it was the Leveller and mortalist Richard Overton who in 1644 asserted the 'Creatureship' of humanity, arguing that beasts partake in the afterlife, and that human dominion over animals was 'not merely a figure of the unnatural political order of England', but 'inseparable from it'.[45] With the waning of the doctrine of similitudes, of correspondences between the human animal and other animals, through which beasts and men might engage in a certain reciprocity, and the beast instruct the man from time to time, rather than the other way around, animals became increasingly both sentimental objects, devoid of agency, and forms of property.[46] Erica Fudge has found that seventeenth-century English law differed from continental European law in emphasizing simultaneously the status of animals as owned, and as individuals whom the owners had to know; animals were both property and had knowable characters.[47]

The new proprietorial relation to animals took at least two forms in the later eighteenth century, one directed against the instrumental use of animals by humans, the other rendering them objects of intensive cultivation. Arguments for compassion toward animals occurred throughout the eighteenth century but with increasing frequency from the 1770s onwards, coinciding with anti-slavery debates.[48] The 1790s witnessed the advocacy of the rights of brutes as well as the rights of man, woman and African slaves.[49] In 1794 the young Samuel Taylor Coleridge would go so far as to address an ass, 'I hail thee *Brother* –,' sympathizing with this 'POOR little foal of an oppressèd race!' and wishing to share his vision of Pantisocracy, or 'Peace and mild Equality', with him.[50] More than an exercise in Laurence Sternean sentiment, though in the spirit of Sterne, this undergraduate effusion sought to encompass even the oppressed of other species within Coleridge's visionary ethics. Today's animal rights movement finds its proper origins in the later eighteenth century. Anti-cruelty sentiment was first directed against lower-class sports such as bull-baiting and became a means of social regulation.[51] Cruelty was 'only applicable to the most stupid, ignorant, and uncivilized part of our countrymen,' announced the Rev. James Granger in 1772: 'Those of higher rank and knowledge are far more humane and benevolent.'[52] Edward Augustus Kendall would make a yellow-hammer declare in 1799, 'The refined part of mankind . . . are always the friends of Nature.'[53] But the anti-cruelty movement would eventually present a serious challenge to hunting.[54]

Experimental husbandry, on the other hand, went hand in hand with hunting. The rising demand for meat among an expanding urban population stimulated changes in sheep and cattle breeding in the interests of commercial meat production, sacrificing quality for quantity and quick growth. What Robert C. Allen calls the second, or landlords', agricultural revolution, culminating in the later eighteenth century, is often exemplified by Robert Bakewell's in-and-in bred cattle and New Leicester sheep, produced at Dishley Grange Farm in Leicestershire, and famous by the 1770s.[55] It appears to have been no accident that Bakewell lived in serious hunting country. Hugo Meynell, the first Master of the Quorn Hunt, lived only 6 miles from Dishley, and was breeding foxhounds on the same principles in the 1760s.[56] This mid-eighteenth-century innovation in cattle and sheep occurred long before any alteration happened to pigs, which fed on waste products and were the mainstay of cottagers' and laborers' larders. The Old English pig breeds, such as the Gloucester Old Spot, which was fed in the Severn valley on the surplus whey from cheesemaking, windfall apples in the autumn and the residues of pulp from the cider press, flourished in non-intensive systems of farming and were slow-maturing, until being crossed with exotic imported pigs from China and elsewhere, beginning in the late eighteenth century.[57] Before being bred as commercial meat-producing machines, pigs were used not only to hunt truffles, but also game. In the early nineteenth century, 'Slut', a black sow from the New Forest owned by a gamekeeper, could point and retrieve partridges as well as any gun dog.[58]

The origins of selective animal breeding in England lie with the production of that other vital piece of equipment for hunting, the thoroughbred horse, beginning in the sixteenth century.[59] The making of the modern throughbred from imported Near Eastern bloodstock, especially the pure-bred Arab, is yet another chapter in the Orientalist romance, the English imperial appropriation of exotic 'Oriental' goods.[60]

The landlords' revolution in agriculture, with its enclosures and farm amalgamations, was directed at maximum profit and drove many people from the land. Rather than increasing output and distributing its benefits widely, as had happened with the first or yeomen's agricultural revolution in the seventeenth century, this revolution concentrated benefits among the elite through higher rents and a reduction in agricultural labor. Allen's conclusion is, he claims, 'unavoidable – most English men and women would have been better off had the landlords' revolution never occurred'.[61] The capitalization of agriculture progressively introduced what some historians have called protoindustrialization. Before the establishment of the factory system, a landless proletariat had already come into being.[62] J.R. Wordie cautiously concludes that even if the English system did not actively drive people off the land through parliamentary enclosure, it nevertheless prevented people from *getting onto* the land as population grew. In this way, the English system, devised and operated by English landlords, *made available* a labour supply for industry, although it did not *create* it.'[63]

Hence urbanization, but also the staging of seemingly empty parklands and fields. The great estate, the landscaped park and the country house may have depended upon the toil of many, but these places were designed to minimize the visibility of labor. Gentility was to be enacted in splendid isolation. The landscape gardener Humphry Repton, criticizing his predecessors in 1816, regretted 'the prevailing custom of placing a House in the middle of a Park, detached from all objects, whether of convenience or magnificence; and thus making a country residence as solitary and unconnected as the Prison on Dartmoor'.[64]

As the geographer Michael Bunce observes, '[T]he aesthetic appreciation of the countryside is inseparable from the social order which created it.'[65] Country houses baldly situated on acres of ostentatiously close-shaven lawns reminded people other than Repton of penal institutions. In early modern England, the 'entrenched hierarchical structure of rural society', as Bunce puts it, had this obvious effect of privatizing property and minimizing the presence of the lower orders both inside and beyond the park pale.[66] But the entrenched order based on land ownership had a less obvious effect as well. Rank denoted leisure, and this meant that 'agrarian objectives were often subordinate to the requirements of gentrification'.[67] Often countering the drive for maximum agricultural productivity and profit was the desire to enhance 'the amenity value of the rural landscape', which for the gentry meant 'the sporting pleasures of hunting, shooting and fishing' and 'a life of genteel ease in carefully landscaped surroundings'.[68]

Maintaining a landscape to admire, walk in, ride, hunt and shoot over meant conservation, not thoroughgoing exploitation of the land. Here is where an ecological sensibility among early modern landowners and their tenants and laborers came into play. When early in the nineteenth century John Clare lamented how 'muse and marvel where we may / Gain mars the landscape every day', he was criticizing the putting of wild, unimproved common grazing land under the plough or enclosing it to make 'Green paddocks', not protesting the parklands of great landowners.[69] Today's farmers and farm workers may well 'maintain a functional aesthetic which is totally at variance with the prevailing urban view', regarding 'a "clean" weed-free field as an object of consummate beauty', feeling 'impatient with the disfiguring clutter of useless hedgerows and trees', and appreciating 'the hygienic design of modern farm buildings'.[70] But early modern farmers and laborers seem to have shared with their landlords a less entirely functional view of the rural landscape than the modern one characterized by Howard Newby. It was their livelihood, but their source of recreation too.

From the seventeenth century to the nineteenth century, country dwellers of all social ranks expected to take pleasure from what we might call the 'amenity' value of rural places. The gentry might have landscaped parks and game-preserving woodlands, from which local people might be officially excluded, but laborers knew how to exploit what remained of former common lands and woods during their brief hours of leisure, and how to infiltrate emparked estates for relaxing, botanizing or poaching. Clare reported how he climbed over the wall into the park at Burghley in order to read James Thomson's *The Seasons* in beautiful surroundings, away from prying eyes.[71] But for Clare a wild common, such as Swordy Well in his native parish of Helpston, could offer as much beauty as a landscaped park. And such places were imperiled by the agricultural improvers' and land agents' promises of enhanced profits, advocating enclosing commons and opening up the parish by means of new roads, putting more acreage under the plough, and making every acre pay.

Stewardship of land and animals, and paternalistic relations with tenants and laborers, sometimes acted as a brake on the devotion to agrarian improvement measured in market terms. And these were most likely to be practiced where older, less socially polarized patterns of behavior held out against urbanized modernity. Hunting with hounds and coursing with greyhounds often fulfilled this democratizing function. In 1821, the radical William Cobbett singled out hunting and coursing as 'those pursuits where all artificial distinctions are lost', where all those 'attached to the soil', whether resident native gentry, farmers, or laborers, were brought together in a leveling enactment of community.[72] Howard Newby allows that fox-hunting 'created a following which embraced all social classes while providing the country gentry and squirearchy with a convenient excuse for coming together and sharing the same consuming mystique.'[73] From the eighteenth century to the present day, upper- and middle-class interests have dominated the landscape not only through owner-ship but also through the movements to conserve and protect the countryside.[74] Nevertheless, many who are not gentry have imitated or aspired to this style of life, and the amenity value of landscape, where it has been preserved, is intermittently accessible even to the propertyless.

For a number of reasons, then – the perceived unjustness of the game laws, urbanization, changes in attitudes toward animals, changes in land use and animal breeding, brought about by agrarian improvement and capitalization – sporting culture by the 1790s had become something of a minority interest, though one of great social, even national, significance.

Inventing modern fox-hunting, inventing the countryside

One further consequence of these changes was that a beast of chase that was neither game nor private property, but vermin, became the most desirable object of pursuit. We will continue to hear of such noble quarry as the hart or stag (male red deer), hind (female red deer), buck (fallow deer) and roe-deer, but scarcity had made deer-hunting a highly restricted sport as early as the seventeenth century, almost entirely reserved for the royal chase in royal forests. As Richard Blome opined in 1686, the pursuit of deer should be confined to 'Persons of *Estate* and *Quality*' who possessed 'the Priviledges and Conveniences of *Forests*, *Chases*, and *Parks*'.[75] By the 1790s, George III was hunting stags carted to and from the meet, who would be spared to run another day, and by the early nineteenth century, hunting carted stags had become a metropolitan as well as royal pastime, popular with City merchants and financiers.[76] Hare chasing, on the other hand, was the traditional recreation of country gentlemen and women, whether it took the form of hunting with a pack of scenting hounds, or coursing with greyhounds or 'gaze hounds', who were stimulated by sight. Hare chasing, according to Blome, afforded 'Delight and recreation to every Man', and even men 'of a lower Rank' might 'sometimes divert themselves with the *Hare*',[77] as plenty of unqualified folk did. Of mounted fox-hunting, which both Charles II and James II pursued enthusiastically for its fast pace, Blome cautiously opined that 'AMongst the Divertisements used by the *Gentry* of this Kingdom, *Fox-hunting* is of no small esteem'.[78] Fox chases required greater speed and a straighter line across country than hare chases, promising to recoup with interest some of the dash of stag-hunting. By the early nineteenth century, riding to hounds on the line of a fox in the changed – improved and enclosed – agricultural landscape, especially in the formerly open-field Midland shires, demanded 'new and exciting skills'.[79] This new leaping at racing pace also made life for women riding side-saddle especially difficult, and though some female thrusters persisted, aided after 1830 by developments in the side-saddle, women in the nineteenth-century hunting field came to be regarded as exceptions, whereas in previous centuries their participation in field sports had been unexceptionable.[80]

Modern mounted fox-hunting had evolved 'by around 1820', and technically has changed very little since.[81] Dedicated to speed and bold riding rather than intimate contemplation of a country, mounted fox-hunting quickly developed its own exclusive rituals and language, constitutive of a distinct national identity for those aspiring to claim it. Reviewing in 1926 the life of Jack Mytton (1796–1834), a famous fox-hunter, Virginia Woolf may not have wished to pursue connections between English poetry and English hunting, but she did intimate how crucial fox-hunting had become for the very texture of English prose, and for Englishness itself as a national identity. Of the English sporting writers (Peter Beckford, Charles St. John, Robert Smith Surtees, Charles James Apperley or 'Nimrod'), Woolf wrote:

> They have had their effect upon the language. This riding and tumbling, this being blown upon and rained upon and splashed from head to heels with mud, have worked themselves into the very texture of English prose and given it that leap and dash, that stripping of images from flying hedge and tossing tree which distinguish it not indeed above the French but so emphatically from it.[82]

In the course of the nineteenth century, mounted fox-hunting had become, and still was in 1926, the sign and physical grounding of Englishness, so emphatically different from anything French. Already in 1868, Anthony Trollope had despaired of conveying to any foreigner 'an adequate idea of the practice' of fox-hunting, so peculiarly English a sport had it become.[83] 'Open to all classes', except, Trollope was careful to note, waged laborers, this country pursuit had the effect 'of making all classes for a time equal in the country':[84]

> [F]oreigners cannot be made to understand that all the world, any one who chooses to put himself on horseback, let him be a lord or a tinker, should have permission to ride where he will, over enclosed fields, across growing crops, crushing down cherished fences, and treating the land as though it were his own, – as long as hounds are running; that this should be done without any payment of any kind exacted from the enjoyer of the sport, that the poorest man may join in it without question asked.[85]

Fox-hunting was the liberty and equality of the freeborn Englishman in action, however poor; but not for the laboring man.

Perhaps the development that most distinguished late nineteenth-century hunting from earlier forms was the introduction in the 1850s and 1860s of wire fencing, and barbed wire in particular.[86] With labor costs for putting up wire a tenth of the price of those for post and rail, wire was a boon for farmers short of capital, but when used to reinforce a hedge, or put on the blind side of a fence, wire was invisible to jumping horses.[87] Wired-up enclosures put paid to the notion of hunting as enacting English liberties. Farmers had to be placated into taking down their wire for the hunting season, and 'ware wire' was added to the hunt followers' litany.

Modern fox-hunting was very much an 'invented' tradition in the sense that Eric Hobsbawm and Terence Ranger intend when they describe cultural forms that seem to date from 'an immemorial past', such as the pageantry surrounding the British monarchy 'in its public ceremonial manifestations', as retrospective and often quite recent inventions.[88] Fox-hunting in its modern form, invented sometime between the 1780s and the 1820s, was very different from older field sports as they had been practiced. As late as 1735, in William Somervile's popular and important poem *The Chace*, which all subsequent hunting writers would cite, we hear hounds called 'dogs', their 'sterns' referred to simply as 'tails', and the fox's brush (tail) described as 'tipp'd' rather than 'tagged' with white. All these would be considered solecisms in the nineteenth- and twentieth-century literature of hunting.[89]

Such changes are difficult to date precisely, but red coats for men, with hunt colors on the collar and club buttons, were a comparatively late invention. Raymond Carr associates the new uniform with subscription hunts, as opposed to private packs of hounds, suggesting that red dress coats were originally worn to hunt club dinners in provincial towns.[90] The Tarporley Hunt Club changed from blue to red coats in 1769 with the substitution of foxes for hares.[91] Red frock coats began appearing in portraits of sportsmen by George Stubbs as early as 1770, most likely implying the sitter's fondness for a fox chase.[92] Although some eighteenth-century private packs had distinctive liveries, such as the Goodwood blue of the Duke of Richmond's Charlton Hunt, portrayed by Stubbs in pictures *circa* 1759–60,[93] hunting attire remained various during the eighteenth century, with fox-hunters coming out in 'long, loose coats and a variety of breeches, boots and hats'.[94] Certainly by the 1850s, 'the tight-fitting scarlet coat – whether its colour came from the Tory red as opposed to the Whig blue is questionable – adorned with five brass buttons was "correct" ', though black coats were also worn.[95] Fox-hunters' uniform had evolved into what was, in effect, the fox's livery: red coat, white bib, and black extremities.[96]

By 1800 in the Midland shires, and in most of the country by 1820, the older notion of hunting a country slowly and patiently with a scratch pack of dogs, beginning very early in the morning, and putting up whatever game was present, or of shooting as simply stalking with a gun, was being dismissed as old-fashioned by members of the rural elite. The naturalist's sensibility had all but disappeared from fashionable sporting culture, where speed and big bags were all the rage. However, another older tradition persisted in upland districts, as it still persists today – in mid-Wales, the Cumbrian Fells and other counties with difficult or remote terrain – the pack of foxhounds usually followed on foot. This sort of hunting remains much closer to fox-hunting's pedestrian origins. It is the one form of hunting with hounds that a recent study has found potentially irreplaceable – because both effective and cost-efficient – as a means of limiting fox predation in hill sheep-farming districts, especially when the hunt employs a 'flexible *modus operandi*' as in mid-Wales, sometimes taking to the horse in lowland country, using terriers for digging out, or carrying shotguns or rifles for shooting foxes in plantations.[97] Foot packs of foxhounds are thus a special case, and not to be equated with modern mounted fox-hunting. Foot packs of beagles which hunt hares preserve the pedestrian practice without the same justification of controlling a predator. Hares 'are simultaneously considered a pest, are a game species culled for their meat and for sport, and are of conservation concern', since they are included in the national Biodiversity Action Plan.[98]

By the 1790s and early nineteenth century, the new exponents of nature were the walkers, and they were often (also) poets. Like so many other British institutions during this time of intense national rivalry and war, English fox-hunting and English walking were considered to be distinguishable most precisely and dramatically from anything French.

'Green Pastoral Landscapes' and the picturesque countryside

Between 1780 and 1830, even as fox-hunting in the Midland shires became the new national sport, and older forms of hunting came increasingly to seem the pastime of poachers and other benighted rural folk, urban tourists began to seek out rural places for their amenity value. In 1712 in the *Spectator*, Joseph Addison had proposed that it was not the possession of broad acres that set one apart from the vulgar mass, but rather the ability to take pleasure in the consumption of landscape as an aesthetic object. This is a sentiment that would be echoed by Wordsworth a hundred years later, appealing to people of 'pure taste' to protect the Lake District as a 'national property', and thus helping to inspire the National Park movement:[99]

> A man of a Polite Imagination, is let into a great many Pleasures that the Vulgar are not capable of receiving . . . He meets with a secret Refreshment in a Description, and often feels a greater Satisfaction in the Prospect of Fields and Meadows, than another does in the Possession. It gives him, indeed, a kind of Property in every thing he sees, and makes the most rude uncultivated Parts of Nature administer to his Pleasures.[100]

The polite man transcends the legalities of mere possession of property to derive a greater pleasure from contemplating landscape. While this aesthetic pleasure may exceed possessive pleasure, it nevertheless resembles it. A man might well, by implication, own land without properly appreciating it, be rendered incapable of satisfaction in the land itself by an untutored imagination. Notice that a 'Description' of a prospect, oral or written, provides its own refreshment. Now one can visit the country imaginatively without even having to go there. The countryside packaged as a literary phenomenon, a reading experience for urban audiences, had come into being by the very beginning of the eighteenth century. Notice also that an educated taste in landscape makes one a connoisseur equally of the cultivated – fields and meadows – and the 'rude, uncultivated', that which will come to be called 'picturesque'. Purely aesthetic pleasure is to be derived from these landscapes, which need not be agriculturally improved. The man of polite imagination can go to the rudest, remotest parts and find them catering to his taste, actively pleasing him.

When William Wordsworth enthused about the scenery of the Wye valley in 'Lines written a few miles above Tintern Abbey, on revisiting the banks of the Wye during a tour, July 13, 1798', he admired the mixed agriculture of this 'green pastoral landscape' (1. 159), anciently enclosed, its foliage so luxuriant that signs of human habitation were hidden and no inhabitants could be seen.[101] The Wye valley's 'pastoral farms, / Green to the very door' (ll. 16–17) were pastoral in the literal sense of being devoted to grazing rather than tillage, with orchards and woodland intermixed.[102] But they were also pastoral in the literary sense of suggesting to his mind the ancient genre of pastoral, as opposed to georgic, verse. The pastoral and the georgic, poetic forms derived from Greek and Roman precedents, offer differing views of country life. In pastoral verse, shepherds live lives of comparative leisure. In a pastoral idyll, no one labors and everyone is nourished by a natural plenitude. Virgil's *Georgics*, by contrast, offered advice to landowners about husbandry. Georgic verse presupposes a need for labor and cultivation to ensure survival. Resources will be consumed, individual people, plants and animals will get used up, but good stewardship should ensure the survival of all species. The georgic imagines what would now be called a sustainable relationship between production and consumption.

The literary historian Michael McKeon finds in Romantic pastoral such as Wordsworth's, in which labor and consumption are most conspicuous by their absence, a contradiction between 'the dream of a direct apprehension of nature' and 'the inevitability of nature's imaginative construction'.[103] Pastoral fantasies have always had a particular appeal for audiences removed from agrarian realities. It takes a knowledgeable eye to recognize the difference between a sweet pasture and a sour one, sheep-cropped old turf and rye-grass monoculture, healthy fields and too great a presence of yellow buttercups or ragwort – pretty but not indicative of good husbandry. McKeon proposes that readers ever since Wordsworth's day have assumed a pastoral attitude toward nature without even noticing it. He calls this phenomenon 'the profound assimilation of pastoral inquiry through its deliberate and increasingly imperceptible application to all experience'.[104]

During the two agricultural revolutions, from the late sixteenth into the early nineteenth centuries, even urbanites would most likely have known something about farming. Manuals of husbandry, no matter how prosaic, continued to legitimate themselves throughout the eighteenth century as a publishably genteel discourse by invoking Virgilian precedent. The 'agricultural improvers and projectors of the period', as Frans De Bruyn observes, justified 'disruptive new modes of agricultural capitalization, organization, and technology by referring these activities to the familiar moral and public imperatives of an ancient georgic tradition'.[105] Husbandry and even the noblest literary genre, poetry, were considered fit companions. Amongst some very stiff competition in Elizabeth I's reign, Thomas Tusser's *Five hundreth points of good husbandry* (1573) became the bestselling book of verse, subsequently appearing in 20 editions between 1573 and 1638 alone.[106] Tusser's popularity endured for 250 years, as evidenced by John Clare's owning a copy of a new edition of *Five Hundred Points* published in 1812.[107] As late as 1800 Bloomfield could praise 'The rich manure that drenching winter made, / Which pil'd near home, grows green with many a weed', because the dungheap offered 'A promis'd nutriment for Autumn's seed' (*The Farmer's Boy*, 'Spring', ll. 188–90). Once the heyday of georgic verse had ended, by about 1789, a good dungheap, though it remained the pillar of early modern husbandry, was less and less likely to figure on the list of tourists' preferred sights.[108]

The fiercest defenders of the 'countryside' – as pastoralized aesthetic object – have, ever since the eighteenth century, tended to be members of the metropolitan middling sort rather than upper- or lower-class country dwellers. Conceiving of the countryside as a source of aesthetic pleasure has usually constituted a retreat from the urban marketplace at the same time that it was a product of it. Today even rural Britain is largely inhabited by non-farming folk, so a predominance of pastoral as opposed to georgic views is understandable.[109] It is significant that when the National Farmers' Union (NFU) came to name its special insurance policy for people who live in the country but do not farm, they chose the name 'Countryside'. Offering mutual insurance to owners of 'more than a garden but less than a farm', NFU rhetoric extended to non-farmers a sense of belonging to the rural landed interest on a *national* level through landownership of a particular kind (Fig. 4).[110] People with incomes not derived from land but who own extensive landholdings are likely to be highly capitalized, non-laboring newcomers to the country, commuters or weekend country dwellers. They are the inheritors of the ideology of the Roman villa, repackaged through the centuries, but always most flourishing during periods of greatest metropolitan growth (such as 'ancient Rome, eighteenth- and nineteenth-century Britain, and the twentieth century throughout the West').[111]

As early as the first two decades of the eighteenth century, such members of the greater London *literati* as Addison were promoting landscape as an object

Figure 4 Advertisement for the National Farmers' Union Mutual Insurance 'Countryside' Policy (August 1993). By courtesy of the NFU. Photograph: Kevin Shea.

of consumption, and Richard Steele, founder of the *Tatler*, and Alexander Pope, the most celebrated poet of his time and an influential figure in garden design thanks to five leased acres at Twickenham, were using the word 'picturesque' to signify aesthetic approval wherever 'irregular and rugged beauty' was to be found (*OED*). A taste for the picturesque had implications for the whole of the countryside, which might be imagined as one great landscape painting, with scant regard for agricultural realities or sporting prospects. The new picturesque taste contributed to a vision of England's green and pleasant land conceived at the expense of both farmers and farm workers, many of whom continued to prize busily populated, productive landscapes and country sports.[112]

Picturesque aesthetics as expounded by the clergyman and schoolmaster William Gilpin was a program for seeing, sketching and otherwise consuming landscape. Between 1760 and 1800, tours of the Wye valley, north Wales, the Lake District, the Scottish Highlands and other remote upland areas had begun to be undertaken, by coach, on horseback or on foot.[113] Thomas West's *A Guide to the Lakes*, first published in 1778, ran to ten editions during the next 30 years.[114] Touring the Scottish Highlands in 1775, Mary Anne Hanway demanded that her coachman drive not as briskly as usual, but *'sentimentally'* through the country, while she rode 'pencil in *hand*' to capture Nature at the trot.[115] Gilpin helped formalize the tourists' itinerary. So popular did Gilpin's views become that lady tourists in the Lake District in 1800 were observed 'reading Gilpin's &c while passing by the very places instead of looking at the places'.[116] Gilpin's picturesque landscape tourism promised both erotic and sporting excitements:[117]

> Every distant horizon promises something new; and with this pleasing expectation we follow nature through all her walks. We pursue her from hill to dale; and hunt after those various beauties, with which she every where abounds.[118]

Clearly, Gilpin took for granted the inherently sporting nature of his English audience. He hoped, however, to capitalize on this sporting spirit in order to recruit people for the picturesque program, which he described as a superior pursuit to hunting:

> The pleasures of the chase are universal. A hare started before dogs is enough to set a whole country in an uproar. The plough, and the spade are deserted. Care is left behind; and every human faculty is dilated with joy.
>
> And shall we suppose it a greater pleasure to the sportsman to pursue a trivial animal, than it is to the man of taste to pursue the beauties of

nature? to follow her through all her recesses? to obtain a sudden glance, as she flits past him in some airy shape? to trace her through the mazes of the cover? to wind after her along the vale? or along the reaches of the river?[119]

We are now entering the world of the walker – or at least he who *doesn't* hunt, but rides in order to view and sketch – a higher form of being above mere blood sports. Gilpin says nothing about the point of view of the animal being hunted; it is only 'a trivial animal', whereas nature is a goddess worthy of hot pursuit. And the final act of this chase? Once cornered, nature will have to yield up her beauties to the seeker after the picturesque. Ovidian rape and metamorphosis, another form of venery, replace hunting in this supposedly bloodless pastime. Alan Liu has identified a moment of arrested violence as characteristic of picturesque aesthetics.[120] I think we can connect this arresting of violence with the displacement of hunting, and the recruiting of its venereal associations, by the picturesque program.

Gilpin's ploy of deliberate substitution – the picturesque tour for 'the pleasures of the chase' – is exposed in William Combe's satire, *Doctor Syntax in Search of the Picturesque* (1809), in which the Gilpin-like Rev. Dr Syntax declines a lord's invitation to hunt, so keen is he on making sketches and writing up his tour of the Lakes in the hopes of earning a fortune:

> Your sport, my Lord, I cannot take,
> For I must go and hunt a Lake;
> And while you chace the flying deer,
> I must fly off to *Windermere*.
> 'Stead of hallooing to a fox,
> I must catch echos from the rocks;
> With curious eye and active scent,
> I on the *picturesque* am bent;
> This is my game, I must pursue it,
> And make it where I cannot view it.
>
> (Canto 13, p. 104)[121]

In 1941 the guild socialist H.J. Massingham deplored the continuing influence of the picturesque as a 'giant worm, trailing its slime over the counties of England'. 'How did the misbegotten thing,' Massingham asked, 'insinuate itself into the language, like the little owl and the grey squirrel, only with incomparably more devastating effect?'[122] To picturesque aesthetics, Massingham assigned responsibility for both urban industrial squalor and modern agribusiness, 'the earth conceived as industrial plant managed by absentee officials'.[123] These sentiments have been recently echoed by Howard

Newby, who links the urban disdain for modern farming, farmers and farm workers to an ignorance of practical husbandry and an enthusiasm for the picturesque.[124]

In its own way, however, late eighteenth-century picturesque theory served a conservationist and even democratic agenda much as old-fashioned hunting could be said to have done. The two could even be combined in painting, mingling two aesthetics that were otherwise so often at odds. In 1792, the year before the outbreak of war with France and the same year in which Gilpin's *Three Essays* were published, George Morland, the most prolific and probably most popular painter of rural English life,[125] endeavored in *The Benevolent Sportsman* to defend the old hunting and shooting culture both picturesquely and patriotically (Col. Pl. 1). Morland, always in debt and often drunk, was far from an elitist. He had 'a complete aversion to anything resembling polite society', and his 'chosen companions were the very lowest he could find'.[126] Never ' "apparently happier than in the midst of peasants" ',[127] Morland in this picture represents a deliberately old-fashioned English squire as personally generous, modest in dress and demeanor, and conspicuously lacking in the new refinement that would soon be repudiated by Bloomfield. This sportsman appears in the very terms John Clare used to describe his patron Lord Radstock, as having a good deal of 'bluntness and openheartedness about him and there is nothing of pride or fashion'; 'he is as plain in manner and dress as an old country squire.'[128]

The benevolent sportsman signifies that this is England, not France, damn it! In England, according to Morland, it was possible to be a landlord and a qualified sportsman and still be a man of the people, perhaps even harbor egalitarian sympathies. Out with his gamekeeper and gun dogs, the squire dispenses charity not to dutiful, industrious tenants, but to gypsies. Gypsies were nothing if not picturesque, for they posed a potential threat to social order. Gilpin had wished to ban 'The spade, / The plough, the patient angler with his rod' from picturesque landscapes and substitute gypsies for a rough and wild effect:

> . . . far other guests invite,
> Wild as those scenes themselves, banditti fierce,
> And gipsey-tribes, not merely to adorn,
> But to impress that sentiment more strong,
> Awak'd already by the savage-scene.
>
> (ll. 574–80)[129]

Morland's painting fuses the sporting and the picturesque aesthetics, suffusing the scene with vaguely radical sentiment.

Yet the sportsman's patriotism is nevertheless assured; it is legible in his very English seat, knees bent and feet well forward, upon his English horse. Gypsies have long been associated with horse-dealing, and the sportsman rides what would still be known in horse-dealing circles as a 'good hairy' horse, a cross-bred cob, up to carrying a gentleman's weight, but nicely shaped and balanced, with a quality head testifying to those refinements of thoroughbred blood – speed, spring-loaded comfort, sensitivity, stamina. He is a good horse, but not a flashy one. Significantly, the sportsman rides with plenty of metal in the horse's mouth, as if already prepared for the new fast-paced fox hunting. The double bridle or 'bit and bradoon' has a curb bit as well as a small-ringed, full-cheek snaffle. The curb is a more powerful bit than the snaffle, used for collection and flexion, for arching the neck and bringing the horse's head down and in, as well as for greater stopping power. We might notice, however, that in the picture the curb rein lies casually on the horse's neck, while our sporting squire holds his horse still with the lightest of finger-touches on the snaffle or bradoon rein. He rides his horse gently and freely, not with iron-fisted collection as in a Continental riding school. As he rides his horse, so will he govern. Should there be any impetuous impulses toward headlong dashes in his mount, he will undoubtedly engage the curb.

This peculiar pastoral, the 'Idea of England'

Between 1671 and 1831, a particular 'Idea of England' was forged, to use Celia Fiennes's phrase, an idea that found material form in landscapes which seem at once closed and open, enclosed and agriculturally improved for maximum productivity, yet permeable as well to accommodate common rights and sport.[130] Between 1671 and 1831, the period of conflict over the game laws, private property was being both legally consolidated and actively contested. Contesting landowning privilege in the name of common rights, by both poaching and trespassing, became widespread, even as the emparking of grounds as well as the enclosure of open fields, common lands and waste lands continued apace. A particular disposition of the bodies of humans and horses encouraging free, forward movement was perfected which facilitated cross-country riding and fox-hunting but also mimed the liberty of the subject so important to English national identity. In a sense, walking replaced field sports for those who abhorred the brutality or bloodiness of the chase or the shoot but still yearned to know a country in the way those who hunted and shot over it, or netted, ferreted or fished it, knew it. The instinct to gather, if not to hunt, remained in place, whether the walker collected botanical specimens or merely picturesque views.

Clare, who claimed that he had 'never shot even so much as a sparrow' in his life,[131] collected wild flowers and observed children hunting for snails in much the same spirit as the gypsies, with whom he socialized and poached:

> I was never easy but when I was in the fields passing my sabbaths and leisures ... or running into the woods to hunt strawberrys or stealing peas in church time when the owners was safe to boil at the gipseys fire who went half shares at our stolen luxury.[132]

The gypsies' 'common thefts' were only 'trifling depredations taking any thing that huswifes forget to secure at night hunting game in the woods with their dogs at night of which all are fish that come [to the] nett except foxes,' Clare reported.[133] His own fieldwork owed much to gypsy 'ways and habits' in which he had been 'initiated', and he was 'often tempted to join them'.[134]

At the turn of the twentieth century, John Buchan's characters, whether great public men fatigued by the cares of office, or shepherds, gamekeepers or poachers, sought solace in the hills and on the moors, and the proper means for doing so remained field sports, listening to 'the song of being and death as uttered by wild living things since the rocks had form'.[135] Even a Scottish ploughman with advanced views about politics finds himself diverted on the day of a big election by an unlooked-for opportunity for trout fishing:

> This was the day which he had looked forward to for so long, when he was to have been busied in deciding doubtful voters, and breathing activity into the ranks of his cause. And lo! the day had come and found his thoughts elsewhere. For all such things are, at the best, of fleeting interest, and do not stir men otherwise than sentimentally; but the old kindly love of field-sports, the joy in the smell of the earth and the living air, lie very close to a man's heart.[136]

Buchan's phrase 'the old kindly love', with its wordplay on 'kin' and 'kindness', suggests that a certain kinship between humans and animals can be expressed only in the sports of field and stream. Isaak Walton's *The Compleat Angler* (1653–76) had exemplified 'how to use nature, to live in intimacy and harmony with its terms, to honor the reciprocity between use and needs in a way that sharply contrasts with landholding and urban power,' as Steven Zwicker puts it.[137] In Buchan, fishing, like hunting, is a form of natural-historical fieldwork and a ritual enactment of human animal being, as much as a recreation.

Even historians of landscape who are hostile to field sports have admitted how fox-hunting and pheasant shooting led to the cultivation of landscape features in defiance of the rationalization of agriculture at all costs.[138] Although 'The Enclosure-Act Myth' – that the modern English landscape is 'almost wholly the artificial creation of the last 250 years' – needs to be laid to rest, it is also the case that Leicestershire and other open-field or champion regions were profoundly affected by the last wave of enclosures. The open-field country was peculiarly susceptible to becoming the Planned Countryside of the agricultural improvers, as Oliver Rackham puts it.[139] Particularly in the Midland shires, whose champion lands were enclosed within hedges according to parliamentary Enclosure Acts between 1750 and 1850, and laid down to pasture for cattle and sheep, the preservation of gorse coverts and the planting of spinneys to encourage foxes produced a landscape quite different from the product of maximum agricultural extraction.[140]

The case is still being made today for field sports as part of a program of conservation of both landscape and wildlife. The zoologist David Macdonald, who has spent thousands of his nocturnal hours freezing under hedges in order to study foxes, and whose proposals should thus carry some weight within the conservation and animal rights lobbies, considers that the consequences of banning fox-hunting are not likely to be straightforward, nor easily predicted. Furthermore, 'fox-hunting is of minor significance to foxes in particular or amongst wildlife issues in general.'[141] In subsequent studies he has suggested that some alterations in hunting practice would also be beneficial (eliminating digging, earthstopping and terrier work wherever possible, and developing drag-hunting to make it more attractive).[142] Most importantly, the fox-hunting lobby should concede that hunting is a sport, a form of recreation, dedicated as much to preserving as to controlling foxes.[143] With the possible exceptions of upland sheep farming and game bird-preserving areas, it is not at all clear that foxes require population control, and where rabbits are an agricultural pest, as in cereal farming, fox predation may sometimes be of economic benefit.[144] In 1987, Macdonald opined that a mixed economy of shooting and hunting could be beneficial for foxes as well as those 'who want to photograph foxes, or to harvest their skins, or to spare them unpleasant deaths [poisoning, strangulation in snares, gassing] at human hands'.[145] More recently, in contributing to Lord Burns's Inquiry into Hunting with Dogs, Macdonald and his fellow researchers stress that 'questions of wildlife management are not answered by biology alone', and must be addressed with 'a level of complexity and inescapable inter-disciplinarity' 'often not adequately considered in discussions of wildlife management'.[146]

Hunting and shooting have historically contributed to the conservation of wildlife and habitats.[147] They could be said to aid in environmental management rather than to detract from or destroy the environment. However, their potential usefulness for a green politics has been neglected even by their proponents, forgotten along with several hundred years of history.

Part I of this book demonstrates how the 'countryside' was invented during the period 1631–1831 in various ways, by various means. Relations between hunting, natural history, and ecology are explored in Chapter 1. Changes in landscape and writing about the land are discussed from the landlords' point of view in Chapter 2, and from the poaching masses' point of view in Chapter 3. Chapter 4 investigates the meaning of 'sports', summarizes changes in field sports other than hunting with packs of hounds, and analyzes the significance of dog-keeping. Chapter 5 argues that vegetarianism and anti-cruelty sentiments caused a radical pedestrianism to emerge in the 1790s which was removed from earlier sporting exploits. Parts II and III endeavor to capture the physical and literary pleasures of hunting (Chapters 6, 7 and 8) and walking (Chapters 9, 10 and 11), respectively. Readers interested in how women figure in this story should turn immediately to Chapter 6. Chapter 11 is a case study of how and when Dartmoor first became a tourist destination, exemplifying the arguments of this book.

About one thing fox-hunters, game fishermen, field zoologists and botanists, walkers, ecologists and landscape conservationists can agree: there should be limits to the rights and entitlements of the owners of private property. The Diggers argued in 1649 that in the beginning the earth was created a 'Common Treasury, to preserve Beasts, Birds, Fishes, and Man', with 'not one word' said 'That one branch of mankind should rule over another'.[148] The countryside as 'an Idea of England' that can only be understood by being hunted-and-gathered is with us still. This book is indeed the place to inquire about how much English poetry depends upon English hunting.

Part I
From Country to Countryside

1
The Greenness of Hunting

> A man cannot be a true sportsman who is not also a true naturalist.
> Attributed to Thomas Smith by C.E. Hare,
> *The Language of Sport* (1939)

Popular green thinking today, often manifested in vegetarianism and support for animal rights, rejects hunting as cruel and argues for minimal interference with the lives of animals, both wild and domestic.[1] On the other hand, field biologists and others who work with animals professionally recognize that to know animals well, to learn about their lives intimately, requires close study of the sort that is bound to interfere in some sense, that might require direct intervention or otherwise make a difference in those lives, and that involves an uneasy combination of scientific skepticism and anthropomorphic thinking.[2] There is an important distinction to be drawn here between empathy, based on knowledge of the other party, and sympathy, based on personal feeling and projection. Here pet-ownership and attitudes toward animals derived from pet-ownership may diverge most sharply from scientific or professional practice. The sympathy felt by the animal lover seems to arise naturally and to require no special knowledge or preparation. Field scientists, professional animal trainers and farmers engaged in traditional husbandry would disagree.[3]

Ecological theorists occupy a third position. Whether deep, social, historical or bioregional ecology is being professed,[4] most ecological theorists today agree that a critique of Enlightenment thinking with regard to the instrumental use of nature by humanity is now urgently called for. Human society must recognize that it too is subject to ecological imperatives, such as limited resources (air, water) and sustainable population size. Some new form of human responsibility toward global ecological wastage, some new mode of reciprocity, is needed. Exactly what shape this should take varies, from deep ecologist Dave Foreman's call for 'a return to the Pleistocene', to the socialist Ted Benton's conclusion that a properly social ecology would recognize the human species as one species among others, at once subject to ecological constraints but, equally, responsible for devising solutions to globally destructive – maladaptive – behavior.[5] Ecological theorists differ from environmentalists in their more radical stance on the need utterly to transform modern capitalist-industrial societies as well as to preserve as much wilderness as possible.[6]

The modern literary and cultural response to the challenge of ecological theory has most often been generically pastoral, not georgic, as if the earth might be saved by a change of heart.[7] Just as we are all more or less tourists now in relation to country matters,[8] we are also all more or less urban pastoralists, more accustomed to looking at nature than to hunting and gathering, and often not quite sure what we are looking at. The georgic ethos of a necessary, but not purely instrumental, *consumption* of nature – not just a reverent contemplation of it – depends upon a sensibility now alien to most citizens of the metropolis. Yet this very georgic ethos of conservationist stewardship and use is one with which we shall need to come to terms once more if we are to negotiate a more realistic – or ecologically sustainable, to use the buzzword – relation to the natural world than our present one.[9] Our present relation is unthinkingly instrumental and unthinkingly pastoral at the same time.

In order to understand how these differing views have come about and what is at stake in each, we need to consider human and animal relations as they have evolved historically, and the histories of animals in relation to human projections – how animals have figured as objects, subjects and agents capable of narratives of history and feeling. The chase once provided the primary English cultural site for exploring these relationships.

Once upon a time hunting, shooting and naturalistic field-work were inseparable, as evidenced by the work of Gilbert White (1720–93).[10] Before taking up natural-historical studies as a middle-aged clergyman, White was a typical 'hearty' undergraduate at Oriel College, Oxford: 'He rode, hunted, shot and fished with almost ferocious vigour. Ironically – given what was to become his consuming passion – he used to gun down small summer migrants to keep his hand in for the winter shoots.'[11] Of the many forms of recreation open to White at Oxford, he 'seems to have preferred riding and shooting', and so 'kept a dog and carried a gun, being especially fond of shooting partridges and hares'.[12] He always rode between Oxford and Selborne because he loved horses, but far from operating as a connoisseur of horse flesh, he found that a 'broken-winded, spavined nag well met his needs, provided that the creature was an old playmate'.[13] Natural history seemed to White a natural outgrowth of sporting culture.

Published in 1788–89, a few months after the founding of the Linnean Society, White's *The Natural History and Antiquities of Selborne, in the County of*

Southampton partakes of a multiplicity of discourses: scientific, literary, aesthetic, sporting, naturalistic. From the seventeenth through the early nineteenth centuries, science and literature were by no means opposed or even separated, as they would subsequently become. As John Wyatt observes, 'Whatever Coleridge in his most dogmatic mood may have determined, men who dealt with material things saw themselves as part of a unified high culture.'[14] Although the 'Men of Science' were evolving new languages for their disciplines of study, they remained unclear about their differences from 'men of letters'.[15] Thus we should not be surprised by the amount of overlap of ideas and tropes between such apparently disparate kinds of writing as zoology and botany, theories of the picturesque, and poetry. Although the aims of the tourist in search of picturesque views, the collector of rare plants, the sportsman- or poacher-naturalist, and the poet seeking new effects, might seem unrelated and even contradictory, there are convergences.

While expounding the principles of picturesque beauty, William Gilpin opined that the way stags met their deaths was common knowledge among what will probably strike the modern reader as three quite different categories of observer: 'The sorrows of the dying stag – his sighs; his tears; and the unfriendly return his distresses find from all his former companions, are circumstances in his history well-known to the naturalist, the forester, and the huntsman.'[16] The divisions of knowledge we take for granted did not yet exist even in the last decade of the eighteenth century. Thomas Pennant, one of Gilbert White's correspondents in *The Natural History of Selborne*, seems to have aimed his *British Zoology* (1766) at least partly at poets, hoping to inspire them to pay closer attention to the animal world:

> Descriptive poetry is still more indebted to natural knowledge, than either painting or sculpture: the poet has the whole creation for his range; nor can his art exist without borrowing metaphors, allusions, or descriptions from the face of nature, which is the only fund of great ideas.[17]

In 1777, John Aikin sought even more explicitly to bring together poetry and natural history, claiming that he would be satisfied if he succeeded in securing for Pennant at least one 'fellow-labourer in your interesting researches into BRITISH ZOOLOGY'.[18] Aikin declared that the purpose of his essay was:

> to add incitements to the study of natural history, by placing in a stronger light than has yet been done, the advantages that may result from it to the most exalted and delightful of all arts, that of poetry.[19]

Close observation was necessary not only for scientific exactness, but for the proper formation and exercise of taste: 'Taste may perhaps be fixed and explained by philosophical investigation; but it can only be *formed* by frequent contemplation of the objects with which it is conversant' (p. 155). Here we find overlaps between science, poetry, theories of the picturesque and the cultivation of polite taste more generally: 'Thus, when a careless eye only beholds an ordinary and indistinct landskip, one accustomed to examine, compare, and discriminate will discern detached figures and groups, which, judiciously brought forwards, may be wrought into the most striking pictures' (p. 156).

Not only poetry but picturesque theory could share knowledge with zoology. Of the ass, Pennant observed:

> The qualities of this animal are so well known, that we need not expatiate on them; its patience and perseverance under labor, and its indifference in respect to food, need not be mentioned; any weed or thistle contents it: if it gives the preference to any vegetable, it is to the *Plantane*; for which we have often seen it neglect every other herb in the pasture.[20]

On the one hand, Pennant assumes his audience has some knowledge of the ass, including knowledge of what we might call the animal's moral as well as physical qualities. The ass is patient and persevering, a humble and modestly heroic animal, as well as an easy beast to keep and feed. Such moral speculation into the ass's character would today be judged unscientific. As William Ashworth puts it, 'Ecosystems are amoral'; they 'are not made of chains of right and wrong, but of chains of cause and effect'.[21] Pennant, however, does add to what he assumes everybody already knows some empirical observation of his own: asses prefer plantanes to other food.

How similar are Gilpin's comments on the ass in the context of picturesque theory in 1791:

> Besides the horse, the forest is much frequented by another animal of his genus, inferior indeed in dignity; but superior in picturesque beauty; I mean the ass. Among all the tribes of animals, scarce one is more ornamental in landscape. In what this picturesque beauty consists, whether in his peculiar character – in his strong lines – in his colouring – in the roughness of his coat – or in the mixture of all – would be difficult perhaps to ascertain. The observation however is undoubtedly true; and every picturesque eye will acknowledge it.[22]

Gilpin, though his object is aesthetics, combines the moral and physical qualities of the ass to capture its essence, as does Pennant. Like Pennant, Gilpin too assumes that the ass's character, especially his humble inferiority to the horse, is so well known as to need no exposition. Expounding the tenets of picturesque taste, Gilpin emphasizes the ass's ornamental value in landscape as its most salient feature. But the animal's 'peculiar character', which everybody knows, is inextricable from its physical appearance in arriving at an explanation for its peculiarly picturesque beauty.

White would have had no difficulty in thinking that the interests of sportsmen and naturalists might coincide, any more than the interests of naturalists and poets, or naturalists and picturesque theorists. Bin's or Bean's Pond on the verge of Wolmer-Forest is 'worthy the attention of a naturalist or a sportsman', because it provides 'such a safe and pleasing shelter to wild ducks, teals, snipes, &c. that they breed there. In the winter this covert is also frequented by foxes, and sometimes by pheasants; and the bogs produce many curious plants.'[23] The wild inhabitants of the pond's vicinity repay close study, whether the object is sport, food for the pot or a greater knowledge of the natural world. Objects of a gun – wild ducks, teals, snipes, pheasants – or of a fox-hunter's obsessive interest, or of a naturalist's close inspection – bog plants – are equally worthy of mention here, because closely linked. Thus does White provide us with a recognition of ecological interdependence without naming it as such. And he does so with his gun barrel at the ready. It is as if the sights of a gun and the magnifying glass or microscope were merely two sorts of lenses for getting closer to the natural world.

Many of White's observations were based on examining and dissecting the dead bodies of animals and birds, a scientific interest but also a sporting one: 'Many times have I had the curiosity to open the stomachs of *woodcocks* and *snipes*; but nothing ever occurred that helped to explain to me what their subsistence might be: all that I could ever find was a soft mucus, among which lay many pellucid small gravels' (p. 125). So many corpses were brought him by his neighbors, including gamekeepers, that the pages of *The Natural History of Selborne* read like a repository of curious remains. We learn that White has 'got an uncommon *calculus aegogropila*, taken out of the stomach of a fat ox; it is perfectly round, and about the size of a large *Seville* orange' (p. 92), or that 'The osprey was shot about a year ago at *Frinsham-pond*, a great lake, at about six miles from hence, while it was sitting on the handle of a plough and devouring a fish' (p. 97), or that 'One of the keepers of *Woolmer-forest*' sent him 'a *peregrine falcon*, which he shot on the verge of that district as it was devouring a wood-pigeon' (p. 278). Studying a live bird may be preferable, but not possible: 'It was not in my power to procure you a black-cap, or a less reed-sparrow, or sedge-bird, alive', he writes to Daines Barrington (p. 124).

Some of White's knowledge of field study, including the close observation of behavior as well as the inspection of entrails, comes directly from his sportsman's past. On the maternal instinct in birds, he reports that 'a partridge will tumble along before a sportsman in order to draw away the dogs from her helpless covey' (p. 150). Sometimes he distances himself from this knowledge while still using it: 'At this distance of years it is not in my power to recollect at what periods woodcocks used to be sluggish or alert when I was a sportsman; but, upon my mentioning this circumstance to a friend, he thinks he has observed them to be remarkably listless against snowy foul weather' (p. 145). Even deer-stealers are comprehensible to White, in spite of his friendly relations with local gamekeepers. Poachers love the pursuit of game as much as lawful or qualified sportsmen. The deer of the Holt 'are much thinned and reduced by the night-hunters', reports White: 'Neither fines nor imprisonments can deter them: so impossible is it to extinguish the spirit of sporting, which seems to be inherent in human nature' (pp. 25–6). Poachers 'harass' the deer (p. 25) as well as kill them; White evidences some social distance here, but he also does the gentlemanly thing of granting 'night-hunters' the status of sportsmen.

In 1878, Richard Jefferies could still describe informal shooting or 'potting', stalking quietly with gun in hand, as both a mild form of sport and an unexceptional, even requisite, part of a naturalist's field training:

> 'Potting' is hardly sport, yet it has an advantage to those who take a pleasure in observing the ways of bird and animal. There is just sufficient interest to induce one to remain quiet and still, which is the prime condition of seeing anything; and in my own case the rabbits so patiently stalked have at last often gone free, either from their own amusing antics, or because the noise of the explosion would disturb something else under observation.[24]

The bag is more often a visual one than a visceral one. But field sports and natural history are part of the same enterprise.

Although White's writing has become a sacred text for British ecologists, the 'spirit of sporting' has become so unfashionable in conservationist or ecological circles that White's sympathy with sportsmen has largely disappeared from view. Consider the fact that the editor of the popular Penguin edition of White's *The Natural History of Selborne* and biographer of White, Richard Mabey, is also the author of *Food for Free*, first published in 1972 and reprinted nine times, brought out in a new color edition in 1989, and again reprinted in 1992. This long-term bestseller might well be described as that most paradoxical thing, a naturalist's guide to nearly vegetarian poaching. Mabey admits, 'There is something akin to hunting here: the search, the gradually acquired wisdom about seasons and habitats, the satisfaction of having proved you can provide for yourself.'[25]

Today, however, there seems to be a prevalent belief that catching something animal is somehow categorically much more violent and problematical than gathering something vegetable. Mabey stresses that this is a book about wild *plant* foods, 'which is', he claims, 'the simple reason (apart from personal qualms) why there is nothing about fish or wildfowl' (p. 15). Presumably, those personal qualms are of a vegetarian sort, and only 'fish and wildfowl', no mammals of any kind, are even permitted to figure as potential objects of search and acquisition. In the very next sentence, however, Mabey risks offending ethical vegetarians because his gathering, if not his hunting, desires have got the better of him: 'But I have included shellfish because from a picker's eye view, they are more like plants than animals. They stay more or less in one place, and are gathered, not caught' (p. 15). The hunting desires of human gatherers would seem not entirely to have vanished, making distinguishing between some plants and some animals in the field a tricky business.

Some green paradoxes in hunting writing

What was the philosophy of hunting that sportsmen-naturalists such as White and Jefferies inherited from previous centuries? Understood in its own terms as ritual or sporting practice, hunting is not simply reducible to animal-killing.[26] Hunting tradition constitutes the chase as both human immersion in the natural world, in animal being, and a meditation upon human responsibilities towards fellow creatures. Death, if it occurs at the end of the chase because the humans' animals, the hounds, have outwitted or outrun the wild quarry, is not to be indulged in lightly. The literature of hunting foregrounds this ambivalence, the human entitlement to enter into the hunt, and urge the hounds to do so, balanced against the necessary proprieties that must be observed if the hunt is to be a properly sporting event. If this paradox has proven increasingly inaccessible to people with modern, urban, post-Enlightenment sensibilities, it was nonetheless accessible to early modern English people.

Within pre-eighteenth-century hunting writing such as George Gascoigne's *The Noble Arte of Venerie or Hunting* (1575), there is often a peculiar merging of man and beast, an identificatory loss of human distinctness in the chase.[27] 'We are the stag or circling hare and we are also, radically, not,' as literary historian Eric Rothstein puts it. Rothstein remarks upon this 'ambivalence' toward hunting (and shooting) as emerging in English poetry with the famous stag hunt in John Denham's *Coopers Hill* (first published in 1642, revised in 1655). Once this ambivalence has lost force, according to Rothstein, distinctly anti-hunting sentiments emerge, as in James Thomson's *The Seasons* (1726–30). 'For the earlier poems, we are the stag or circling hare and we are also, radically, not. From this tension arises a poetically fruitful ambivalence,'

Rothstein argues, which vanishes once the new taxonomy, in which animals have their own status within a classificatory grid of the natural world, takes hold.[28] What Rothstein overlooks, besides the differences between hunting and shooting, are the inherent paradoxes or 'ambivalences' of hunting literature, displayed in works earlier than *Coopers Hill*, such as Michael Drayton's *Poly-Olbion* (1622), and Gascoigne's *Noble Arte*, which equates the chase with monarchy and nobility, portraying in woodcut Elizabeth I about to take the knife from a kneeling courtier in order to break up a deer.[29] A certain ambivalence about taking the lives of animals can be traced back to before the early modern period, but hunting discourse itself is built upon this ambivalence, with its shifting identifications between humans and animals.[30]

What is most easily overlooked in the representation of hunting and other field sports in England is the simultaneously scientific and critical dimension of the discourse, in both how-to manuals and more literary kinds of writing. Gascoigne's is not the first hunting treatise in English, but it was influential and contains original practical advice as well as original verse. It followed upon *The Art of Venery* by William Twici, huntsman to Edward II, *The Master of Game* (1406–13) by Edward Plantagenet, second Duke of York, based upon Gaston Phoebus's *Livre de chasse*, and *The Boke of St. Albans*, published in 1486 from mainly French manuscript sources, which proved enormously popular, appearing in some 15 editions in the sixteenth century.[31] Like most of its predecessors, Gascoigne's book is largely a translation from the French, in this case of Jacques du Fouilloux's *La Vénerie*, taken from the fifth edition of 1573, which had been enlarged by inclusions from the *Maitre de la Chasse*, begun in 1387 by Count Gaston de Foix (Gaston Phoebus).[32] The protocol for the royal pursuit of deer had first been elaborated in France, and the English continued to look to French texts for guidance, though English practices and terms were beginning to differ, as Gascoigne himself indicates. For instance, the French and English methods of breaking up a deer differ, with the French huntsman beginning by cutting off the right forefoot and presenting it to the prince, and the English having the chief personage cut a slit along the brisket and down towards the belly, 'to see the good-nesse of the flesh, and howe thicke it is'.[33]

Gascoigne advocates the utility (food, clothing, medicine) and pleasures of hunting (the chance to show one's skill in the art and be rewarded by the great), proving the importance of woodcraft. When the nobles enjoy the hunting feast, they rehearse 'whiche hounde hunted best, and which huntesman hunted moste like a woodman' (p. 128). Gascoigne demonstrates that considerable naturalistic knowledge was necessary for successful hunting, and that the language of hunting was not only about excluding the socially inferior, who would not be versed in the correct terminology, but also about having a systematic way of classifying the knowledge obtained through fieldwork. Who, apart from a huntsman, a field naturalist or biologist, would know how to identify, as well as name, the 'fewmet or fewmyshings' of deer and 'all beastes which live of browse', the 'Lesses' of boar or bear, the 'Croteys' of hares and coneys (rabbits), the 'feance' of 'vermyne or stinking chases, as Foxes, Badgers, and such like', and the otter's 'Sprayntes' (pp. 97–8)? Thus the treatise gives details of the kinds of ordure characteristic of each beast of chase as well as a poem advising how to make a proper report to the Queen on the detailed findings of a 'ryngwalke' (p. 77) that 'harboured' (p. 78) – both located and deterred from moving off because of the hound's scent left behind – a hart before the hunting of it. The deer must be sufficiently old (at least six years) and carry ten points or branches of horn on his head in order to qualify as a warrantable stag or hart royal:

>An Hart of tenne, I hope he harbord bee.
>For if you marke his fewmets every poynt,
>You shall them finde, long, round, and well
>   annoynt,
>Knottie and great, withouten prickes or cares,
>The moystnesse shewes, what venysone he beares.
>. . .
>Then if she aske, what Slot or view I found,
>I say, the Slot, or view, was long on ground,
>The toes were great, the joyntbones round and short,
>The shinne bones large, the dewclawes close in port:
>Short joynted was he, hollow footed eke,
>An Hart to hunt, as any man can seeke.
>
>   ('The report of a Huntesman upon the sight of an
>   Hart, in pride of greace', pp. 96–7)

Knowing how to read excrement and 'slots' or hoof marks for the condition as well as age of the animal was crucial to success as a huntsman.

Gascoigne's book also questions mankind's right to take animal lives without consideration of their interests and point of view. A dialogue between hunting folk and their animal prey ensues, with the hart, hare, fox and otter pleading for mercy in a series of embedded poems. Most of these poems are Gascoigne's original interpolations,[34] as are many asides pertaining to his distinctly English experiences in the hunting field, such as his section on coursing with greyhounds (pp. 246–50) and his opinion that hares make the best first beast of chase for 'entryng', or initiating young hounds into hunting (p. 37). The animals' pleas often take the form of trying to direct attention away from their own species as desirable beasts of chase and substituting other beasts instead. Naturally, there are moral allegorical implications in this scheme. But empirically based knowledge of the natural world is present too. There is a recognition of the competition between species – almost a Darwinian concept of species-specific struggle for survival, what Tennyson would call 'Nature, red in tooth and claw'.[35] The hart asks piteously, why he must be sacrificed to feed mankind, who begrudges him the merest blade of grass or corn, even in times of starvation:

> Must I with mine owne fleshe, his hatefull fleshe so feede,
> Whiche me disdaynes one bitte of grasse, or corne in tyme of neede?
> Alas (*Man*) do not so, some other beastes go kill,
> Whiche worke thy harme by sundrie meanes: and so content thy will.
>
> ('The wofull words of the Hart to the Hunter', p. 139)

The hart claims that other beasts are more destructive of humankind's interests than red deer, who only feed on the tender shoots of planted crops when nothing else is available. Commoners living near royal forests and deer parks would have known better, since red deer particularly favor young corn or vegetable crops and, in winter, the bark of young trees; to timber crops they can be almost as damaging as goats.[36] Gascoigne himself reports that 'if it snow, they feede on the tops of the mosse, and pill the trees even as a Goate will doe' (pp. 72–3). Such special pleading on the hart's part is explicitly self-interested and rhetorically fashioned at the expense of other species. The hart goes so far as to wish war would break out in order to divert humans from the chase. And should there be a leader who insisted on deer-hunting anyway, the hart craves that that man, like Acteon, might see Diana naked and be similarly punished by being turned into a hart and torn apart by his own hounds:

> Untill his houndes may teare, that hart of his in twayne,
> Which thus torments us harmelesse *Harts*, and puttes our hartes to payne.
>
> (p. 140)

The hare goes even further by protesting that unlike the red deer (hart), fallow deer (buck) or roe-deer, she does no harm at all to humankind:

> The Harte doth hurte (I must a trueth confesse)
> He spoyleth Corne, and beares the hedge adowne:
> So doth the Bucke, and though the Rowe seeme lesse,
> Yet doth he harme in many a field and Towne:
> . . .
> But I poore Beast, whose feeding is not seene,

> Who break no hedge, who pill no pleasant plant:
> Who stroye no fruite, who can turne up no greene,
> Who spoyle no corne, to make the Plowman want:
> Am yet pursewed with hounde, horse, might and mayne
> By murdring men, untill they have me slayne.
>
> ('The Hare, to the Hunter', p. 177)

To add insult to injury, the hare concludes, she has no use value for mankind and thus must be hunted for the sporting thrill alone. If there is a lesson to be learned from hare-hunting, therefore, it is: '*Grevous is the glee / Which endes in Bloud*, that lesson learne of me' (p. 178). Pursuit ending in a kill is intrinsic to the natural world, there is glee to be derived from it, but humans should have a care. The glee derived from blood sport is 'grevous' and not to be indulged in lightly. The hare's plea, like the hart's, is somewhat disingenuous. The delectability of a hare properly cooked was well known, and must still be appreciable by English palates or hares would not continue to hang in such fashionable shopping venues as Oxford's Covered Market. The hare protests too much: she *is* food, and she provides the best sport.

As in fables, so also in the field: the devious hare, though defenseless except for speed, could instruct as well as delight her pursuers. Gervase Markham ranked hare-hunting high among 'Countrey Contentments' as 'euerie honest man, and good mans chase', and 'the freest, readiest, and most enduring pastime.'[37] Nicholas Cox announced in 1674, 'As of all Chaces the Hare makes the greatest pastime and pleasure; so it is a great delight and satisfaction to see the Craft of this little poor Beast in her own self-preservation.'[38] In medieval tradition, the hare's duplicity had been coupled with gender ambiguity and even sexual deviance.[39] Gascoigne's harmless hare is female; by the eighteenth century, the hunted hare would be generically feminized as 'bold puss', a tradition that has continued to this day, so that Shakespeare's male Wat in *Venus and Adonis* (1593; ll. 679–708) and Margaret Cavendish's in 'The Hunting of the Hare' (see Chapter 6) now strike an odd note.[40]

In Gascoigne's poem, the hare's harmlessness emblematizes the requisite humility of the hunter, who should not gloat at his conquest but rather ask pardon of his prey while administering the *coup de grâce*. By 1735, the heart-rending, because human-sounding, shrieks of 'puss' when caught were no longer so paradoxically grievous but gratifying to the hunting field as they had once been.[41] By 1750, hare-hunters were on the defensive:

> A Lover of Hunting almost every Man *is*, or *would be thought*; but twenty in the Field after a Hare, my Lord, find more Delight and sincere Enjoyment, than *one* in twenty in a Fox Chace, the former consisting of an endless

Variety of *accidental* Delights, the latter of little more than *hard Riding*, the Pleasure of clearing some *dangerous Leap*, the Pride of striding the *best Nag*, and shewing somewhat of the bold Horseman; and (equal to any thing) of being *first in at the Death*, after a Chace frequently from County to County, and perhaps above half the Way, out of *Sight or Hearing* of the Hounds. So that, but for the *Name* of Fox-hunting, a Man might as well mount at his Stable-Door, and determine to gallop twenty Miles an End into another County.⁴²

It was otherwise with hare-hunting, its devotees insisted, because of the hare's stratagems, which provided many 'accidental Delights' while keeping the hunt within a comparatively small compass so that hounds could be both heard and seen, and elderly as well as youthful followers might enjoy the sport:

How are both *young* and *old* lost in delightful Enchantment, when Puss has *baulked* the Dogs, dropt the Pack, and on some rising Hillock plays, *in Sight*, her little Tricks; *leaps* here, *doubles* there, now *sits on End and listens*; then *crouched*, (as if sunk in the Earth) deceives the unexperienced Eye, and *creeps* to a Squat.⁴³

Hare-hunting as a soft option for the elderly had been the theme of Eustace Budgell's essay in the *Spectator* (1711), which had related how Sir Roger de Coverley, once renowned in his neighborhood for his 'remarkable Enmity towards Foxes', but now too old for fox-hunting, had taken up hare-hunting.⁴⁴

Ever faster hounds and horses led to impatience with the mazy line of hare-hunting, and coupled with softer hearts and an occasional scarcity of hares, contributed to the new preference for the 'lesse cunning' but more physically demanding hunting of the 'much hotter' or 'stinking sent' of the fox.⁴⁵ Gascoigne himself accounted fox-hunting, 'especially within the grounde', as but 'small pastime' (p. 186). Earlier hunting treatises describe foxes being hunted in their earths with terriers, like rats, and dug out, like badgers. Upland foot packs in sheep farming areas such as mid-Wales are the modern representatives of this tradition. Even today informal or 'rat-catcher' clothing is worn to mounted meets outside the official fox-hunting season. Chaucer describes Daun Russell, the fox, being chased by dogs and shouting villagers with staves and trumpets ('bemes') in 'The Nun's Priest's Tale':

> Of bras they broghten bemes, and of box,
> Of horn, of boon, in whiche they blewe and powped,
> And therwithal they shriked and they howped.⁴⁶

As with the displacement of hare-hunting by fox-hunting, when mounted hunting of the fox began is difficult to ascertain with certainty, but its earliest advocacy is usually attributed to Sir Thomas Cockaine of Derbyshire, who declared in *A Short Treatise of Hunting* (1591):

> And this tast I will give you of the flying of this chase, that the Author hereof hath killed a Foxe distant from the Covert were hee was found, foureteene miles aloft the ground with Hounds.
>
> By that time either Nobleman or Gentleman hath hunted two yeares with one packe of Hounds, the same will hunt neither Hare nor Conie, nor any other chase save a vermine.[47]

It would be the 'flying' of this chase that would attract bold horsemen, including Charles II and his brother James II, after the Restoration and the decimation of deer parks during the Civil Wars.

In 'The Foxe to the Huntesman', Gascoigne's fox lectures humanity on human vanity and luxury, protesting that humans 'can spoyle, more profit in an houre, / Than Raynard rifles in a yere, when he doth most devoure' (p. 199). As in the Aesopian fable, so also in Gascoigne's verse, the fox is a clever interpreter, close to human in his self-serving logic. From the fox's point of view, human devotion to the chase appears as necessary, yet objectionable, as their 'costly clothes' and 'deintie fare' or 'couches stuft with doune' (p. 199):

> But wherto serve these sundry sports, these chases manyfold? Forfoth to feede their thoughts, with drags of vaine delight, Whereon most men do muse by day, wheron they dream by night. (p. 199)

The fox cannot conceive that humans will give up hunting any more than he will. But he wishes they would be less sanctimonious about human superiority and vulpine culpability than they tend to be.

'The Otters oration' (accompanying illustration, Fig. 5)[48] clinches the argument that a proper knowledge of the literature of hunting and hunting's meanings will transform human behavior from unthinking instrumentality toward the animal world into a relation more profound, a kind of reciprocity. By protesting that humans criticize beasts for precisely their own vices of gluttony, bloodthirstiness and wasteful behavior, the otter reveals the duplicity at the heart of the human language of 'beastliness'. It takes a beast of chase to set the record straight. The rituals of hunting should teach the followers of the chase not to waste their fellow creatures but to pursue them with reverence. To follow the chase is to seek for grace. If hounds and

Figure 5 Anonymous illustration for 'The Otters oration', opposite page 202 from George Gascoigne's *The Noble Arte of Venerie or Hunting* (1575). Reference shelfmark, Douce T 247(1). By courtesy of the Bodleian Library, University of Oxford.

humans win the day, 'grevous glee' and sober joy should result, for no animal's life is to be taken lightly.

> When men their time and treasure not mispende,
> But follow grace, which is with paines ygot,
>
> (p. 205)

then and only then will the sporting ideal in its fully religious-ritual sense have been realized. We have here not so much an ambivalence about the practices of field sports as a practice of ambivalence, elevating ambiguity about the relation between human and animal to an art-form. The scientific and the sacramental, the instrumental and the empathetic are fused.

Writers within the tradition of hunting literature often shift their point of view, their identifications and investments, from human to animal vantage points and back again. Michael Drayton, producing his chorography of Great Britain in 1622, represented field sports as among the greatest pleasures of the British Isles. The poet figures himself intermittently as both hunter and hunted. Animal identification begins with the animals closest to humans in the chase, hounds and horses. Britain is supreme in sport because the best horses and hounds are to be found there, and some regions, such as the West Country, are particularly favored for the strains they produce:

> What pleasures hath this Ile, of us esteem'd most deere,
> In any place, but poore unto the plentie heere?
> ...
> A horse of greater speed, nor yet a righter hound,
> Nor any where twixt *Kent* and *Calidon* is found.
> ...

> And as the Westerne soyle as sound a Horse doth breed,
> As doth the land that lies betwixt the *Trent* and *Tweed*:
> No Hunter, so, but finds the breeding of the West,
> The onely kind of Hounds, for mouth and nostrill best;
> That cold doth sildome fret, nor heat doth over-haile;
> As standing in the Flight, as pleasant on the Traile;
> Free hunting, easely checkt, and loving every Chase;
> Straight running, hard, and tough, of reasonable pase:
> Not heavie, as that hound which *Lancashire* doth breed;
> Nor as the Northerne kind, so light and hot of speed,
> Upon the cleerer Chase, or on the soyled Traine,
> Doth make the sweetest cry, in Wood-land, or on Plaine.
>
> ('The third Song', p. 40)

This hymn to hounds as sporting dogs gives us the distinct impression of a sporting Drayton. He may be writing poetry, but he embeds a brief treatise on hunting. The taxonomy of kinds of hound represented here will be repeated in almost exactly these terms until the nineteenth century.[49] A hound with a keen nose, up to difficult scenting conditions, which will produce 'the sweetest cry' and travel neither too fast nor too slow for a moderately mounted field accompanied by foot followers, will remain the ideal for nearly two centuries – plenty of hound music and a kind of hunting that would last all day. (Only when mounted fox-hunting came to supplant all other forms of hunting in the English imagination would pace be everything.)

But when Drayton merges his poetic practice with hare coursing later in the poem, we sense his doubled empathy, his simultaneous identification with the hare as well as with the greyhounds:

> When each man runnes his Horse, with fixed eyes, and notes
> Which Dog first turnes the Hare, which first the other coats,
> . . .
> But with a nimble turne shee casts them both arrere:
> Till oft for want of breath, to fall to ground they make her,
> The Greyhounds both so spent, that they want breath to take her.
> Here leave I whilst the Muse more serious things attends,
> And with my Course at Hare, my *Canto* likewise ends.
>
> ('The three and twentieth Song', 2: 73)

To 'coat' means to outstrip another in the chase. Not trusting to his audience's command of such technical vocabulary, Drayton appends a note defining 'coat'. Hunting is thus simultaneously assumed to be popularly appealing and arcane.

The practice is well known, but the parlance is more elusive. Drayton both identifies his own poetic muse with hunting, or rather coursing, and has her withdraw her investment in it for 'more serious things'. Coursing is not serious; it is a source of delight, a diversion or recreation provocative of mirth. In keeping with this jocund mood, Drayton has the coursing match end in a draw between hare and greyhounds. All exuberance spent, all fall down in exhaustion together, including the poet, whose canto ends abruptly with the course. This bloodless anticlimax casts coursing in a salubrious, not a sanguine light. Drayton expects his readers to be delighted with this representation. No conflict arises between intellectual or artistic production and sporting pursuits.

So powerful does this tradition prove that in later periods, what might at first look like straightforwardly anti-hunting poems often turn out to be peculiarly indebted to the tradition of hunting or sporting verse. Elizabeth Moody is very much of her time in staking out an anti-cruelty position in 'To a Gentleman Who Invited Me to Go A-Fishing' (1798), declaring 'No sports are guiltless to the feeling mind' (l. 4).[50] But her conclusion by means of rhetorical questions opens up the possibility that, as in the world of Walton's *Compleat Angler*, fishing and writing, rather than being diametrically opposed activities, in fact share some strange congruences:

> Shall I, who cultivate the Muse's lays,
> And pay my homage at Apollo's shrine,
> Shall I to torpid angling give my days,
> And change poetic wreaths for fishing-line?
>
> Sit like a statue by the placid lake,
> My mind suspended on a gudgeon's fate;
> Transported if the silly fish I take,
> Chagrined and weary, if it shuns the bait?
>
> (ll. 9–16)

This description of fishing sounds uncannily like a description of waiting for poetic inspiration to strike, rather than a contrast to it.

Hunting and social ecology

William Cobbett, the most famous radical of his day, was surprisingly attached to hunting. He defended the rights of cottagers and laborers to take game and protested the harshness of the punishments inflicted for transgression of the game laws.[51] Cobbett's endorsement of rural sports as ' "manly" recreations' that encouraged 'social and cultural interaction between élite and non-élite' was crucial for his radical envisaging of the nation.[52] As Ian Dyck explains,

As he crossed the threshold to radicalism, Cobbett distilled from the ranks of English society a vertical configuration of countrymen – including landlord, farmer, labourer and village tradesman – which he defined, not according to relationships with the means of production, but according to rural residence, a love of the land and a shared opposition to the culture and ideology of the city.[53]

Leonora Nattrass characterizes Cobbett's writing as '*constitutive* of a Radical nation' in that he 'refuses to exclude the centre entirely from his work' and holds 'hostile readerships together in his texts'.[54] In fact, Cobbett's *Rural Rides* might well be modeled on his vision of the hunting field.

Cobbett's views on agriculture and sporting culture, and his construction in his writing of a radical, class-divided, but socially cohesive community, instance an ethic and an aesthetic of stewardship of land, animals and people that even modern farmers and agricultural workers might be able to endorse. This ethic had georgic roots; it was grounded in the literature of husbandry and improvement, but was as concerned with the welfare of the laboring poor as with profits for landowners, and with the beauty or amenity value of landscape as well as its usefulness. In the Island of Thanet in Kent, Cobbett remarked:

> Invariably have I observed that the richer the soil, and the more destitute of woods; that is to say, the more purely a corn country, the more miserable the laborers . . . No hedges, no ditches, no commons, no grassy lanes: a country divided into great farms; a few trees surround the great farm-house. All the rest is bare of trees; and the wretched laborer has not a stick of wood, and has no place for a pig or cow to graze, or even to lie down upon. The rabbit countries are the countries for labouring men.[55]

Cobbett's attention to the health of agricultural land, but also to woods, commons, heaths and their wild inhabitants – his concern for what we would now call the ecosystem – was inseparable from his concern for the people who lived in a place, especially the poorest classes. Cobbett, like the social ecologist Murray Bookchin, regarded human liberation as an ecological project, and defending the countryside, conversely, as a social project.[56] What connection might there be between these views and Cobbett's being a keen follower of hounds?

Visiting Herefordshire in November 1821, Cobbett enjoyed hare-hunting with Philip and Walter Palmer as a change from inspecting their pastures, orchards, coppices, timber trees and successful crops of swedes. First there is the bonding with the horse, the bold, cooperative and clever hunter:

> They put me upon a horse that seemed to have been made on purpose for me, strong, tall, gentle and bold; and that carried me either over or through

everything. I, who am just the weight of a four-bushel sack of good wheat, actually sat on his back from daylight in the morning to dusk (about nine hours), without once setting my foot on the ground.[57]

This long day's hunting was over the steepest ground he had ever ridden: 'A steep and naked ridge, lying between two flat valleys, having a mixture of pretty large fields and small woods, formed our ground. The hares crossed the ridge forward and backward, and gave us numerous views and very fine sport' (1: 32). Here 'views', even in this picturesque woodland pastoral country, meant views of hares, and of hounds following hares. Cobbett, like Celia Fiennes some 100 years earlier, as we shall see in the next chapter, is not ready to give up hunting a country or sporting in the country for simply consuming landscape by looking at picturesque views.

Being able to claim 'A whole day most delightfully passed a hare-hunting, with a pretty pack of hounds' (1: 32) meant for Cobbett that he shared not only the pleasures of the landlords but of their tenants and laborers too. For Cobbett hare-hunting and fox-hunting were democratizing, levelling pursuits where 'all artificial distinctions are lost' (1: 38). Only a 'shallow fool', according to Cobbett, could fail duly to

> estimate the difference between a resident *native* gentry, attached to the soil, known to every farmer and labourer from their childhood, frequently mixing with them in those pursuits where all artificial distinctions are lost, practising hospitality without ceremony, from habit and not on calculation; and a gentry, only now-and-then residing at all, having no relish for country-delights, foreign in their manners, distant and haughty in their behaviour, looking to the soil only for its rents, viewing it as a mere object of speculation, unacquainted with its cultivators, despising them and their pursuits, and relying for influence, not upon the good will of the vicinage, but upon the dread of their power. (1: 38)

Cobbett thus accurately predicted the conditions in which hunting would come to be viewed as an emblem of class privilege, not of the levelling of class distinctions or a responsible form of stewardship.

It is important to note that the hunting of which Cobbett approved was provincial, and not the most fashionable hunting of his day. During the early decades of the nineteenth century, the great grasslands of the Midland counties of Leicestershire, Northamptonshire and Rutland, but more particularly the packs of hounds that hunted there – the Quorn pre-eminently, but also the Belvoir, Cottesmore and Pytchley, and later, the Billesdon or Fernie, a country that split off from the Quorn in the mid-nineteenth century – began to be referred to by hunting people as 'the shires'. David Itzkowitz suggests that some would include such packs as the Atherstone and the Warwickshire, 'but strictly speaking,' he notes, 'only the first group are "shire" packs. All others, from the greatest to the smallest, are "provincial" packs.'[58]

Compared with the great grasslands of the Midland shires, Herefordshire was not only remote from London, but it had never been an open-field or champion country, and it remained a prime example of the Ancient as opposed to the Planned Countryside, in Oliver Rackham's terms. This was one of the landscapes most prized by theorists of the picturesque: William Gilpin, and the Herefordshire landowners Uvedale Price (who lived at Foxley) and Richard Payne Knight (who lived at Downton), and by William Wordsworth in 'Lines written a few miles above Tintern Abbey' (1797). Herefordshire was the home of the English georgic poem, according to John Barrell.[59] 'Here were broad valleys, orchards, hop-fields, woodlands, parkland and hedgerows, scattered with small hamlets and threaded by deep lanes. This was an old enclosed county of modest properties, of minor gentry, smallholders and cottagers', as the geographers Stephen Daniels and Charles Watkins point out.[60]

There were limits to the democratizing effects of hunting in Cobbett's day, especially in more agriculturally innovative parts of the country.[61] Cobbett himself regrets the disappearance of private packs of hounds in many counties such as Berkshire, observing that 'forty years ago, there were *five* packs of *fox-hounds* and *ten* packs of *harriers* kept within *ten miles* of Newbury', whereas in 1821 there was only one foxhound pack, maintained by subscription, and no harriers except his friend Mr. Budd's.[62] Regretting the disappearance of private packs (especially of old-fashioned harriers) is, for Cobbett, equivalent to mourning the passing of personal generosity and hospitality among the better-off, the essence of democratic spirit within the rural community. The new subscription hunt clubs were democratic in that anyone could join who could afford to, but they did not replace the older cross-class linkages that private packs had sometimes encouraged, or their equating of hunting a circling hare with collective perambulation of a country, 'hunting the borough'.

The occasion of a fox chase could lift the spirits of even poor country dwellers, but foot-following was hard work. In 'Autumn', in *The Farmer's Boy*, Robert Bloomfield memorialized the country around Euston Hall in Suffolk, of which the Duke of Grafton was lord of the manor, by commemorating the Duke's great foxhound, Trouncer. 'Inscribed on a stone in Euston Park wall', Bloomfield reports, is ' "*Foxes, rejoice! here buried lies your foe*" ' (l. 332 and n.). But there was nevertheless a drawback for the humble foot-follower:

> O'er heath far stretch'd, or down, or valley low,
> The stiff-limb'd peasant, glorying in the show,
> Pursues in vain; where youth itself soon tires,

>   Spite of the transports that the chace inspires;
>   For who unmounted long can charm the eye,
>   Or hear the music of the leading cry?
>
>   (ll. 297–302)

However transporting the music of the chase, without a horse to keep pace the would-be follower was so disadvantaged as to be right out of it. To hear the leading hounds in the pack and to obtain a properly picturesque view of hunting – one to 'charm the eye' – one must be mounted. And to be mounted, it would seem, one must be better off than the 'stiff-limb'd peasant' in Giles's country. Cobbett's sort of hunting was not a means to pleasure equally accessible to everyone, and it was fast disappearing except in provincial countries in the West and the North. In the most remote and difficult terrain – mid-Wales and Cumbria – foot packs persisted.

'A man cannot be a true sportsman who is not also a true naturalist', an adage attributed to Thomas Smith,[63] carried more weight before field sports became fully commercialized in the course of the nineteenth century (see Chapters 4, 7 and 8). In 1838 Smith himself wrote, more prosaically, that 'Few things are so necessary for a huntsman to acquire as a thorough knowledge of the country he hunts.'[64] Smith also wrote *The Life of a Fox* (1843), ostensibly by Wily, a New Forest fox, who invites for supper and a swapping of stories a convocation of foxes from different parts of the country. This is an empathetic as well as empirically knowledgeable text, advocating the preservation of foxes through the building of artificial fox earths, criticizing gamekeepers for trapping and maiming foxes, condemning absentee landlords who let their shooting or sell their game at any price, and protesting against the digging out of foxes who have escaped fairly from hounds and gone to ground.[65] Smith was a huntsman who could get hounds to perform extraordinarily well 'because he could do the work good hounds do'; his 'infinite patience, decision and speed' earned him the highest praise from Robert Smith Surtees, who regarded Smith as the greatest master of his day.[66] Interested in 'contemporary science and progressive agriculture', he was both an inventor and an innovator in hunting practice, but his contemporaries found his 'passionate interest in natural history' a 'mere eccentricity'.[67] By the 1820s and 1830s hunting and natural history were parting company, and Smith appears to have maintained some of the eighteenth-century connections between them, alongside his interests in science and progressive agriculture.

A modern coda

Hunting allegories continue to influence many professionals who work with animals, even those who object to hunting's 'undeniable, if unquantifiable' cruelty to foxes and other quarry.[68] Unlike strict behaviorists, who have been criticized from an animal rights point of view for their mechanistic treatment of animals,[69] the zoologist Macdonald admits that field study involves both scientific skepticism and residually anthropomorphic thinking. Like Gilbert White before him, Macdonald has had to look to gamekeepers, farmers and shepherds for reliable data about fox behavior, not to mention specimens for postmortem analysis. These people Macdonald calls 'countrymen of the old school', who are possessed of a 'paradoxical admiration, almost affection', for foxes 'characteristic of many who spend a lifetime trying to kill them'.[70] There is the Cambridgeshire gamekeeper Mr. Spring, for example, who sent Macdonald parcels containing the anal glands of foxes killed on his beat, and also supplied him with eight orphaned fox cubs for hand-rearing (p. 132).[71] When Macdonald and his wife Jenny watched a hunt with the Blencathra hounds, a foot pack, they were jubilant that the fox escaped, apparently as a consequence of vulpine quick-wittedness. 'A scientist's burden of scepticism' made Macdonald uncertain that the fox actually outwitted hounds by hiding and running back along the line of his scent, but he confesses 'that is what it looked like' (p. 174). Like many before them, the Macdonalds saw for themselves the fox's 'legendary ingenuity' and apparently 'uncanny' ability to 'evade doom' when chased by hounds (p. 172), enacted as it can only be in the hunting field.

As described in *Running with the Fox*, Macdonald's modern version of White's natural-historical study combines such techniques as radio-tracking with hand-rearing of fox cubs to establish a close working relationship with individual foxes. Macdonald engages in what we might call intersubjectivity, or mutual exchange with his fox subjects. Not many self-professed animal lovers would endure the violation of property, let alone personal space, Macdonald experienced while sharing his bed with Niff, the prototype hand-reared fox, enduring sleepless nights with her warbling in his ear, licking him and digging holes in his scalp. He observed the wholesale destruction of his property, including books, as she chewed objects and buried caches of food throughout the house (pp. 56–60). Having risked the condemnation of the scientific establishment by his closeness to his fox subjects, Macdonald writes: 'Mercifully, I was wrong in my fear that Niff would grow up de-foxed by my companionship. In fact, ours was to become a thrilling, professional relationship, with her the active partner and I the passive onlooker' (p. 57). The risk of anthropomorphic projection and human intervention in natural processes – the human contamination of foxiness – has, it would seem, been repaid in this case by a rich psychic, but also scientific, dynamic.

Macdonald's sometimes painfully, but always empirically, gathered knowledge of fox behavior has put him in a position to empathize with how a fox thinks. This knowledge is akin to the huntsman's or field naturalist's woodcraft, and it does not entitle Macdonald to intervene heroically, beyond administering the occasional antibiotic, to save the lives of foxes when they become ill or injured, or to prevent circumstances from affecting them unpleasantly. Nor does it entitle him to moralize about fox behavior, either to defend or blame them for their actions. When foxes engage in surplus killing, for example, in hen coops, pheasant pens and other enclosures in which prey do not run, this represents a maladaptation, 'the misfiring of a trait that is useful in another context (like modern man's tendency to take too much sugar)', as Macdonald puts it (p. 164). The fox's normal behavior becomes inappropriate when the prey's behavior is abnormal. It has been this very tendency toward surplus killing that has given the fox the reputation of bloodthirsty felon. But scientific skepticism dictates a critical distance here between anthropomorphic projection and acceptable explanation. As Macdonald puts it, 'The fact that we can never know what, if anything, a fox feels about killing does not affect the fact that it is nonsensical to judge them on human cultural values' (p. 164). Such words as 'evil, vengeful and wicked', Macdonald claims, whatever 'occasional relevance' they may have to discussions of human morality, have 'none whatsoever to fox behaviour' (p. 164).

Macdonald doesn't sympathize with the fox by identifying his own feelings with what a fox must feel. Sympathy contains a consuming, appropriating element. We sympathize with those we think are like ourselves in some way. Empathy, on the other hand, is a relation across difference, an attempt to understand another being in its own terms. From the professional animal worker's point of view, the urban pet-owner's usual fantasy of non-interference with pets, the sympathetic wish that they be 'free' to behave as they like, may constitute unthinking cruelty. People who work with animals have to know empirically, even if they do not use the scientific language of behaviorism, the behavioral patterns of the animals with which they work. As Richard Jefferies reported in 1878, gamekeepers and shepherds, having closely associated with dogs since childhood, have no doubt that dogs 'think and reason in the same way as human beings, though of course in a limited degree', and that 'the labourers who wait on and feed cattle are fully persuaded of their intelligence, which, however, in no way prevents them throwing the milking-stool at their heads when unruly'.[72]

Another way of putting it would be to say that the allegorical stories people who work with animals tell themselves about their animal partners must bear some workable relation to the animals' capacities. Vicki Hearne, for instance, argues from a trainer's perspective:

> It won't do to suggest that the dog can just live peacefully around the house while we refrain from giving any commands that might deprive him

of his 'freedom', for that simply doesn't happen. We are in charge already, like it or not . . . A refusal to give commands or to notice that commands are being given is often a refusal to acknowledge a relationship, just as is a refusal to obey.[73]

Following the dog trainer William Koehler, who argues that 'the dog has the "right to the consequences of his actions" ' (p. 44) and distinguishing between 'kindness and cruelty, or perhaps between rightness and cruelty' (p. 45), Hearne observes: 'A sharp, two-handed, decisive upward jerk on the training lead, performed as impersonally as possible, is a correction. Irritable, nagging, coaxing tugs and jerks are punishments, as beatings are. The self-esteem of the handler gets into them' (p. 45). Corrections, on the other hand, 'are administered out of a deep respect for the dog's moral and intellectual capacities' (p. 45).

If this is also the rhetoric of the socialization of children, involving a good deal of anthropomorphic projection, the usefulness of such allegory as a mode of interpreting the dog is constantly being tested in practice, in a direct way. And it is definitely a form of allegory derived from hunting, not from pet ownership. In William Koehler's terms, according to Hearne, to get 'absolute obedience from a dog – and he means absolute – confers nobility, character and dignity on the dog' (p. 43). The allegories appropriate to great dogs or great horses are allegories of heroic dedication to what they know – the knowledge of hunting, tracking, jumping, dressage, and so on. In the case of a pointer, this means 'to be staunch to wing and shot, to hunt wholeheartedly where lesser dogs lose spirit, to work with full courtliness and gallantry' (p. 50). The value of the foxhound to the fox-hunter lies in the allegorical story the fox-hunter tells about the hound, about 'the value and greatness of fox hunting' (p. 261). There is nothing here of allegories of human freedom disguised as fantasies about how we should interfere with our pets as little as possible.

Empathy for animals, based upon the recognition of their inherent value despite their differences from human beings, ideally demands some knowledge of their mental capacities and behavior, their distinctness as separate species beings. Hunting is undoubtedly a stressor of the hunted, detrimental to the welfare of the quarry, yet findings regarding the stress suffered by hunted animals are difficult to interpret, as Macdonald, Tattersall, Johnson and their co-authors suggest. They remain unsure:

> whether to be more impressed by the observation that, at 385bpm [beats per minute], the heart rate of foxes being chased was 18% higher than that of foxes running voluntarily, or by the observation that the heart rate of those running voluntarily was only an equivalent amount less than those being chased.[74]

Interpretation is further complicated by the fact that these findings were obtained by chasing foxes in an enclosure, where they had no escape route. Might this experiment have been more stressful, therefore, than a cross-country chase? Scientific measurement of stress, precondition for calculating possible suffering and human cruelty, remains a difficult task.

Horses, which are herd animals, and hounds, which hunt in packs, are both highly social kinds of animals with a dominance hierarchy. Both horses and hounds thus usually respond enthusiastically to the group dynamics of hunting.[75] Stephen Budiansky contrasts the naturalness of hunting with the unnaturalness of dressage, which is usually perceived by sophisticated urbanites as the height of civility and refinement in horse training:

> To suggest that there is something unnatural about dressage is to risk bringing upon oneself a torrent of abuse from its devotees, but the truth of the matter is that these collected gaits are not ones that a freely moving horse, or a green horse, or a wild horse, will ever select on its own – unless it is highly emotionally aroused . . . Through steady schooling a horse can be taught to move at a collected gait perfectly calmly, just as a police horse may be taught to overcome its natural nervousness at the sound of a gun shot. But the enshrinement of the high school as an ideal continues to do great mischief to our understanding of the natural movements of the horse.[76]

S.G. Goldschmidt declared categorically in 1927, when hunting was much less likely to be criticized than it is today:

> THE only work to which a horse can be put from which it derives any pleasure, and in which it takes any personal interest, is fox-hunting . . . The instinct of self-preservation in the wild ancestor would account for the joy in hunting and slaying a carnivorous animal. I can explain in no other way the keenness in a hunt and satisfaction at a kill some horses display, even though they may never have seen hounds before.[77]

Budiansky gives a physiological explanation for this behavior: galloping with the herd out hunting stimulates the fear-and-fight-or-flight response, which releases endorphins, 'natural opioids'.[78] Hence horses accustomed to going hunting, whenever they hear a hunting horn or observe what looks like hunting going on, often appear eager to follow hounds even without human encouragement.

The well-being of humans and all other species depends upon preserving natural levels of biodiversity. Species compete with one another, but they also connect. Deep ecologist Dave Foreman has championed a return to gathering and hunting societies, while from a social-ecological point of view, 'it is sheer distortion to associate all pain with suffering, all predation with cruelty'.[79] '[T]he mutually supportive interplay between photosynthetic life forms and herbivores, far from providing evidence of the simplest form of "predation" ... is in fact indispensable to soil fertility from animal wastes, seed distribution, and the return (via death) of bulky organisms to an ever-enriched ecosystem', Murray Bookchin argues.[80] Predation is not only natural, but it may prevent the very suffering that animal rights activists deplore.[81]

Hunting privileges the preservation of particular species. Individuals may be killed, but hunted species are managed – sometimes by being fed and housed, sometimes merely by being allowed to feed and reproduce themselves unmolested – so as to sustain them as a resource for future hunting. As William Taplin put it in 1772: 'It is most certain if there were more Shooters, there would be more Birds; this is as strictly true as the antient Adage that has been so often experienced, "The more Hounds, the more Hares."'[82] To this end, Taplin proposed that 'The Qualification should be extended to every Freeholder of Five Pounds *per Annum* in *England*', so that the game would 'have as many Protectors, as it has Enemies now'.[83] Taplin's end has been achieved by different means. In spite of pressure from human population and intensive agriculture since the Second World War – especially mechanized and inorganically regulated agriculture – and in spite of the erosion of food supplies and habitats for wildlife at a pace that was simply unthinkable in the eighteenth century, there is no shortage of foxes or pheasants in Britain today.

Whether the fox-hunting lobby might wish to take up a more thoroughly conservationist position, returning to its roots in natural history and naturalistic field study, is an open question. What we can most usefully recover from Gascoigne's, White's, Cobbett's and Smith's legacy is the combining of empathetic and scientific endeavors, still practiced in naturalistic field study such as Macdonald's. They illustrate in different ways how to begin to reconcile an ethical rejection of instrumental thinking about nature with the continuing need of humans as a species to consume at least some natural resources, and how to balance the need for human intervention in nature with the recognition that humans too are subject to ecological forces.

2
Land, and Writing about Land

> My love of the country's abidin',
>   And Nature I'm always salutin',
> For when I'm not shootin' or ridin'
>   I'm huntin' or fishin' or shootin'.
>
> Charles, Viscount Harkaway, in A.P. Herbert,
>   *Tantivy Towers: A Light Opera in*
>   *Three Acts* (1931)

There is no doubt that between 1671 and 1831 the landed classes wished to derive maximum benefit from forests, commons and their own estates. Like crops and timber, game animals might be regarded as merely so much plenty to be harvested. It has thus seemed highly unlikely to previous surveyors of this field that a conservationist sensibility might be found among landowners or fanciers of field sports. But the seeking of recreational pleasure as well as profit from the land has a long, if neglected, history, and the perpetuation of field sports could be said to have countered the drive for agricultural improvement in some respects. Pursuing the 'Countrey Contentments' of hunting, coursing and shooting, landlords produced a landscape less single-mindedly devoted to intensive agricultural production than it would otherwise have been. Preserving deer, foxes, hares, pheasants and their habitats was a form of conservation, of environmental management not devoted to maximum extraction. Hunting a country, in all its scents and stinks, its blood and other effluvia, meant taking an interest in and intervening in the ecological balance of a piece of country on behalf of what would today be called its leisure amenities. And what were the pleasures of hunting, precisely? Might they not have included a knowledge of and sensitivity to local topography, flora and fauna, even an obsession with their preservation, that modern conservationists would approve of, if not envy?

This chapter examines the landlords' contribution to the invention of the English countryside as a sporting Elysium, and alternatively, as a painterly landscape. Both these amenity uses of the land ran counter to the drive for productivity and profit known as the Agricultural Revolution. We will investigate changes in land use and the English landscape itself, but also shifts in representation of and writing about land, for the two are so intimately bound as to be nearly inseparable. When husbandry and poetry, the most elite genre, were considered fit companions, it was possible for landowners, farmers and laborers to share a literary aesthetic, and there is some evidence for this until late in the eighteenth century.[1] Urbanization and the influence of metropolitan taste changed both the aesthetics of landscape and writing about land. Dung disappeared from the literary landscape, as did the scents and odors of hunting and sporting culture.

The invention of the countryside has been marked by progressive exclusivity. Like the pleasures of field sports, the benefits of agrarian improvement were more widely and equably distributed in the seventeenth century than they would be in the eighteenth or nineteenth centuries. As Nigel Everett opines, 'Emparkment seems to have had some of the same emotional appeal as the game laws in the pursuit of privileges and those "marks of opulence" noted by [Adam] Smith, which others do not and generally can not enjoy.'[2] But that quintessential signature of English landownership – the landscaped park devoted to pleasure rather than use – also signifies a quintessentially English art[3] and a public resource which the culture of rambling has made potentially available to anyone.

The poetry of agrarian improvement

Throughout the late sixteenth, seventeenth and eighteenth centuries, population pressures dictated, and an ethic of experiment and industry enabled, keen attention to agricultural improvement. In the sixteenth and early seventeenth centuries, 'men made war upon the forests, moors, and fens with a zeal which they had not felt for some three hundred years.'[4] Deforestation occurred at such a pace that as early as 1575 George Gascoigne found it unnecessary to say much of deer-hunting in what he called the 'high woods' made up of 'great trees' because 'men take such order for high woodes nowe adayes, that before many yeres passe, a huntesman shall not be combered with seeking or harboring an Hart in highe woodes'.[5] In 1785 in *The Task*, William Cowper mourned tree-felling on country estates, grateful that a local landowner, John Courtney Throckmorton of Weston Underwood, allowed him access to woodland:

> Ye fallen avenues! once more I mourn
> Your fate unmerited, once more rejoice
> That yet a remnant of your race survives.
>
> (Book 1, 'The Sofa', ll. 338–40)[6]

In Book 6, 'The Winter Walk at Noon', Cowper sympathized with wild animals whose only refuge from their human predators was the fast-vanishing remnant of ancient forests: 'The wilderness is theirs with all its caves, / Its hollow glenns, its thickets, and its plains / Unvisited by man. There they are free' (6: 401–3). The identification with animals articulated by Cowper and others in the last two decades of the eighteenth century (see Chapter 5) might profitably be connected with a general anxiety regarding deforestation and the loss of wild commons and other habitats.[7]

Before 1750, the population of England seemed unable to expand beyond between five and six million, a maximum reached in the early fourteenth century and again in the mid-seventeenth century.[8] The sheer length of the period of agrarian capitalization, from the late sixteenth to the mid-nineteenth century, which tackled this problem of food supply, can best be grasped by means of Robert C. Allen's identification of two agricultural revolutions in English history, the yeomen's and the landlords'. The yeomen's is the radically early one for which Eric Kerridge has argued.[9] Although its legal basis was laid in the sixteenth century, it was mainly a seventeenth-century technological revolution brought about by yeomen and small farmers, the benefits of which were widely distributed. The landlords' revolution, consisting of enclosure and farm amalgamation, began in the fifteenth century but happened mainly in the eighteenth, according to Allen, and brought profits to the large landowners but hardships to small farmers, cottagers and the mass of the laboring population.[10] Although during the eighteenth century the number of small farmers actually increased, the landlords' revolution nevertheless succeeded in bringing about 'the economic and social marginalization of the small cultivators in English rural society', while pauperizing the most vulnerable.[11]

The mid-seventeenth century, the period of the Civil War and the political experiments of the Commonwealth and Protectorate under Oliver Cromwell, should be understood as a period of intense population pressure – a crisis of resources as well as of religion and politics. The rhetoric of plenitude – of crops, woods, game – to be found in so much seventeenth-century English writing should be carefully scrutinized. Early in the second decade of the seventeenth century, when Ben Jonson celebrated lordly entitlement at Penshurst, the domination of master and mistress over the natural bounty of their estate appeared complete. Yet, as James Turner reports, '[I]t is widely recognized that the condition of the rural poor was more terrible in the years 1620–50 than at any other time.'[12] The genre of the country-house poem had already begun to gather force, even as the social order it celebrated was vulnerable to attack. We have here a good instance of what Raymond Williams has characterized as 'the inherent dominative mode':[13]

> Each bank doth yield thee conies, and the tops,
>   Fertile of wood, Ashour and Sidney's copse,
> To crown thy open table, doth provide
>   The purpled pheasant with the speckled side;
> The painted partridge lies in every field,
>   And for thy mess is willing to be killed.
>
>   (ll. 25–30)[14]

This slavish devotion of the natural world to its human owners both typifies the pastoral poetry of rural retirement and foreshadows many an imperial adventure. Rabbits, trees, pheasants and partridges deliver themselves to the Sidneys, the owners of Penshurst, as produce ready and willing to be consumed. The estate represents a great repository of plenty to be harvested for the asking. Such a rhetoric of natural bounty and bloodless consumption should be read alongside early seventeenth-century anxieties about the ratio of population to agricultural productivity and the consequent drive for improvement. A painless harvest gathered through sport is the ultimate wish-fulfillment.

The georgic, unlike the pastoral, recognizes the need for both human labor and consumption of the natural world. The georgic is the pre-eminent genre of the landlords' revolution, though its rise to favor pre-dated the mid-eighteenth century, and georgic verse was popular with yeomen and laborers.[15] In 1697, John Dryden announced in his translation of Virgil's *Georgics*, the originary poem of advice about husbandry, that it was 'the best Poem by the best Poet'.[16] As late as 1784, when discussing her favorite authors, Ann Yearsley, the 'milk-woman of Bristol', would cannily claim that 'Among the Heathens', she had 'met with no such Composition as Virgil's Georgics', thus simultaneously demonstrating her piety and her professional good sense.[17] The georgic middle style, with its mixture of high and low content and sublime and mock-heroic language, was located on the *scala Virgilii*, the Virgilian hierarchy of genres, above pastoral eclogue and below epic. This was the genre that best answered the needs of a mercantile society and an expanding British empire until late in the eighteenth century. As Karen O'Brien puts it, 'Georgic presented poets with an adaptable middle style that could rise to national prophecy and rapture or descend to technical detail without breaching generic decorum.'[18] Virgilian georgic could make the most mundane, even filthy, matters poetically viable, at least until *circa* 1789.[19] As Addison remarked, Virgil even 'breaks the clods and tosses the dung about with an air of gracefulness', and English poets were encouraged to do likewise.[20]

One of the pleasures of georgic verse seventeenth- and eighteenth-century readers undoubtedly experienced was recognizing the witty mixing of

Virgilian precedent with a knowledge of new farming procedures specific to English agricultural improvement. John Dyer's description in *The Fleece* (1757) of sheep feeding upon turnips, for instance, has been found risibly pretentious by modern readers for its use of periphrasis – elevated, latinate poetic diction appropriate for the georgic:

> ... with busy mouths
> They scoop white turnips: little care is yours:
> Only at morning hour to interpose
> Dry food of oats, or hay, or brittle straw,
> The wat'ry juices of the bossy root
> Absorbing; or from noxious air to screen
> Your heavy teeming ewes with wattled fence
> Of furze or copse-wood in the lofty field,
> Which bleak ascends among the whistling winds.
>
> (1: 483–91)[21]

John Goodridge has recently vindicated Dyer by pointing out how useful technical advice is contained within the periphrastic phrasing. Sheep require a balanced diet of roots, which have a high water content, and dry, bulky foods such as oats, hay and straw. 'What we now call lactic acid poisoning, overeating disease, enterotoxaemia and other such dietary disorders may all be caused by excessive root or concentrate eating,' Goodridge claims.[22] The best prevention of digestive illness in sheep lies in what Dyer describes as the shepherd's 'interposing' of dry foods to absorb the turnip's 'wat'ry juices'. Dyer's calling the turnip a 'bossy root' is, for Goodridge, a calculated gesture of concentrating meaning in a single phrase as only poetry can do, combining the 'turnip's physical substantiality (bossy, contrasting and alliterating with the "brittle straw"), and its discovered quality (root)' (p. 165). Modern readers should accommodate themselves to the difference of past taste:

> For those practically or intellectually involved in agricultural developments in the early modern period (including georgic writers and their readers), the turnip is no less than the silicone chip of the New Farming ... We may believe that the turnip is 'low', but Dyer reasonably supposed it to be in the very vanguard of cultural progress. (pp. 165–6)

For Cowper in *The Task* (1785), greenhouse-grown cucumbers were a comparable sign of that progress. Bringing home the georgic discourse of improvement to metropolitan suburban gardens as well as country estates, Cowper hoped to persuade his audience in Book 3, 'The Garden', that 'Who loves a garden, loves a green-house too' (3: 566). The imperial implications of Cowper's poem are subtle.

Beginning in the 1660s, greenhouses enabled exotic imported plants such as orange-trees to be domiciled in English gardens (*OED*). The very Englishness of the mixture of plants in garden and greenhouse was itself an imperial construction.[23] More successful than Dyer's *The Fleece*, *The Task* made Cowper phenomenally popular among a wide range of readers: gentry, professional middle and commercial lower-middle classes, tradespeople and artisans. Not only country gentlemen, but the families of 'bankers, shopkeepers, manufacturers, farmers, tanners, brewers, millers and clergymen' found 'succour and inspiration in Cowper, whether their politics were radical or conservative'.[24] Cowper's hymn to compost and cucumbers appealed to all:

> The stable yields a stercorarious heap,
> Impregnated with quick fermenting salts,
> And potent to resist the freezing blast.
> For 'ere the beech and elm have cast their leaf
> Decidu'ous, and when now November dark
> Checks vegetation in the torpid plant
> Exposed to his cold breath, the task begins.
> . . .
> The seed selected wisely, plump and smooth
> And glossy, he commits to pots of size
> Diminutive, well fill'd with well prepar'd
> And fruitful soil, that has been treasur'd long,
> And drunk no moisture from the dripping clouds.
> These on the warm and genial earth that hides
> The smoking manure and o'erspreads it all,
> He places lightly, and as time subdues
> The rage of fermentation, plunges deep
> In the soft medium, 'till they stand immers'd.
> Then rise the tender germs upstarting quick . . .
>
> (3: 463–9, 511–21)

For farming folk, even as recently as the 1930s, the coming spring was likely to be named, however unpoetically, ' "next dung-hauling" '.[25] Cowper gives us a good sense of the steamy presence of stable and farmyard muck: 'saturated straw' and 'smoking manure'. 'Stercorarious' means 'Of, produced by, living in, dung and faeces' (*OED*). Cowper's 'stercorarious heap' originates in dung; it is a great heap of straw mixed with dung and urine to form a rich manure or compost, that lodestone of the gardener's art from which all vegetable life springs. This word is both a characteristic instance of periphrasis, and an appeal to vegetable lovers and gardeners alike, who can now find their favorite activity memorialized according to neoclassical precedent. Thus some of the less salubrious realities of country life are miniaturized rather preciously for suburban consumption.

So long as georgic precedents sustained an aesthetic of agrarian productivity, poetry and husbandry continued a limited partnership. The tidying influence of metropolitan taste might occasionally be felt, especially in the presence of urban ladies, for whom dung had no obvious virtues and fewer charms. But commonly until 1789, and in a few instances after that, dung was not only 'the grand pillar of your husbandry', as Arthur Young put it, but a feature in the poetic landscape.[26] In *Five hundreth points of good husbandry* (1573), Thomas Tusser, voice of the yeomen's revolution, had advised managing the dungheap with an eye toward aesthetics as well as practicality. As Goodridge observes, Tusser uses 'short, anapestic tetrameters with a range of alliterative echoes and rhymes, to give pith and energy to his advice'.[27] The manure in the making was daily to be gathered up and stored, but handsomely, in a nicely rounded hill:

> Lay compasse up handsomely, round on a hill,
>   to walke in thy yard, at thy pleasure and will:
> More compasse it maketh, and handsome the plot,
>   if horse keeper daily forgetteth it not.[28]

In 1744, no longer trusting his audience's ability to deal with old-fashioned terms for fertilizing excrement, Tusser's editor added: 'Compass is Dung, of which the Yard should often be clean'd, that the more may be made; and whatsoever a Lady may think, a Farmer thinks Heaps of Dung a very good Ornament to his Dwelling' (p. 151). This obvious gendering, as well as plebeianizing, of dung and farmyard stench aligns femininity with the new polite urban culture, and implies that women of any social pretension will be ignorant outsiders in agrarian scenes. The same ideology will edge women out of both science and the hunting field by the end of the eighteenth century (see Chapter 6).

In 1782, the Quaker poet John Scott echoed Tusser in an eclogue on 'Rural Business; or, The Agriculturalists', in which the first bard advises:

> In vacant corners, on the hamlet waste,
> The ample dunghill's steaming heap be plac'd;
> There many a month fermenting to remain,
> Ere thy slow team disperse it o'er the plain.

The second bard chimes in by echoing Alexander Pope's description in the *Essay on Man*'s first epistle (1733) of how, to possessive and individualistic mankind, the universe seems to have been designed for humans alone:

> The prudent farmer all manure provides,
> The mire of roads, the mould of hedge-row sides;
> For him their mud the stagnant ponds supply;
> For him their soil, the stable and the sty.
>
> (ll. 21–8)[29]

Speaking from the point of view of human 'Pride', Pope had answered the question of 'Earth for whose use?' with such lines as: 'For me, the mine a thousand treasures brings; / For me, health gushes from a thousand springs' (ll. 132, 137–8). Scott narrows the compass to the farmer's possession of all he surveys as potential manure. There is richness and abundance to be found in the country, so long as one seeks fertilizer. Nothing is to be wasted, not even stagnant mud or hedgerow mould or stable and pigsty muck.

Bloomfield's *The Farmer's Boy* (composed 1798; published 1800) represents the death-knell of this tradition. When farm worker Giles first ploughs, then harrows, fields in 'Spring', a split between the ploughman and his audience is already discernible:

> From ridge to ridge the ponderous harrow guides;
> His heels deep sinking every step he goes,
> Till dirt usurp the empire of his shoes.
>
> (ll. 80–2)

Having dirt usurp the 'empire' of their shoes was precisely what self-styled ladies were to avoid when treading in country districts. The tidiness of the farmyard after dung-hauling has occurred in Bloomfield's poem would have pleased Tusser, but it was going out of fashion as poetry:

> At home, the yard affords a grateful scene,
> For Spring makes e'en a miry cow-yard clean.
> Thence from its chalky bed behold convey'd
> The rich manure that drenching winter made,
> Which pil'd near home, grows green with many a weed,
> A promis'd nutriment for Autumn's seed.
>
> (ll. 185–90)

There is still a sense in Bloomfield of the labor required to produce this image of fruitful tidiness. Even the rich smells of the farmyard might be experienced here through the image of lifting the 'rich manure that drenching winter made' from its 'chalky bed', dripping and ripe with odors as well as nutrients. Good stewardship of the land is visible in every green shoot.

The science of agriculture and the science of scent

Hunting with hounds, or the science of 'scent', like the science of horticulture or sheep feeding or manure production, was very much a science of which it could be said that 'the answer lies in the soil'.[30] In *The Chace* (1735), an entire georgic devoted to hunting, 'scent', or the odorous traces left behind by animals which could be perceived and pursued by other animals, was the technical basis of William Somervile's explanation for how modern hunting differed from ancient hunting. Proper improved hounds hunted by sense of smell, not sight. Somervile's passage on scent will be much cited by subsequent writers, constituting the core of later English hunting literature. Beginning to bring Book 1 of *The Chace* to a close, Somervile describes as a 'sagacious brute' (1: 341) the 'lime', or bloodhound used for harboring or tracking deer, and for detecting human predators along the Scottish borders. The keen-nosed hound hunts by scent, defined as follows:

> The blood that from the heart incessant rolls
> In many a crimson tide, then here and there,
> In smaller rills disparted, as it flows,
> Propell'd, the serous particles evade
> Through the open pores, and, with the ambient air
> Entangling, mix: as fuming vapours rise,
> And hang upon the gently-purling brook,
> There by the incumbent atmosphere compress'd.
> The panting chase grows warmer as he flies,
> And through the net-work of the skin perspires;
> Leaves a long streaming trail behind, which, by
> The cooler air condensed, remains, unless
> By some rude storm dispersed, or rarefied
> By the meridian sun's intenser heat:
> To every shrub the warm effluvia cling,
> Hang on the grass, impregnate earth and skies:
> With nostrils opening wide, o'er hill, o'er dale,
> The vigorous hounds pursue, with every breath
> Inhale their grateful stream; quick pleasures sting
> Their tingling nerves, while they their thanks repay,
> And in triumphant melody confess
> The titillating joy. Thus on the air
> Depend the hunter's hopes.
>
> (1: 364–86)

As in Dyer's passage on the turnip, or Cowper's on cucumbers, so also here: every aspiration of the English georgic is present in it.[31] Hunting is represented as a science, as well as an art. A proper attention to the land, and to the nation in all its glories, requires a knowledge of hunting. Hunting is both a natural and a highly developed and methodized practice. As in so much georgic writing of the period, the heart of this natural, but modern and improved, practice lies in the pursuit of 'effluvia'. As with Cowper's 'stercorarious heap' and Bloomfield's 'rich manure that drenching winter made', from which vegetable life springs, so too with scent, from which hunting draws its being. The country is meant to be apprehended with all five senses, and the sense of smell is a particularly crucial one.

In 1781 Peter Beckford (like Somervile, a New College man), revised Somervile's theory of scent. As with Cowper's 'English' garden, which was home to imperial imports, so also with Beckford's version of the typical sporting English squire. It was said that Beckford 'would bag a fox in Greek, find a hare in Latin, inspect his kennels in Italian, and direct the economy of his stables in exquisite French'.[32] He was the cousin of William Beckford of Fonthill, the eccentric author of *Vathek* (1786); both inherited Jamaican sugar fortunes.[33] In *Thoughts upon Hunting*, written while recovering at Bristol Hotwells from a bad fall,[34] Beckford gave it as his opinion that the soil, as well as the temperature and moisture of the air, was always crucial for determining whether there would be scent, or whether scent would lie so that hounds could follow it: 'Experience tells us, that difference of soil, occasions difference of scent; and on the richness of soil and the moderate moisture of it, does scent also depend, I think, as well as on the air.'[35] By arguing that 'the answer lies in the soil', Beckford brought Somervile even more fully in line with georgic precedent than the poet of *The Chace* had himself managed to do.

Today scent is explained by the differential *between* the temperature of the ground and that of the air; scent carries best when warm air rises and carries odors with it, and when there is fog, as opposed to when spiders' webs on the ground are coated in hoar frost.[36] Foxes in particular are a rich source of scents. There are at least six sources of fox odor. The one most commonly perceived by humans, according to David Macdonald, is the fox's urine, which fulfills many functions, from marking a territory – 'the odourous equivalent of birdsong' – to ' "bookkeeping" ' – sprinkling urine on empty food cache sites as a reminder not to waste time exploring them subsequently.[37] But there are five other sources of fox perfume as well: the tail gland known as the 'violet' gland, since its odor is reminiscent of those flowers, which is activated when the tail is raised or lashed in greetings; the anal sacs, which produce an acrid-smelling milky fluid sometimes left on fox droppings; fox droppings or 'billets' themselves; a set of glands in the skin around the chin and angle of the jaw, whose function has not been determined but is likely to be social; and finally, the scent glands 'in the pinkish skin between their toes and pads', which give off a 'pleasant, sweet smell' (p. 125). This is the scent that hounds follow.

Hunting a country in the sense that Somervile or Beckford intended requires the use of all five senses (not merely sight) and an intimate local knowledge of topography, plants and animals that would do a naturalist credit. Before the advent of modern fox-hunting, country gentlemen's hunting was not so much species-specific as a matter of going out at daybreak with hounds, looking to put up whatever game there might be. Hares, foxes, rabbits, deer: any of these might provide sport, and finding them required a knowledge of the country. 'Whereas the great aristocrats might hanker after deer, the country gentleman hunted anything that jumped up in front of his hounds and in England the most abundant quarry consisted of hares and foxes.'[38] The naturalistic knowledge required to hunt was never confined to the upper classes, though the gentry tried to reserve it for themselves, especially after 1671. Where country gentlemen hunted, lesser mortals followed. Between 1671 and 1750, they often did so with permission. After 1750, where once informal arrangements had eased the harsh legalities of the Game Act, these ceased.[39] During the century after 1750, the landlords' revolution hardened class divisions in country districts in concert with the drive for privatization of the land and maximum extraction from it.

Although enclosure wrought changes in hunting and shooting practices, coursing, hunting with hounds and poaching nevertheless preserved something of the sense of openness, of unboundedness, in the landscape that predated enclosure. Most profoundly and paradoxically, the hunted landscape remained one in which private property boundaries were blurred or overridden. This could mean the depredations of sporting landowners at the expense of their own tenants or other farmers. In 1573 Tusser protested against footpaths and 'hunting and hauking' as destructive of agriculture: 'With horse and with cattel what rode, / is made through every man's corne?' (p. 92). In *Agriculture* (1753), Robert Dodsley lamented how a farmer's ripened crop might be trampled by stag-hunters:

> In vain, unheard, the wretched hind exclaims;
> The ruin of his crop in vain laments:
> Deaf to his cries, they traverse the ripe field
> In cruel exultation; trampling down
> Beneath their feet, in one short moment's sport,
> The peace, the comfort of his future year.
>
> (3: 75–80)[40]

Members of the sporting culture often literally rode over other people's interests and livelihood. But the blurring of private property boundaries in hunting could also mean that the characteristic English landscape had to appear both open and closed, both champion for sporting and enclosed for agriculture, both open to the freeborn, liberty-loving Englishman exercising his rights of way and common, and closed by a park pale against threats to social order.

The closed yet open landscape, and the ideology of the landscaped park

Celia Fiennes began her manuscript of travels in 1682 by noting that Newton Toney in Wiltshire, where her mother lived, was 'all on the downs a fine Champion Country pleasant for all sports – Rideing, Hunting, Courseing, Setting and Shooteing'.[41] The field sports of the seventeenth century, Fiennes tells us, were most enjoyably pursued in an open, unenclosed, not very heavily wooded country, a champion or champaign country. Downlands signifies high grasslands, ancient pasture, not plowed fields. There could be good gallops here, and an uninterrupted view of hounds or greyhounds going about their work, as in John Wootton's *Hare Hunting on Salisbury Plain*, painted for Henry, the third Duke of Beaufort in the late seventeenth or early eighteenth century (Plate 1). Field sports served both as an index to the richness and plenty of a place, answering to human utility, and as a sign of liberty to indulge in recreation or diversion: a little allegory of English freedom.

The champion is the very landscape of medieval open-field mixed agriculture which would be most dramatically altered during the sixteenth- and seventeenth-century enclosures, and then again during Parliamentary Enclosure Acts between 1750 and 1850, when the extent of new hedges planted equaled all those planted in the previous 500 years.[42] Other kinds of country had already been enclosed for centuries,[43] especially the grass-growing woodland and upland areas of the North and West of England, where a pastoral economy survived with fewer noticeable changes to the landscape than occurred in the champion, mixed corn- and grass-growing lowlands of the South and East.

In 1573 Tusser advocated the 'several', or enclosed, over the champion landscape, and promoted enclosure as conducive both to increased productivity and to the delights of private property:

> More profit is quieter found,
>   (where pastures in several be:
> Of one silly aker of ground,

then champion maketh of three.
Againe what a joy is it knowne:
when men may be bold with their owne.

(p. 93)

As Andrew McRae notes, Tusser's individualism, directed at the industrious and thrifty husbandman or smallholder, is taken 'to an extreme at which any of the traditional claims that impinge upon the farmer's sovereignty are depicted as a form of theft'.[44] Communal access to common land, exercise of common rights or the collective labor of fellow commoners Tusser represents as unprofitable. As Tusser's editor of 1744 stated categorically:

> [F]or where there is a great deal, what is every Body's Care, is no Body's Care; for it is not only the Shepherd, the Ox-boy and the Poor, but Farmers and Gentlemen will filch from one another, form pretended Privileges out of bad Customs, such as Foot-paths, Sheep-drifts, Privilege of Hunting and Hawking. (p. 92)

Guaranteeing private property rights guaranteed the prosperity of all, while hunting and hawking, and the champion country so amenable to these recreations, were distinctly disruptive of individualistic good husbandry.

Fiennes's preferred landscape of open rolling downlands over which to ride, hunt, course and shoot was the very one condemned by Tusser. It was also the landscape that Horace Walpole would dismiss as old-fashioned in 1771. The dominance of the eye over all the other senses, and the elevation of pictorial criteria above other criteria, such as hunting and gathering possibilities, marked a revolution in land use packaged as a change in landscape aesthetics:

> Since we have been familiarized to the study of landscape, we hear less of what delighted our sportsmen-ancestors, *a fine open country*. Wiltshire, Dorsetshire, and such ocean-like extents were formerly preferred to the rich blue prospects of Kent, to the Thames-watered views in Berkshire, and to the magnificent scale of nature in Yorkshire. An open country is but a canvas on which a landscape might be designed.[45]

Thus speaks the voice of the Town, of metropolitan London taste. By the 1770s, 'ocean-like' open prospects were out of favor with pictorially-minded critics who were not sportsmen and women. By the early nineteenth century, 'a fine open country' would be less prized by the proponents of field sports themselves, as hunting with hounds and shooting underwent profound changes. And the drive toward agricultural efficiency, begun in the sixteenth and seventeenth centuries but reaching new levels of intensification in the later eighteenth and nineteenth, would seal the fate of the old open-field agriculture and its 'ocean-like extents' once and for all. The little allegory of English freedom that was hunting, coursing, and so on, would not perish, however. It would be transformed.

When Walpole described William Kent leaping the fence to institute a natural, informal kind of English gardening, he captured the spirit of English sporting culture in an age of enclosure. For Walpole, the 'capital stroke, the leading step to all that has followed', was the finessing of boundaries by making them appear to disappear:[46]

> [Kent] leaped the fence, and saw that all nature was a garden. He felt the delicious contrast of hill and valley changing imperceptibly into each other, tasted the beauty of the gentle swell, or concave scoop, and remarked how loose groves crowned an easy eminence with happy ornament, and while they called in the distant view between their graceful stems, removed and extended the perspective by delusive comparison. (4: 138)

Walpole had attributed the first stage in this innovation to Charles Bridgeman, who created a new form of boundary between the grounds immediately surrounding the house and the land beyond those grounds. As Walpole described it, the 'common people' called the new 'sunk fences' 'Ha! Ha's! to express their surprize at finding a sudden and unperceived check to their walk' (4: 137). Once again the seemingly natural is highly ideological. The estate grounds were bounded and exclusive, yet did not look it. If Bridgeman had introduced the ha-ha, Kent fully exploited its possibilities. Leaping the fence himself, Kent encouraged the surrounding country, however wild or worked, to be incorporated into his design, the whole of the countryside to be regarded as a visual pleasure-ground.

Like the attack upon the open champion landscape, once favored by sporting ancestors, as looking too glaringly open, the description of landscape here is attributable to criteria derived from painting, especially the landscapes of the Poussins, Salvator Rosa and Claude Lorrain.[47] Viewing landscapes pictorially could literally mean seeing them through Claudean eyes. By looking through a Claude Glass, a convex mirror, the viewer miniaturized the reflected landscape and minimized its details or possible deformities.[48] We might, however, also read in Walpole's language a rebellious relation between a younger generation of intellectual 'familiarized to the study of landscape' and previous generations of sportsmen. We need look no further back than Horace Walpole's father, Sir Robert Walpole, Britain's first prime minister, to find a 'sportsman-ancestor' whose interest in a 'fine open country' would have been considerable, whatever his devotion to enclosure as a mode of agricultural improvement. At Houghton, the family estate in Norfolk upon which he lavished the rewards of office, Sir Robert had a ha-ha installed in the Bridgeman manner (4: 137), kept deer and 'engaged in extensive emparkment (shifting the local village in the process)'.⁴⁹ The *DNB* reports that he opened letters from his huntsman before any other correspondence, kept a pack of harriers at Houghton and a pack of beagles at his house in the New Park, Richmond, where he hunted one day in the middle of the week and on Saturdays, thus initiating the modern Parliamentary weekly holiday. In short: 'His recreation was in field sports' (*DNB*).

The younger Walpole subordinated opportunities for sporting to pictorial criteria. Everything was to be tailored for the eye of the spectator on foot, so that he might have the experience of walking in a landscape composed as the greatest painters composed their landscapes. Or might that image of Kent leaping the fence carry within it a trace of the sporting mentality? Are we not invited to ride the landscape in our mind's eye? To feel and taste the undulations of the land as a gentle swell or concave scoop experienced on the back of a horse capable of leaping any likely obstacle? The best writing on the relations between painting, landscape and poetry has assumed a walking viewer only, but given that landscaped parks were so often transformed deer parks, that landowners were most often riders as well as walkers of their estates, that field sports were so crucial to the gentry's self-conception and identity, why not admit into the calculus that landscapes were ridden in imagination as well as walked?

Ecology and the landlords

What might hunting landlords have contributed to rural ecology? Certainly, they have helped shape the rural landscape. In 1825, Clare wrote, 'Went to see the Fox cover in Etton field sown with furze some years ago which now present a novel appearance & thrive better then on their native heath tho the place is low ground.'⁵⁰ Both the thinly spaced trees in the hedgerows of the east Midlands and the gorse and wooded fox coverts of Leicestershire and Northamptonshire are attributable to fox-hunting. As W.G. Hoskins notes:

> In Leicestershire organized fox-hunting developed during the 1770s, in time to enjoy the exhilaration of galloping over miles of unfenced country. Enclosure made things more difficult, or perhaps we should say necessitated new and exciting skills, but at least there were no close ranks of trees to make the fences impossible.⁵¹

So those sparsely wooded hedgerows, with well-spaced ash and sometimes elm trees, are a legacy of field sports. We might conclude that 'but for enclosure, foxhunting would never have become as popular a sport', as John Patten puts it. Where once Salisbury Plain 'was reckoned by some to be as good a country as it was open' – certainly the opinion of Celia Fiennes – the new landscape of hedges and ditches 'became the idealised one for sport',[52] and flying fences at racing pace became all the rage.

Good arable farmers grubbed up natural gorse patches, says Hoskins, and it was only hunting landlords who took over, fenced and preserved 'odd pieces of common land, old cow-pastures that had been allowed to get out of hand', in return for a cash payment to the holders of common rights.[53] Examples of these include Ashby Pastures, Thorpe Trussels and Cossington Gorse in the Quorn country in east Leicestershire.[54] 'These "gorses" filled up the odd corners of parishes, and may be quickly spotted on the Ordnance map,' according to Hoskins.[55] Then there are the planted coverts, 'fenced around by large fox-hunting landlords like Aylesford or Sir Francis Burdett, the radical politician, who hunted in Leicestershire in the 1820s' (p. 197). One-inch maps of the east Midlands 'are splashed all over with these shreds of green, usually distinguishable from true, ancient woodland by their small size and their regular shape' (pp. 197–8). Hoskins cannot help observing that these coverts are a 'very noticeable feature' of the actual landscape, 'for they are often the only clump of trees anywhere in sight over thousands of acres' (p. 198).

One need not be a field sports enthusiast to recognize how fox-hunting and pheasant shooting led to the cultivation of landscape features that would otherwise have been swallowed up in the drive for agricultural profits. Tom Williamson and Liz Bellamy write:

> Whatever we may feel about the morality of killing animals for pleasure and about the psychology of the people who do it, hunting has clearly had a considerable impact on the landscape of rural England. Fox covers, pheasant covers, deer parks, rides, forests and warrens were introduced for the specific purpose of fostering rural sports and many of these features are still extremely common. Indeed, the continued importance of shooting is directly responsible for the survival of vast quantities of woodland in southern England and of moorland in the north.[56]

Elsewhere Williamson speculates that 'shooting was arguably *the* most important influence on the style of the landscape park'.[57]

When improvements in guns made shooting birds on the wing practicable, pheasants became an especially desirable quarry. Believed to have been brought over by the Romans from its habitat on the River Rioni (*Phasis*) in Georgia, near the Black Sea, to adorn their English villas *Phasianus colchicus* (South Caucasian black-neck pheasant or Old English black-neck) 'shot up over the tree-tops like a rocket, its long tail flaunting, its cocketting cry an incitement to the sportsmen below'.[58] By the eighteenth century, the Chinese ring-necked pheasant, *Phasianus colchicus torquatus*, had also been introduced; by 1766 it appeared in Thomas Pennant's *British Zoology*.[59] The habitat needs of pheasants largely explain the layout and planting of the landscaped park, which was itself usually a transformed deer park. The prevailing look of eighteenth-century English country estates – houses commanding landscaped parks – can be understood as a transformation of medieval deer parks from functional places into prospects satisfying to the eye.[60] Launcelot 'Capability' Brown – ' "I see your park has *great capabilities*", he would say to a prospective customer'[61] – is the landscape gardener with whom modern historians of gardening most strongly associate the levelling, earth-moving emparking of estates. Walpole describes Brown as a 'very able master' and the successor to Bridgeman and Kent.[62] According to Nigel Everett, Brown signifies the Whig approach to land management, in which 'Little is left to nature, accident, or discretion; the end of improvement is the clear distinction of personal property from the common, the rustic, the public.'[63]

Brown's designs for landscape parks, still visible at such places as Blenheim and Burghley,[64] emphasize the domination of the country house over the surrounding prospect, though, like his predecessor Kent, he attempted to naturalize this domination. Walpole thought that the modern English taste in gardening had indeed achieved this look of unadorned naturalness. English superiority to other nations was therefore as natural – and indisputable – as the greenness of English verdure:

> We have discovered the point of perfection. We have given the true model of gardening to the world; let other countries mimic or corrupt our taste; but let it reign here on its verdant throne, original by its elegant simplicity, and proud of no other art than that of softening nature's harshness and copying her graceful touch. (4: 147)

Improvement here is rendered transparent; it is a mere copying of the natural, an informal introduction of a few creature comforts to what was already there.

Conifer plantations of between 15 and 20 years old were considered particularly attractive; that is to say, they were attractive to pheasants, especially if they contained the larch, 'on account of its branches growing nearly at right angles from the stem: this renders the sitting position of the birds very easy'.[65] Once mature, these plantations required underplanting with shade-resistant species to achieve a shrub layer of cover for the birds – 'hence,' according to Williamson, 'the vast quantities of bird cherry which appear in many eighteenth-century accounts, and the great masses of rhododendron planted in belts and clumps in the nineteenth century ("the crowning plant for game cover", in the words of Alexander Forsyth)'.[66] No less exotic than the pheasant, the rhododendron, imported from the Himalayas, owed its rapid spread to the 'Pheasant Imperative'.[67] Since pheasants were creatures of the woodland edge, small clumps or long, thin strips of woodland with sinuous or scalloped edges were preferable to large plantations.

This trace of functionality within the design of the landscaped park should not be slighted, since it suggests that considerations other than purely aesthetic ones continued to operate within the new landscape gardening. The famous serpentine line, popularized by Pope at Twickenham and imitated everywhere, may have begun as an affront to Continental geometry, but it had other uses as well. The sinuous or scalloped edges of belts of woodland may owe as much to pheasants as to aesthetics. That look of shaved lawns gradually giving way to wilder verdure, so distinctively English according to Walpole, is the mark of gamekeeping as well as picturesque taste. Williamson concludes that once we have described the laying out of an estate's grounds as a shooting reserve suitable for pheasants, convenient for owners and guests, and mindful of anti-poaching security measures, we will have described the essential design of the eighteenth-century English landscaped park:

> One logical arrangement would be to have a number of small blocks – let us call them *clumps* – scattered around the landscape. A more extensive area of woodland, planted in the form of a thin linear strip (perhaps with wavy or scalloped edges) could be placed on the periphery of the reserve: let us call it a *belt*. Where drives ran in from the outside world it would be sensible to erect a lodge, to control access and to act – in effect – as a base for security guards engaged in permanent surveillance against the incursions of poachers.[68]

Demonstrations of landlordly power through exercising sporting rights thus effected environmental conservation. Preserved woodlands and heaths, gorse coverts and hedgerows, parkland clumps and belts of trees: all these features of the rural landscape we take for granted today can be attributed to the mania for field sports.

It is no coincidence that recent complaints from environmental conservationists about alien species of plants they wish to see eradicated from the British countryside target precisely those plants, such as rhododendrons, that were widely planted to provide habitats for game. A recent photograph of volunteers 'rhodie-bashing' for the Woodland Trust is captioned, 'Rhododendrons, originally introduced as cover for game birds, stifle the growth of native species'.[69] Although only certain species were selectively preserved, and those included imported aliens, promoting wildlife and habitats remained high on the sportsman's agenda, even when such preservation of game conflicted with rigorously productive arable farming practices. Those seeking to rid the landscape of all traces of field sports will have a formidable task ahead of them. As Oliver Rackham observes, the fox, having arrived in Britain by natural processes in prehistoric times, is as native as the oak tree.[70]

Between 1671 and 1831 it became increasingly obvious that not only landed gentlemen could evaluate a prospect. Although these new observers might not own any acres themselves, their landlessness did not prevent them from becoming 'familiarized to the study of landscape', as Walpole put it. Readers of the *Spectator* had been assured of this since 1712. Gilpin and other theorists of the picturesque consolidated the tradition. Domestic tourism meant that picturesque theory could have a democratizing effect, with tourists invading landscaped parks literally as well as figuratively. The movement meant that the new walkers in search of botanical specimens or picturesque views were often mistaken for poachers in search of illegal game.

3
Game and the Poacher

> And I give you my word
> That a sensitive bird –
> A point for our foolish reproachers –
> Prefers its career
> To be stopped by a peer
> And not by unmannerly poachers.
>
> The Earl of Tantivy, in A. P. Herbert,
> *Tantivy Towers: A Light Opera in
> Three Acts* (1931)

Landlords divided society into themselves – qualified sportsmen – and everybody else – poachers to a man, or woman. Most English people were thus inclined not to regard poaching as a crime. Except for the organized, heavily armed poaching gangs of the late eighteenth and early nineteenth centuries who poached exclusively for the blackmarket, poachers were not even thought to be 'unmannerly'. The greatest field naturalists were often poachers. Poaching was a national pastime.

Game legislation described

Between 1671 and 1831, game legislation generated more printer's ink and more outbreaks of violence than did reform of the franchise.[1] As a 'Gentleman of Lincolns-Inn, A Freeholder of Middlesex' remarked in 1771, 'One would almost be tempted to think, from the peculiar attention the Legislature hath in all ages paid to the preservation of the Game, that there was some sovereign medicinal quality in the blood and juices of these animals.'[2] What were the legal technicalities? The Game Act of 1671 raised the property qualification to lords of the manor and those who had a substantial income from landed property: £100 yearly from a freehold estate, or leases of 99 years worth £150; one could also qualify by being the son and heir of an esquire, or by owning parks, warrens, chases or free fisheries (22 & 23 Chas. II c. 25). As William Blackstone observed with some asperity, this qualification meant that 'fifty times' the property was required 'to enable a man to kill a partridge, as to vote for a knight of the shire'.[3] This remained the basic qualification act until that of 1831 (1 & 2 Wm. IV. c. 32).[4] The Game Reform Bill of 1831 qualified all holders of game certificates to kill game, subject to the law of trespass. A game certificate could be purchased for stamp duty of £3.13s. 6d. This Act made game for the first time the property of the owner of the land on which it was found. Sporting privileges remained the prerogative of landlords, rather than of occupiers or tenants.[5] Not until the Ground Game Act of 1880 was passed were tenant-farmers given the right to shoot rabbits and hares over their land.

According to the strict definition that applied between 1671 and 1831, not all beasts of chase were game, only hare, partridge, pheasant and moor fowl. Deer and rabbits were not game, but still far from being trivial objects of pursuit. Ever since the Middle Ages, there had been a property qualification for hunting deer and rabbits; under the Game Act of 1605, no person was permitted to take them unless he had at least £40 a year from land, or goods worth at least £200. But during the course of the seventeenth century, deer and rabbits began to be more and more enclosed in deer parks and rabbit warrens, and came to be viewed as private property under the law. The Game Act of 1671 thus omitted deer from its list of protected species, though it still included rabbits; but rabbits too were dropped from the list in 1692. They might still be pursued for sport or meat, or in the case of rabbits, the trade in skins, but they were no longer game.[6] Those who went after them without the permission of the landowner on whose land they were found were not so much poachers as thieves, and subject to much harsher punishments – say, transportation for seven years for deer stealing, as opposed to a £5 fine or three months in prison for game poaching.[7] Other beasts of chase – foxes, otters and badgers – were vermin. Dispatching them was regarded as a community service and required no property qualification.

The politics of game legislation

'Game stood for land' because it

> could be excluded from the legal marketplace of moveable property much more rigorously than land itself, than the great estates that were ideally transferred through inheritance but that in practice were often bought and sold. Although there was a thriving black market in game, it could legally circulate only as gift.[8]

By limiting access to sporting privileges, the game laws succeeded in privileging inherited rank and land over mere financial clout, which was becoming increasingly a matter of capital investments other than landed property as the eighteenth century wore on.

By legislating that game belonged to them as a class and not to the Crown, the Game Act of 1671 consolidated the interests of the landed gentry in at least two senses.[9] Before the Act, only the monarch had the prerogative to hunt where he pleased and to take whatever measures he saw fit to preserve the game. 'After 1671, he still had those rights – but now the gentry had them also,' as Munsche comments. The Act of 1671 allowed lords of manors to appoint their own gamekeepers; however, unless those gamekeepers were themselves qualified, they were not permitted to hunt simply by dint of the gentry's writ, not until 1707 when this last vestige of the royal prerogative was surrendered (5 Anne c. 14 s. 4).[10] We could thus understand the Game Act as part of the Restoration settlement agreed upon by the landowning classes that limited the monarch's power.

By reserving the taking of game for the wealthier gentry, the Game Act not only usurped royal prerogative on their behalf, but also extorted a certain revenge against forms of wealth other than land. The qualification for hunting during James I's reign had varied, depending on the game (for partridges and pheasants, freeholds worth £40 a year, leaseholds worth twice that amount, or goods and chattels valued at £400 or more; for hunting deer and rabbits, the qualification was higher, and for possessing hunting dogs and nets, slightly lower). This qualification represented an increase over the Qualification Act of 1389, passed in the wake of the Peasants' Revolt, which aimed to prevent those with lands worth less than 40 shillings a year from keeping hunting dogs, or using ferrets, nets, hare-pipes or other 'engines' to take deer, rabbits, hare or other game. But between 1610 and 1671, income from trade, stocks or offices, if ample enough, remained sufficient to qualify a man to course a hare or shoot a partridge. After 1671 this was no longer so.[11] A final battle of the English revolution had been fought and won by the men of landed property against commercially successful tradesmen and artisans, stock-brokers and financiers, political office-holders and great urban merchants.

Outside the legal edifice lay the world in which, 'whatever the law said, most rural men hunted, hunted because they needed to or because they believed that it was their right'.[12]

Game, politics and ecology

Both the revolutionary experiments of the mid-seventeenth century and the gentry's Restoration backlash against them could be understood as responses to ecological as well as political crisis. Within agrarian history, a flash-point had been reached: 'Hitherto the theory about gentlemen's exclusive rights to game, and the practice whereby all classes hunted it, had remained happily divergent; now they were converging uncomfortably.'[13] Many developments contributed to this phenomenon: the Royalist gentry's determination to enjoy themselves after years of deprivation during the Civil War and Interregnum; the knowledge they had acquired in exile on the Continent regarding the pleasures of horsemanship and the advantages of strict regulation of game preservation; the new economy of leisure which made pleasure a serious business; new grain surpluses for feeding game and horses for hunting and racing; and increasing pressure, thanks to enclosure and improvement, on uncultivated lands – waste and woodland – formerly available for hunting.

The game laws both signaled ecological crisis and provided strategies for dealing with it. The scarcity of game animals meant that the gentry wished to reserve them for themselves, while punishing those upstarts with income fron non-landed property who had exerted newfound power during the Civil War and Interregnum – merchants, tradesmen, artisans. The proto-communist experiment of the Diggers on St George's Hill still inspired fear. In terms of social ecology, or the analysis of the interrelations of human with animal needs, the deprivations of the poor increased exponentially as improving enclosers and engrossers of common land included wild animals among their privatized perquisites.

The mutual entanglements of country and city

The view that 'members of the landed interest did not share in the spoils of the fiscal-military state is nonsense', according to John Brewer; the gentry and aristocracy invested in public funds and securities and held lucrative state offices.[14] Nevertheless, this diversification of income went hand in hand with an increasing snobbery toward finance and trade. 'It is no coincidence,' Brewer observes, 'that, in a period when office-holding became a vital support for those who were genteel but disadvantaged either by birth (because they were a younger son) or by poverty, the snobbery towards "trade" grew apace. By the late eighteenth century the landed classes had fully developed that exquisite condescension towards trade which was hardly present a hundred years before.'[15] Hardly present? We can locate a form of that condescension in the institution of the Game Act of 1671. Adherence to the game laws figured largely in the construction of an anti-trade, self-consciously landed identity among the gentry, often in direct contradiction of the facts of a family's wealth having derived from involvement in financial or commercial ventures.

1 George Morland (1763–1804), *The Benevolent Sportsman* (1792). By courtesy of the Syndics of the Fitzwilliam Museum, University of Cambridge.

2 Thomas Gainsborough (1727–88), *Mr and Mrs Robert Andrews* (c. 1748–50). By courtesy of the National Gallery, London.

3 George Stubbs (1724–1806), *Laetitia, Lady Lade* (1793). By permission of The Royal Collection (copyright) 2000, Her Majesty the Queen Elizabeth II.

4 John Wootton (1682/3–1764), *The Bloody Shoulder'd Arabian* (1724). By courtesy of private collection.

5 Jacques-Laurent Agasse (1767–1849), *Sleeping Fox* (1794). By courtesy of the Oskar Reinhart Collection, Winterthur, Switzerland.

Thomas Gainsborough's painting *Mr and Mrs Robert Andrews* (c. 1748–50) (Col. Pl. 2) represents just such an assertion of identity as the second agricultural revolution, the landlords' revolution, was getting underway. Michael Rosenthal observes that Robert Andrews having himself painted with his farm in the background, rather than the more traditional landscaped park, marks him as one of the new breed of gentry improvers: '[W]e are meant to admire his progressive farming, lines of stubble showing that he has used a seed-drill, as well as his possessions: his dog and his wife.'[16] Rosenthal could have added to this list of possessions Andrews's flintlock gun, and the suggestion of potentially bagged game on his wife's lap. Noting that Frances Andrews wears a 'round-ear' cap under a milkmaid's straw hat, the height of fashion during the 1730s and 1740s, and that Robert Andrews's slouching pose and shooting jacket mark him as even more fashionably informal and rustic, Ann Bermingham comments, 'Quite simply, the Andrewses can afford to be themselves.'[17]

But this look of studied rusticity, this naturalizing of country landownership, becomes more complicated in the light of certain financial arrangements. Read in its historical context, Gainsborough's painting depicts not so much the representative of a country gentry dating from time immemorial as the scion of 'gentlemanly capitalism', in P.J. Cain and A.G. Hopkins's phrase.[18] As Cain and Hopkins argue, 'The peculiar character of the modern British aristocracy was shaped by merging its pre-capitalist heritage with incomes derived from commercial agriculture.'[19] Between 1688 and 1850, City of London financiers and providers of services, great overseas merchants and merchant bankers in effect apprenticed themselves to the landed interest, thus creating 'a form of capitalism headed by improving landlords in association with improving financiers who served as their junior partners'.[20] Robert Andrews's father, Robert senior, was a rich Scottish moneylender, who often lent money to landowners, and counted among his debtors Frederick, Prince of Wales.[21]

Metropolitan money owned the countryside. This social fact produced certain social antagonisms, rather than rural harmony. As Bermingham points out, by choosing to paint his subjects in a farm rather than landscaped park, Gainsborough risked exposing the very social conditions the Andrewses presumably wished to mask with their look of studied rurality. The painting holds in tension their landlordly leisure with the rural labor by which it was supported. The picture threatens to dissolve into a series of contradictions: 'the owner opposed to his property, his leisure to others' production, and his pseudo-hunting to their real gathering.'[22] We can be more precise than Bermingham and rephrase this last as the injustice of his shooting in relation to the laborers' real poaching. Some social unease or anxiety about the status of game literally erupts into an unpainted space in the picture, where bagged birds should have been in Mrs Andrews's lap.

Robert Andrews, himself a qualified sportsman, may have been entirely entitled by law to shoot or otherwise take game on his estate, but that privilege and the restrictions of older common rights it represents had become so socially contested that the sight of dead game was now nearly not respectable. It was approaching unrepresentability. 'Gainsborough may have intended to paint a pheasant shot by her husband in Frances's lap: this part of the picture is unfinished, only a shape being outlined,' observes John Hayes.[23] Bermingham concurs with Hayes, adding in a footnote, 'Since birds often carried erotic connotations, the dead pheasant might have seemed inappropriate and thus been left unfinished.'[24] Malcolm Cormack, describing Mrs Andrews 'nervously' clutching 'to her best silk dress what Gainsborough may have intended to be a pheasant', thinks it was a 'flash of common sense' on Gainsborough's part to leave the bagged bird unpainted.[25] Stephen Deuchar goes even further in terms of the aesthetic advantages of the unfinished space, claiming that the 'omission of game itself from the formula' of the couple's property ownership is particularly appropriate because it 'promotes the sense of potential yet to be fulfilled (for game has yet to be caught), sharpening rather than diminishing the spectator's admiration of their ownership of land and the benefits accruing from their legal right to exploit it'.[26] Anthony Vandervell and Charles Coles, however, prefer to see unpainted partridges. They report, 'It is said that the artist painted out a brace of partridges originally laid in Mrs. Andrews's lap. The detail shows the mixed farming of wheat and sheep, making for good partridge country.'[27] Pheasant or brace of partridges? The unpainted space is undecidable. It is literally a gap in the picture, open to interpretation. And it marks nothing so much as the endlessly debated place of the game laws in eighteenth-century English society.

The gentry's monopolizing of field sports, and a new anathematization of field sports by some intellectuals, can be seen as power plays within the gradual shift taking place in English society from making social distinctions within a minutely graduated system of rank to a broadly differentiated class system, in which economic and political power would become ever more inseparably interfused and detached from the country estate, except in so far as the estate remained an important means for displaying wealth, a continuing venue for conspicuous consumption.

Rights of common: a long history of struggle

The importance of rights of common for the laboring classes has remained a persistent theme throughout the histories of agricultural improvement and game legislation. In addition to grazing rights, rights to fuel, fertilizer, and building materials, the wild birds and animals of the common were shared by the poor. In many parishes, fowling, snaring rabbits, or even poaching the king's deer were common practices among the poor. Commoners appropriated:

> hares, fish, wood-pigeons, and birds' eggs; together with beech-mast from the copses, for their pigs; crab-apples and cobnuts from the hedgerows; brambles, whortles, and juniper berries from the heaths; and mint, thyme, balm, tansy, and other wild herbs from any little patch of waste.[28]

This subsistence economy was the one most vulnerable to disruption and destitution through privatized enclosure by engrossing landlords.

Where common fields and pastures, some of which were communally enclosed, persisted, so too did the common rights that accompanied them – well into the late eighteenth and early nineteenth centuries. According to J.M. Neeson, both sides of the published debate agreed that this final phase of Parliamentary enclosure 'would ensure labourers' complete dependence on a wage, and encourage the proletarianization of small farmers'.[29] This assumption of the effects of enclosure 'was so thoroughly worked into the social vision of both defenders and critics as to be beyond dispute'.[30] The only argument was whether to welcome or denounce this change.

Was the early nineteenth century the worst period for English smallholders and agricultural laborers? Remember Turner's assessment that the 'condition of the rural poor was more terrible in the years 1620–50 than at any other time'.[31] Andrew McRae has discovered a crisis of hardship even earlier, describing the 'years of dearth and famine around 1596–8' as 'one of the most intense rural crises of the early modern era'.[32] The prospect of agrarian suffering and unrest stretches back at least to William Langland's late fourteenth-century *Piers Plowman*, the last poem for some centuries in which we are allowed to see the country as a 'field full of folk' – lest they be about to riot against some injustice or other (though they seldom did).[33] The same receding horizon applies to any attempt to assign a single point of origin to controversies over hunting and other field sports, although there is good reason to concentrate on the years between 1671 and 1831. Tusser, intent on exhorting his readers to favor enclosure as early as 1557, was far from keen on hunting. The gentry of his time would probably have agreed, since many abandoned their deer parks and turned their land over to more profitable cattle grazing.[34] But not all agricultural writers took this attitude. In *A Way to get Wealth*, a collected edition of his works published in 1623, Gervase Markham published his guide to improved husbandry, *The English Husbandman*, alongside its recreational companion, *Countrey Contentments*. Like Tusser, Markham was one of the most popular authors of the seventeenth and eighteenth centuries. *Countrey Contentments* went through 14 editions during the seventeenth century alone.[35] Markham turns up as ' "the most celebrated author on farriery" ', and ' "the Alcoran" ' of northern and Scottish squires, according to the horsewoman heroine Diana Vernon, of Walter Scott's *Rob Roy* (1817).[36] As Roger B. Manning notes, Markham, who had 'scant regard for the Game Laws', never raises the question of who was legally qualified to hunt. He ignored hunting seasons and 'saw nothing wrong with hunting red deer the year round'.[37] Markham represents a yeoman's or smallholder's pro-field sports position that is as powerful a current in the writing of the land as Tusser's pro-enclosure opposition to it, and that became especially controversial after 1660.

The poacher in art and literature

Late sixteenth- and early seventeenth-century cony-catchers of broadside ballad fame are comic or sympathetic figures, who snare fair maids as well as rabbits, or only mildly sinister vagrants and cozeners like Shakespeare's Autolycus, 'a Rogue', in *The Winter's Tale* (1610–11).[38] The whole business becomes more serious after 1671, and even more so after 1750. Eighteenth-century and early nineteenth-century poachers stand on their dignity; they thieve because they have themselves been robbed. Like freeholders inside the purlieus of forests maintaining that they defended the Ancient Constitution when they 'resisted the efforts of the early Stuart monarchs to deny them hunting rights, which the Tudor kings had respected',[39] these later poachers haunt gentlemen sportsmen; they are shadowy doubles of the qualified. As James Hawker, the Victorian poacher, observed, 'If I Had been Born an idiot and unfit to carry a gun – though with Plenty of Cash – they would have called me a Grand Sportsman. Being Born Poor, I am called a Poacher.'[40] Gamekeepers represent a further ambiguity, poised between their landed employers and their fellow servants, licensed to do what in their own persons they could not do.[41] As Charles Kingsley remarked in 1862, 'a keeper is only a poacher turned outside in, and a poacher a keeper turned inside out.'[42]

George Morland's painting of 1790, *Tavern Interior with a Sportsman Refreshing* (Plate 2), captures some of this lower-class hostility toward the privileges of the qualified sportsman. As Stephen Deuchar comments, 'Tellingly, the covetous gaze of the "unqualified" men towards the tantalizing dead game is made behind its owner's back; they poach it, mentally – and . . . there is no doubt at all that they do so without right or consent, let alone the landowner's knowledge. Morland's painting, perhaps understandably, was not reproduced as an engraving for the open market.'[43] Deuchar objects that Morland's picture was not 'beneficial' as an intervention on behalf of reform of the Game Laws because it reintroduced the 'resentment' and 'connotations of class conflict' which reformers of the Game Laws were keenest to displace from their movement.[44] On the contrary, I think these connotations and this resentment were so well known as to be unavoidable, and the painting captures powerfully the talismanic appeal of game across this class divide. It problematizes the class divide itself by making the sportsman the sort of fellow who prefers common, down-to-earth, country pleasures to exclusivity or luxury. What separates the unqualified men from the qualified? There is hardly any difference in their dress or taste in entertainment. Yet we have to imagine that a gulf of entitlement through landed property separates them. This sportsman is downwardly mobile, fraternizing with the plebs, which might well have irked the unqualified men more than an attitude of exclusivity. Rubbing shoulders with them in the tavern, he rubs in their faces the minimal sartorial, but maximal social and economic, difference between them – metonymized by the dead body of a hare.

The game that Morland bags for us in this picture is evidence that game continued to exert a powerful appeal among the rural population regardless of rank. Notice the covetous longing with which the unqualified men look upon the sportsman's bag. Freely available meat for the table, or for a quick, lucrative turn-around on the blackmarket, caught in the traditional fashion or cleanly shot, remained a potent lure for the unqualified, from the very poor to the small landowner or tenant farmer. Game as protein caught according to common right rather than comparatively recent statute also had the added cachet of being one of the gentry's privileges. Aping the gentry's privileges while resenting and resisting the exclusivity of their right to have them is part of what made the English an 'ungovernable people' in David Underdown's sense of popular uprisings and rebellious disturbances throughout the early modern period.

In the eighteenth-century game laws and debates about them, we can see at work two notions of property – real and commercial – the one derived from common law, and the other from newer notions of contract law. We can also see the argument for private property working in two ways, at odds. By insisting on the property qualification for hunting, the game laws endorsed the system of private property in land. But they also undermined it, in that game, unlike the enclosed species of rabbits and deer, which were regarded as private property, remained mobile across private boundaries. No law could prevent hares from running through the hedge into someone else's fields. Thus game animals came to impersonate commoners, exercising their rights to perambulate and gather among the commons and open fields. And human commoners, when they exercised traditional rights of common, were branded as poachers, if not thieves. Poachers emerged in the historical record as the demotic doubles of the gentry who obsessively feared and prosecuted them.

If in earlier periods, as Roger Manning has claimed, 'unlawful deer-hunting was a symbolic substitute for war; following the suppression of the mid-Tudor rebellions between 1536 and 1569, it may also have become a symbolic enactment of rebellion',[45] this relation between poaching and political defiance continued until 1831 and beyond. Although born in 1836, five years after the repeal of the old Game Act, James Hawker found himself up against much the same social forces as had bedevilled poachers since the late seventeenth century. Looking back from the perspective of the later nineteenth century, he asserted:

> We Had no voice in making the Game Laws. If we Had i would submit to the majority for I am a Constitutionalist. But I am not going to be a Serf. They not only Stole the land from the People but they Stocked it with Game for Sport, Employed Policemen to Look after it . . . All my Life I Have Poached. If I am able, I will Poach till I Die.[46]

Cobbett reported seeing in 1823, in the village of Up-street near Canterbury, the following sign posted in a garden: 'PARADISE PLACE. *Spring guns and steel traps are set here.*' 'A pretty idea it must give us of Paradise to know that spring guns and steel traps are set in it!' he remarked. Cobbett was convinced the sign spelled new money – 'doubtless some stock-jobber's place' – because 'whenever any of them go to the country, they look upon it that they are to begin a sort of warfare against everything around them. They invariably look upon every labourer as a thief.'[47] Cobbett described the hanging of James Turner and Charles Smith at Winchester in 1822 as 'a thing never to be forgotten by me', memorializing and making popular heroes of these men who had been involved in armed battles with gamekeepers. 'POOR CHARLES SMITH had better have been hunting after *shares* than after *hares*,' he observed bitterly.[48]

A certain social radicalism combined with the knowledge of a naturalist has often distinguished the poacher. The association between poachers and gypsies in the eighteenth century was so powerful that even being seen consorting with gypsies regularly could bring about charges of night poaching, theft or damage to property – all capital offenses. As the Gentleman of Lincolns-Inn queried:

> [W]ho will believe that in the land of liberty it is at this instant a capital crime, maliciously to break down the mound of a fish-pond, whereby any fish shall escape, or to cut down a cherry tree in an orchard, or even to be seen for one month in the company of persons who call themselves or are called Egyptians.[49]

Here gypsies are an unstable, but seemingly self-evident, term: it doesn't matter within the legal discourse if these so-called Egyptians acknowledge themselves to be Egyptians, claim a gypsy identity, or if they are simply so-called by ignorant outsiders.[50] To be seen regularly in the company of vagrants thus classified as a separate race, those foreign dark-skinned brethren, those 'vagrant dwellers in the houseless woods' (1. 21) of Wordsworth's 'Lines written a few miles above Tintern Abbey', was itself a property violation, indicating one was up to no good, caught bloody-handed, deep in filth by association. Walk like an Egyptian, or with Egyptians, and you might find yourself accused of a capital offense.

Going about the woods and fields writing poetry set John Clare sufficiently apart from his fellow villagers that he was the victim of speculation, with some believing him 'crazd', and others putting 'some more criminal interpretations' to his rambles – that he was a 'night walking associate with the gipseys robbing the woods of the hares and pheasants'.[51] Clare became sufficiently 'initiated' into gypsy 'ways and habits' that he was 'often tempted to join them'.[52] Clare reported, having often been in their company, that he 'found them far more honest then their callumniators whom I knew to be of that description', and he learned from them how to live off the land:

> their common thefts are trifling depredations taking any thing that huswifes forget to secure at night hunting game in the woods with their dogs at night of which all are fish that come [to the] nett except foxes but some of them are honest they eat the flesh of Badgers and hedge hogs which are far from bad food for I have eaten of it in my evening merry makings with them.[53]

Clare worried constantly that he would be taken for a poacher, but these early associations were the basis of his field knowledge.[54]

> Took a walk in the field a birds nesting & botanizing & had like to have been taken up as a poacher in Hillywood by a meddlesome consieted keeper belonging to Sir John Trollop he swore that he had seen me in act more then once of shooting game when I never shot even so much as a sparrow in my life – what terryfying rascals these wood keepers & game-keepers are they make a prison of the forrests & are its joalers.[55]

Laborers loitering in rural places were assumed to be loitering with intent, and only poaching seemed a likely intention, not naturalizing or botanizing or writing verse. The 'bat-maker', or maker of children's shoes, Richard Buxton, who published a botanical guide, and his friend James Crowther, who was born in a Manchester cellar and never earned more than a pound a week, but became a leading figure in the Manchester Botanical Society,[56] rambled as much as 15, 20 or even 30 miles a day, looking for botanical specimens. They evaded gamekeepers assiduously, since if they had been caught, they would very likely have been tried for trespassing in pursuit of game or for fish poaching.[57]

Poaching sometimes formed part of a radical politics. Naturalistic fieldwork could be combined with social protest. As a modern self-styled poacher has phrased it, the application of 'natural cunning' to hunting a wild creature surreptitiously entails the double appeal of 'outwitting both the quarry and the representative of authority'.[58] Hawker supported the radical MP Charles Bradlaugh, who 'was not a Poacher of game but a Poacher on the Privileges of the rich Class', and predicted that the Labour Party would 'Rule the world one day'.[59] Hawker himself claimed that he poached 'more for Revenge than Gain. Because the Class poached upon my liberty when I was not able to defend myself' (p. 95). Hawker's journal is replete with a field naturalist's observations gathered from the country round Oadby in the heart of the most fashionable Leicestershire fox-hunting country:

> A Hare is very Sensitive to Smell or Sound. It don't Depend on sight so much. If you can see one coming straight towards you and you stood still, she would go straight by you, but if you only moved or drew your Breath Heavily, she would hear you in an instant. (p. 55)

This was the kind of knowledge indispensable for observing wildlife, and obtainable only from many an hour quiet in the field, not from zoological textbooks. 'The fox and stoat are curious Little Chaps. One eats a lot he Don't kill, the other kills a Lot he Don't eat,' Hawker reports (p. 56). 'Nothing on four legs kills as many head of game as the little Stoat' (p. 58). Natural history contains many lessons from which humans should learn, according to Hawker:

> A Pheasant, like Many Men, is very fond of Drink, but it must be Clean or He will not have it. What a Beautiful Example the Pheasant sets, for many Men will Drink any Filth if it only bears a certain Name. But if there is no Clean Water you won't find Pheasants. They will Roam till they find it. (p. 52)

Reports of poaching on 'shiney' or moonlit winter nights, as celebrated in the ballad of 'The Lincolnshire Poacher', made him laugh: 'For a man who knows his trade will select the Darkest night possible for his work. I have known men leave off when the stars appeared, for Rabbits are best killed by Feeling. A Dark Night, and a Dry wind that is not too strong, that is the night for killing

Rabbits' (p. 48). The ballad of the bold apprentice who took up poaching as a recreation looks highly fanciful from Hawker's point of view:

> When I was bound apprentice, in famous Lincolnshire
> Full well I served my master for more than seven year,
> Till I took up with poaching, as you shall quickly hear:
> Oh! 'tis my delight of a shiney night, in the season of the year.
>
> (1–4)[60]

Going poaching on a 'shiney' night was an adventure more likely to result in a battle with gamekeepers than a full bag.

So much for autobiographical testimony, which in the case of poachers was unsurprisingly limited. The literary poacher of the early nineteenth century was more romantic, but equally upright. In *Our Village* (1824), Mary Russell Mitford memorialized Tom Cordery as the very image of the bold but sensitive poacher. Officially a 'rat-catcher, hare-finder, and broom-maker', Cordery had poached with the best of them until 'the bursting of an over-loaded gun unluckily shot off his left hand'.[61] With his athleticism and knowledge of the wild, Cordery epitomized the independence of the heath, forest and hill-country people of north Hampshire:

> Never did any human being look more like that sort of sportsman commonly called a poacher. He was a tall, finely-built man, with a prodigious stride, that cleared the ground like a horse, and a power of continuing his slow and steady speed, that seemed nothing less than miraculous. Neither man, nor horse, nor dog, could out-tire him. (p. 167)

These are the very characteristics that distinguished Wordsworth and other notable walkers of the 1790s and early 1800s. Like proponents of footpaths and other rights of way, poachers continued to assert rights of common. Cordery striding across the landscape reminded landlords that a fine figure of a sportsman and a fine figure of a gentleman need not be synonymous. But Cordery went one better in being both knowledgeable and tender regarding animals. Mitford considered this trait common among poachers. At his 'uncouth and shapeless' cottage (p. 173), Cordery kept a menagerie of 'pheasants, partridges, rabbits, tame wild ducks, half tame hares, and their enemies by nature and education', 'ferrets, terriers, and mongrels' (p. 174). By erecting 'hutches, cages, fences, kennels, and half a dozen little hurdled inclosures' (p. 174), which to Mitford suggested in their picturesqueness Italy and the landscapes of Salvator Rosa rather than England (p. 173), Cordery managed to keep the peace among these warring species. He evinced that 'remarkable tenderness for animals when domesticated, which is so often found in those whose sole vocation seems to be their destruction in the field' (p. 174). Cordery represents those 'countrymen of the old school', who feel a 'paradoxical admiration, almost affection' for foxes 'characteristic of many who spend a lifetime trying to kill them'.[62] Tom Cordery is as alive to the natural beauty of a landscape as Mitford's narrator herself. One spot in particular, at the edge of the common, 'was scarcely less his admiration than mine', a 'high hill, half covered with furze, and heath, and broom, and sinking abruptly down to a large pond, almost a lake, covered with wild water fowl' (p. 172). The lake's being 'covered' with water fowl suggests a plenitude of wild species. The presence of abundant wildlife was crucial for this paradoxical ability to hunt or poach game in the field – it was there for the taking – while making pets of individual animals of the same species at home. Tom would exclaim with delight at the ' "pleasantness" ' of this scene, conveying the same feeling as would a poet or a painter (p. 172).

John Clare, Robert Bloomfield and the science of the greenwood

We can only regret the limited written evidence left by poachers themselves. Between 1671 and 1831, they are figures of fiction and ballad and legal document. Between 1800 and 1830, a sympathetic audience for rural laboring-class testimony solicited poems and memoirs from writers like Bloomfield and Clare, with the genre of autobiography edging out poetry.[63] We should not be surprised that poets who were agricultural laborers wrote sportingly about hunting and regarded the woods and fields with poachers' eyes. As William Cobbett observed in 1825, when poachers were being hanged unjustly and fox-hunting dominated the sporting press, 'The great business of life, in the country, appertains, in some way or other, to the *game*.'[64] Cobbett was sure that if he were to measure, within a circle of 12 miles, the amount of time spent by everyone over the age of ten talking during the game season 'about the game and about sporting exploits', 'it would amount, upon an average, to six times as much as all the other talk put together; and, as to the anger, the satisfaction, the scolding, the commendation, the chagrin, the exultation, the envy, the emulation, where are there any of these in the country, unconnected with *the game*?' (1: 281).

At the beginning of his publishing career, Clare twice signed himself 'John Clare A Northamptonshire Pheasant', a spelling suggestive of the close connection between 'pheasant' and 'peasant' in some country districts.[65] Butchers with game licenses in my part of Devon still crack Clare's joke in reverse. In game-preserving areas, the pheasant 'became as indigenous as the "peasant", whatever the social cost to the latter'.[66] But it was hunting rather than shooting that made its way into Clare's verse. Evidencing the intimate connection between the music of hounds, hunting and English poetry, some of Clare's earliest and latest poems were hunting songs. Indeed during Clare's lifetime the hunting verse of old ballads had undergone a comparatively recent revival through the tremendous and continuing popularity of Thomas Percy's *Reliques of Ancient English Poetry* (1765) and the sensation of Thomas Chatterton's (1752–70) medieval forgeries and suicide at the age of 17. Mindful of Percy's *Reliques*, which he had seen for the first time in 1820 and found 'the most pleasing book' he had 'ever happend on',[67] Clare indulged in a moment of aristocratic *faux* medievalism in 'Sports of the Field' (1820): 'Then mount on your saddle in doublet of scarlet / & bring in the hounds to the field' (ll. 7–8).[68] The rhythms of English songs and the cadence of hoof beats are inseparable. By composing 'The Milton Hunt', Clare sought to please one of his patrons, Charles Fitzwilliam, Lord Milton, but he was also pleasing himself:

> The Milton hunt again begun
>   Break[s] autumns dappld skys
> While yon red east its blushing sun
>   Awakens in supprise
> The bugle sounds away away
>   The chevy chace begins
> The praise the honours of the day
>   The hardiest hunter wins
>
> For blood bred steeds no reigns can check
>   & true scent nosed hound
> For sportsmen fearless of a neck
>   No chace is more renownd
> The echoing woods are all alive
>   The hounds are on the run
> Oer hedge & gate see how they drive
>   The daring routs begun
>
> The cracking whip & scarlet coat
>   Draws all eyes round em now
> Een startld giles puts in his note
>   & scampers from his plough
> The bugle sounds away away
>   The chevy chase begins
> The praise the honours of the day
>   The hardiest hunter wins.
>
>   ('The Milton Hunt')[69]

'It reads capital dont it,' wrote Clare in a letter to John Taylor in 1821, in which he had proposed a few revisions to achieve the version above. After cutting the original last four lines, which had been about his 'muse', and substituting a repetition of the last four lines of the first stanza, Clare wrote, 'I think it woud make a good hunting Song then & free me of being fond of the barberous sport.'[70] Clare is at best ambivalent about hunting and other field sports, usually taking the side of the hunted animal and identifying with animal pain. The most vivid lines in an earlier hunting song, 'To Day the Fox Must Dye', concern 'Old Reynolds' or 'Reynard' the fox, who 'reals & staggers as he goes / & drops his brush wi' fear' (l. 25, ll. 35–6).[71] The much later 'Hunting Song' from the Northampton Asylum years combines a gaily galloping rhythm – 'Tis the birthday of nature the foxhunters joy / So away to the fields, hark, forward, away –' (ll. 23–4) – with a similar instance of naturalistic observation and empathy with the fox:

> Hark-away to the woods, hark-away do but look
>   The fox has broke cover, and by the bent [spray]
> Drops his tail but a moment, to lap at the brook
> Then flies o'er the red russet fallows away
>
> (ll. 13–16)[72]

The fox's superb confidence at pausing to drink as he breaks cover introduces a casual, light-hearted note into the chase. Like Ann Mallalieu, Clare conveys a sense of privilege at being able to view the fox in such an intimate way.

As a laborer pursuing his 'Careless Rambles',[73] Clare was more than once accused of poaching. He might unthinkingly 'prog' a ball of grass in a hayfield, disturbing a mouse with 'all her young ones hanging at her teats' ([The Mouse's Nest], ll. 1–6), but he was a naturalist without being a sportsman. Like Cobbett, however, he regarded blood sports, from fox-hunting to badger-baiting, as part of rural life, neither totally condemning nor celebrating these pursuits.[74] In the battle between the shepherd and the fox, as represented by Clare, the fox who finds a badger's den to hole up in, 'safe from dangers way', turns the ritual of hunting upside down: 'He lived to chase the hounds another day' ([The Fox], ll. 27–8).

Clare recognized a kinship with Robert Bloomfield in class and taste, and commemorated him as a poet of the 'green', with more lasting value than could be appreciated by the literary marketplace: 'Thy green memorials these and they surpass / The cobweb praise of fashion' ([Bloomfield II], ll. 6–7). Bloomfield celebrated in his 'Hunting-Song' (1802) how fox-hunting opened up the workday agricultural landscape to an enactment of liberty through the music and speed of the chase:

> 3
> Ye Meadows, hail the coming throng;
> Ye peaceful Streams that wind along,
>   Repeat the Hark-away:
> Far o'er the Downs, ye Gales that sweep,
> The daring Oak that crowns the steep,
>   The roaring peal convey.
> 4
> The chiming notes of chearful Hounds,
> Hark! how the hollow Dale resounds;
>   The sunny Hills how gay.
>
> (ll. 13–21)[75]

Even laborers might exult in this outbreak of sport that reclaimed the landscape as a place not of labor solely, but of recreation too.

In 'The Forester', a poem from *May Day with The Muses* (1822), Bloomfield showed 'science' and the woodman's or forester's craft to be one and the same. The forester is at one with his environment, which is both a naturalist's and a sportsman's paradise:

> BORN in a dark wood's lonely dell,
>   Where echoes roar'd, and tendrils curl'd
> Round a low cot, like hermit's cell,
>   Old Salcey Forest was my world.
> I felt no bonds, no shackles then,
>   For life in freedom was begun;
> I gloried in th' exploits of men,
>   And learn'd to lift my father's gun.
>
> O what a joy it gave my heart!
>   Wild as a woodbine up I grew;
> Soon in his feats I bore a part,
>   And counted all the game he slew.
> I learn'd the wiles, the shifts, the calls,
>   The language of each living thing;
> I mark'd the hawk that darting falls,
>   Or station'd spreads the trembling wing.
>
> I mark'd the owl that silent flits,

> The hare that feeds at eventide,
> The upright rabbit, when he sits
>   And mocks you, ere he deigns to hide.
> I heard the fox bark through the night,
>   I saw the rooks depart at morn,
> I saw the wild deer dancing light,
>   And heard the hunter's cheering horn.
>
>   (ll. 1–24)[76]

Like the poacher, the forester acknowledges no law other than his own. His language is the language of the field naturalist. He learns the habits and habitats of different species intimately. Although he kills game, the forester is not only a predator. He consumes, but he also empathizes. Hunting cheers as well as satisfies human needs for survival in the wild wood. The poem's plot turns on the felling of a great oak in a storm. The deer come out to view the damage and to mourn the tree's passing. The forester thus claims for the deer, and for all 'forest-tribes', the power of 'language', of which humans are unjustly proud: 'Who then of language will be proud? / Who arrogate that gift of heaven? / To wild herds when they bellow loud, / To all the forest-tribes 'tis given' (ll. 81–4). He knows the wild animals' language intimately and empirically, whether subtle 'note' or terrorizing 'scream', and it is a more permanent language, and knowledge of it a more scientific knowledge, than the knowledge of human empires, nations and religions:

> Empires may fall, and nations groan,
>   Pride be thrown down, and power decay;
> Dark bigotry may rear her throne,
>   But science is the light of day.
> Yet, while so low my lot is cast,
>   Through wilds and forests let me range;
> My joys shall pomp and power outlast –
>   The voice of nature cannot change.
>
>   (ll. 89–96)

'[S]cience is the light of day', casting permanent illumination. The forester's science of the greenwood, of animal nature, might seem to sound too static here, as if nature had no history. 'All humane things are subject to decay,' Dryden opined in the first line of *Mac Flecknoe* (1682), and within a scheme of time far shorter-term and given to more rapid changes than evolutionary time. Ecosystems, though subject to change, are closer than human social forms to being the signs of permanence that Bloomfield's forester would have them be. As the scientific constructs of human observers, however, ecosystems – and humanity's proper relations with them – are also partly products of culture. Bloomfield's equation of the forester's green science with animal nature, acceptable when poaching and naturalistic knowledge were considered to be synonymous, would surely be disputed today.

One appeal of the new fox-hunting was that the pursuit of sly Reynard, that poacher of poultry and young lambs, mimicked the persecution of human poachers. Plunderer of hen houses, the fox in Somervile's *The Chace* is a 'felon' (3: 37). In Machiavelli's *The Prince* (1532) and in Ben Jonson's play *Volpone* (1605), the fox signified a clever, if devious, chap. '[L]iving by his wits', for centuries medieval Reynard 'was a sympathetic figure'.[77] In the 1790s, Reynard became generally known as 'Charlie Fox', named after Charles James Fox, the Whig leader with radical sympathies and sporting predilections.[78] Once again the beast of chase had political associations. The consciences of the landed interest were clear.

4
The Sporting Life

> Ther is a saying emong hunters that he cannot be a gentlemen whyche loveth not hawkyng and hunting, which I have hard old woodmen wel allow as an approved sentence among them. The like sayinge is that hee cannot be a gentleman whych loveth not a dogge . . .
>
> Anon., *The Institucion of a Gentleman* (1568)[1]

By the 1830s the tourist in search of the picturesque was likely to be a metropolitan man out of sympathy with sporting culture. In Melesina Bowen's poem *Ystradffin* (1839), a tourist meets a Welsh guide 'of a grade beyond the peasants, who usually attend on such occasions, both in rank and intelligence'.[2] The guide represents the poorer rural classes, and the tourist, metropolitan middle-class mobility. The picturesque had become by 1839 what it is today, 'an aesthetic for the increasingly affluent town and city dwellers'.[3] A difference of opinion soon arises between them regarding country sports. During their ride to Ystradffin from Llandovery, there is the sound of hunting in the distance, and the guide's melancholy, induced by memories of the late-lamented owner of the estate of Neuadd, vanishes when he hears hounds:

> . . . a distant sound
> Of Sportsman's shout, and yelping hound;
> Quickly dispers'd the unwonted sigh,
> Uplifts his head, and lights his eye.
>
> Hark to the dogs! I see them now
> Upon Penlifau's darksome brow;
> Oh! how I love to hear that cry!
> My spirit seems with them to fly;
> As o'er the hills their course we trace,

> Eager I feel to join the race;
> In Summer, Autumn, Winter drear,
> We find a healthful pleasure here;
> And tho' the Seasons change the Game,
> The jocund glee is still the same.
> Ah! there they go! and now they're gone!
> The Stranger spake with calmer tone.
>
> 'Doubtless such sports with some agree,
> With them I feel no sympathy;
> Nor with a Sportsman's eye behold,
> These lovely Vales, and mountains bold;
> Yet surely with a zest as true,
> As ever Sportsman met the view.'
>
> (pp. 27–8)

The tourist wishes to be thought a manly man, as full of 'zest' in his hunting after views as any sportsman, but his eye for a country is quite different from that of followers of hounds. Without actually mentioning bloodlust, the tourist manages to convey disdain for those who take pleasure in field sports. He implies that his 'sympathy' lies with hunted creatures rather than with hunters and hounds. In short, he devotes himself to the picturesque, to Nature and Art, not Sport, an attitude that began to be fashionable in the 1780s and 1790s, and had by the 1830s been embraced by a substantial middle-class and gentry minority, as well as some artisans and rural laborers. Sporting culture still flourished, but now there were alternatives to it. Charles James Apperley, the sporting journalist who wrote as 'Nimrod', recognized in 1835 that although he himself had 'but little taste for the picturesque', other gentlemen travelers were its devotees:

> Sportsmen are apt to look at a country with merely a sportsman's eye, as a friend of mine did on his road to Doncaster. 'What a beautiful country!' said one of his fellow travellers. 'Aye,' said he, ' 'tis a pretty country enough, but how the devil do they ride over it?' This I confess is my own case, having but little taste for the picturesque.[4]

Bowen's guide, on the other hand, is a countryman whose spirit 'seems to fly' when he hears hounds. Whether the season be summer, autumn or winter, there will be hunting of the appropriate beast of chase. Spring, as we know, is both dung-hauling time and lambing, calving and foaling time. As Thomas Tusser insisted, in spring farmers have too much work to do to pleasure themselves with hunting. Coming upon a hunt is for the guide, a man above the rank of peasant but still close to the rhythms of farming, a moment of exhilaration, of the unbinding or freeing of his subjectivity into 'jocund glee'. Tracing the hillsides in his imagination as he listens to hounds, he longs to follow them. Coming upon a hunt in progress gives this man intense pleasure. Hunting with hounds is the ultimate diversion from everyday cares, the ultimate recreation or pastime.

Even the Rev. Richard Warner, a pedestrian traveler and no fan of field sports, found himself impressed by a local man's description of hunting in the Welsh hills in 1797. Mr. David Pughe:

> gave us an account of those desperate chaces which dogs and men follow, through regions that no *lowlander* can behold without terror. The sport, however, to those who enjoy it divested of fear, must be most glorious and animating. The rocks and precipices re-echo the united sounds of huntsman, dogs, and horns, and a chorus is formed singular, striking, and indescribable.[5]

Rural sports

The eighteenth-century equivalent of 'recreation' was 'diversion', and Samuel Johnson defines diversion as 'Sport; something that unbends the mind by turning it off from care'.[6] Popular pastimes once known as 'rural sports' were associated with seasonal holidays such as May Day, Whitsuntide, Michaelmas and Christmas. They included a range of activities from church ales, Maypole dances, mummers' plays, football, stool-ball, wrestling, cudgelling and women's smock races, to sports involving animals, such as bull-, bear- and badger-baiting and dog- and cock-fighting. This calendar of 'rural sports' constituted what some historians call 'the old festive culture' of farming and village life before various forms of modernization rendered it obsolete.[7] In this traditional culture, work and recreation might be 'so closely related that they were almost indistinguishable', according to Robert Malcolmson. Poaching, for instance, 'was endemic in many places, and it blended together both profit and pleasure'.[8]

These sports were a matter of controversy throughout the seventeenth and eighteenth centuries. James I and Charles I encouraged traditional holiday pastimes by publishing the *Book of Sports* in 1618 and 1633 as an incitement to 'public mirth'. Milton's citified voyeur-exhibitionist in 'L'Allegro' goes in search of 'Mirth' in the country (l. 38), disdains the 'hairy strength' of the rural laborer (l. 112), and is content to hear hounds at a distance:

> Oft list'ning how the Hounds and horn
> Cheerly rouse the slumb'ring morn,

> From the side of some Hoar Hill,
> Through the high wood echoing shrill[.]
>
> (ll. 53–6)⁹

Rather than deploring hunting as a blood sport, Milton's persona treats it as background music, just another phenomenon of country life. The mirthful wanderer is not a thruster wishing to make a figure in the hunting field. He prefers instead to spend 'Some time walking not unseen / By Hedgerow Elms, on Hillocks green' (ll. 57–8). Not unseen by whom? Presumably by the 'Plowman near at hand', the Milkmaid, Mower and Shepherds (ll. 63–7). Milton's persona desires to be observed by working folk pursuing nothing more nor less than leisure.

There was a 'politics of mirth', as Leah S. Marcus has shown.[10] Royal, aristocratic and gentry patronage was crucial to the survival of popular pastimes. Seen from the top of the social pyramid looking down, they were designed to represent a benevolent paternalism in action, which was not without its self-interested side.[11] Seen from below, they represented a break in the yearly round of labor and care. Far from instancing social harmony, however, they may well have presented opportunities for staging symbolic forms of rebellion, or been forms of carnivalesque rebellion in themselves. From the 1590s onwards the cony-catcher as likeable rogue was a figure in popular as well as more elite literary forms.[12] Certainly poaching, and the refusal to give up what were perceived as traditional common rights to game, are evidence for what E.P. Thompson has called a *'rebellious traditional culture'*.[13]

The language of sport

The language of sport and its derivatives, 'sporting' and 'sportive', was thus rich in connotation and double meanings. Like 'venery', which denoted both the pursuit of beasts of chase, and the pursuit of sexual pleasure (*OED*), 'sport' and its derivatives signaled pleasant pastimes, whether 'amorous dalliance' or the pastime 'afforded by the endeavour to take or kill wild animals, game, or fish' (*OED*). Cony-catching and hare-hunting were particularly susceptible to erotic allegorization. Between the 1660s and the 1830s, however, a narrowing of usage occurred. The language of 'rural sports' shifted from referring broadly to the games and holiday pastimes of the Stuart *Book of Sports*. There was an increasing focus on the sports of field and stream at the expense of other rural diversions, which began to be classified as games, not sports. We can glimpse this split in the titles of two of William Somervile's poems, *Field-Sports* (1742) and *Hobbinol, or the Rural Games* (1760). The older sense of sport as producing 'mirth', in Leah Marcus's or Milton's terms, never disappeared. But by the end of the eighteenth century it had become almost impossible to speak of sport, sports or sportive without some trace of field or blood sports hovering near.

The *OED* gives William Wordsworth's lines from the 1798 poem, 'Lines written a few miles above Tintern Abbey' – 'Hedge-rows, hardly hedge-rows / Little lines of sportive wood run wild' (ll. 16–17) – as an instance of a version of the older meaning: 'disposed to be playful or frolicsome'.[14] The *OED* slightly misleads here, in that if to Wordsworth these hedgerows looked sportive, as in playful or frolicsome, they did so in close, even metonymic relation to the wild creatures who inhabited them, creatures who were very much the object of a sportsman's pursuit. The Wye valley hedgerows, which were 'hardly hedge-rows' at all in that they were so overgrown they hardly seemed to delimit field or property boundaries, not only promised shelter for sheep and thick cover for game and other animals, but metonymically substituted for the animals themselves.

Through metonymic substitution, the hedgerow habitat itself provides the poet with the vicarious thrill of the chase. Wordsworth cannot keep at bay the language of sport. Wordsworth's usage here precisely combines the mirthful or frolicsome with the more gentlemanly connotations of sportive. Never was a poet more haunted by the language and music of the chase who was also so determined to remove blood sports from his ideal countryside and his radically humanized theory of Nature.[15] In *The Prelude* (1805), Wordsworth recalls that during winter ice-skating in the Lake District, he and other local boys imitated 'the chace', 'the resounding horn, / The Pack loud bellowing, and the hunted hare' (1: 462–4). In capturing the 'awful voice' of a shepherd calling his sheep at nightfall in 'Home at Grasmere', Wordsworth compares the 'reiterated whoop' to a bird's call and to the voice of 'a hound / Single at chace among the lonely woods –' (l. 407, ll. 410–13).

That Wordsworth should have represented the music of hounds as characteristic of his native fells is hardly surprising, for his audience would have expected the sounds and sights of hunting there. English rural poetry and hunting had been deeply interfused for several hundred years when Wordsworth came to write. 'No musick can be more ravishingly-delightfull, then a Pack of Dogs in full cry to such a man whose heart and ears are so happy to be set to the tune of such charming instruments,' opined Nicholas Cox in 1674.[16] As Richard Blome put it in 1686, 'that no *Musick* is more charming to the *Ears* of *Man*, than a *Pack* of *Hounds* in full *Cry* is to him that delights in *Hunting*, is to tell you that which experimentally is known, and what hath been sufficiently treated of by others.'[17] The relationship between dogs and humans in English culture – a constant preoccupation – indexes important shifts between 1671 and 1831. This book concentrates upon hunting with hounds, and the challenge of anti-hunting polemics, but other field sports underwent related changes during this period.

Changes in field sports, 1671–1831: fishing, shooting, coursing

Until well into the seventeenth century, most fishermen thought that it made more sense to catch fish with nets, spears or osier baskets than with rod and line.[18] However, as early as 1615, Gervase Markham's popular *Countrey Contentments* advocated angling, anticipating Isaak Walton's royalist *The Compleat Angler* (1655) and Richard Franck's republican *Northern Memoirs* (1694). As with other field sports, the political meaning of fishing, substituting the 'piscatorial' for the shepherds' pastoral, could be equally appropriated by either side.[19]

During the eighteenth century, hawking and netting were largely replaced by shooting. Every sport's fan should have his day within a general sporting culture, recommended George Markland in the poem *Pteryplegia: Or, the Art of Shooting-Flying* (1717):

> Halloo – Halloo – See, see from yonder Furze
> The Lurchers have alarm'd and started Puss!
> Hold! What d'ye do? Sure you don't mean to Fire!
> Constrain that base, ungenerous Desire,
> And let the Courser and the Huntsman share
> Their just and proper Title to the Hare.
> Let the poor Creature pass, and have fair Play,
> And fight the Prize of Life out her own way.
> The tracing Hound by Nature was design'd
> Both for the Use and Pleasure of Mankind;
> Form'd for the Hare, the Hare too for the Hound:
> In Enmity each to each other bound.[20]

Markland came close to coining the phrasing of that first rule of modern shooting, 'Never, never let your gun / Pointed be at anyone.'[21] Markland's version is rather wordier: 'Th' unheeded Muzzle pointed at a Friend, / May instantly unthought Destruction send.'[22]

Shooting itself was transformed by improvements in the flintlock gun, and by the introduction of beaters forming a 'battue' to drive game, that it might be shot flying faster than could be achieved by putting it up with a dog in the traditional fashion.[23] Innovations in firearms technology began to pose the 'severest threat' to wildlife.[24] Alexander Forsyth's patenting in 1807 of the detonating or percussion principle to replace the unreliable steel spark and flint was 'generally regarded as the most important innovation in fire-arms since the original discovery of gunpowder'.[25] By the end of the Napoleonic Wars the copper percussion cap had been introduced; by the mid-1820s sportsmen finally had a weapon with powder that could not be dampened by rain. In

1832 Edward Newman, author of *A History of British Ferns and Allied Plants* (1840), enjoyed trying out this novelty, 'apparently with good effect, on the sea-birds of the Isle of Wight'.[26] Not until the mid-nineteenth century would breech-loaders, at first imported from France, begin to replace muzzle-loaders, which continued to be used by the 'less well-off'.[27]

Shooting game birds or hares was an activity explicitly restricted by the game laws. Cobbett opined in 1825 that there was 'an important distinction to be made between *hunters* (including coursers) and *shooters*'.[28] Avid game preservers were often anti-hunting. Cobbett was clearly of the hunting party, as we have seen, and especially keen on its democratizing effects. He found the shooting (and game-preserving) fraternity 'a disagreeable class, compared with the former; and the reason of this is, their doings are almost wholly their own; while, in the case of the others, the achievements are the property of the dogs' (1: 281), and the 'praises, if any are called for, are bestowed on the greyhounds, the hounds, the fox, the hare, or the horses':

> There is a little rivalship in the riding, or in the behaviour of the horses; but this has so little to do with the personal merit of the sportsmen, that it never produces a want of good fellowship in the evening of the day. A shooter who has been *missing* all day, must have an uncommon share of good sense, not to feel mortified while the slaughterers are relating the adventures of that day . . . (1: 281–2)

Hunting and coursing were animal-centered, according to Cobbett, and that was socially preferable to the human competition of shooting.

Hare coursing with 'gaze hounds', which hunted by sight rather than scent, was an ancient but also very English sport, according to George Gascoigne. Although the bulk of *The Noble Arte* (1575) was a translation of Jacques du Fouilloux's *La Vénerie*, Gascoigne added his own section on coursing because 'they have not that kynd of Venerie so much in estimation in France, as we do it here in England'.[29] Like hunting, coursing focused on animal achievements, but had its human uses. The landscape gardener Humphry Repton advised that landlordly paternalism could display itself effectively through encouraging coursing across the class divide. In recommending improvements to the grounds at Sherringham in Norfolk, Repton suggested that the local poor be admitted to the grounds during at least one month to collect firewood, and that coursing matches be held on the estate to promote goodwill among the villagers. In 1812, after two decades of war with France, Luddism was widespread in industrial districts, the price of wheat was high, and while Norfolk farmers prospered, farm laborers were scarcely able to buy bread.[30] In some districts, Repton had heard 'complaints that the neighbours are all idle thieves and poachers':[31]

With respect to the Game, which is every where, and particularly in Norfolk, the perpetual source of suspicion and temptation, I foresee that at Sherringham it will be one source of conferring happiness; for, there is a great difference betwixt shooting and coursing; one is a selfish, the other a social enjoyment. The villagers will occasionally partake in the sport like those where games of cricket or prison-bars are celebrated; thus promoting a mutual endearment betwixt the Landlord, the Tenant, and the Labourer.[32]

Repton, like Cobbett, found shooting 'selfish' and coursing 'social'. He was confident that Sherringham's owner, the Norfolk squire Abbot Upcher, who agreed with Repton's views on paternalist improvement, would agree that staging occasions for the display of sporting sociality was a small price to pay for 'a happy medium betwixt licentious equality and oppressive tyranny'.[33] Coursing could be said to have maintained something of its democratic or downmarket appeal right through the twentieth century. Greyhound racing in pursuit of electric hares is coursing's urban legacy in the present.

Although coursing underwent some criticism during the eighteenth century, in the contexts of changing attitudes towards animals, changes in hunting practices and controversy over the injustice of the game laws, it re-emerged by the end of the century as a popular spectator sport and betting venue. Celebrating hunting with packs of scenting hounds as a superior, more methodized sport than coursing, William Somervile objected to the overmatching of fast greyhounds to hares in *The Chace* (1735): 'nor the timorous hare / O'ermatch'd destroy, but leave that vile offence / To the mean, murd'rous, coursing crew, intent / On blood and spoil' (1: 226–9). Taken out of context these lines might seem to suggest that the hare was a problematical quarry for any hounds, but Somervile, himself a keen hare-hunter, was by no means critical of the hare as a beast of chase. He objected only to the kind of racing chase she was being subjected to when coursed. The science of scent was, according to him, such a tricky matter that hunting the hare with a pack of harriers gave the hare a much more sporting chance of escape.

Not long after the publication of Somervile's poem, James Seymour's *A Coursing Party* of 1738 (Plate 3) implied there was something rotten in the sport of coursing, period. In the context of the game laws, a certain squeamishness about taking game in this way had become thinkable, even within the sporting fraternity. The very way the hare hangs, so limp and dead, from the servant's hand, coupled with the servant's attitude of 'dutiful submission', as noted by Stephen Deuchar, may strike us with a feeling of unease.[34] The servant submits dutifully to a representative of the class which has both banned him from coursing himself and given him leave to handle game solely for the benefit of his qualified employer. And when we notice the expression on the face of the gentleman sportsman's female companion, we can be even more confident that a note of mute protest has been introduced. 'Look at all you've accomplished – pathetic,' she might be saying, with what we could call a speaking look.

By 1831, these paradoxes have been resolved in a more modern version of coursing than Seymour's, represented in W.H. Davis's *Colonel Newport Charlett's favourite Greyhounds at Exercise* (Plate 4). The greyhounds are depicted with their groom, Tom Bayliss, in the grounds of Hanley Court, Worcestershire, according to Elspeth Montcrieff, who also notes that by the mid-nineteenth century, there were over 350 coursing clubs in existence, with the national Coursing Club being founded in 1858.[35] Here coursing amounts to a display of the colonel's possessions, including his gentlemanly groom and elegant parklands, as well as the greyhounds themselves. As serious a betting venue as horse racing, coursing had ceased to be an informal pastime and become part of the entertainment industry.[36]

Thus coursing, like fox-hunting, became an increasingly public affair between 1671 and 1831, unlike shooting, which remained a private activity even when commercialized through the letting of shooting rights. It has been argued, however, that in another sense both coursing and shooting underwent comparable changes during this period. According to P.B. Munsche, both shooting and coursing ceased to be informal pursuits matching humans against animals, and became competitions between sportsmen, often held before an audience of spectators. No longer contests between hunters and their prey, by the late eighteenth century, coursing matches and shooting parties had reduced the game to 'little more than a moving target', while testing gentlemen's greyhounds and marksmanship, respectively.[37] Munsche speculates that this shift, to a more competitive and commercialized form of sport, might also be linked to limiting the participation of women:

> In the late seventeenth and early eighteenth centuries, it was not unusual for the wives and daughters of the gentry to hunt, hawk and net game . . . Later in the eighteenth century some women still rode to the hounds, but at coursing matches and battues they seem to have been spectators only . . . Perhaps . . . contests between women and game were acceptable in the eighteenth century, but contests between women and men were not.[38]

Coursing seems to have particularly appealed to women, perhaps because coursing hounds perform with comparatively little human intervention. In the Lake District, near Ullswater in 1698, Celia Fiennes decided to pause in her travels to have 'a little Course', proving that she was a sportswoman of the old freeborn hunting and gathering, not the privatizing, legislating school, defying the game laws shamelessly:

> [F]or 3 or 4 miles I rode through a fine forest or parke where was deer skipping about and haires, which by meanes of a good Greyhound I had a little Course, but we being strangers could not so fast pursue it in the grounds full of hillocks and furse and soe she escaped us.[39]

As her editor Christopher Morris remarks, 'Only someone of the most assured social position could have risked this defiance of the new and savage game laws.'[40]

Coursing remained a sport in which women took an intense interest in spite of pressure to function as spectators rather than participants. *The Sportsman's Cabinet* of 1803–4, a sumptuously produced and illustrated encyclopedia of dog breeds, was sponsored by several coursing societies in which peeresses featured prominently.[41] In 1824, Mary Mitford could still wax knowledgeable about coursing, discussing greyhound conformation and performance in graphic detail. But the following comments by George Proctor in *The Quarterly Review* suggest how tastes regarding female sportswomen were changing in the third decade of the nineteenth century. In reviewing *Our Village*, Proctor opined that he should have been better satisfied:

> to have found Miss Mitford less ambitious of astonishing us male creatures by her acquaintance with the mysteries of cricketing and coursing: it is very difficult for a lady to descant gracefully upon the athletic qualities of blacksmiths and ploughmen, the merits of batters and bowlers, of long stops and fielders, and the arithmetic of notches and innings. But it is against the unnatural amalgamation of the craft of the kennel with the light and tasteful pursuits of her sex, that we especially protest. The worrying of the poor timid hare should excite any emotion but pleasure in a female breast; and such technical jargon as the following passage, on the good points of a greyhound, is strangely unbecoming a female mouth ... We would earnestly recommend our fair friend to leave the qualities of the 'little bitches,' and the gross technicalities of the sports of the field, to her coursing acquaintance, the gentleman farmer.[42]

Coursing was coarsening, it would seem. Proctor appears to have been upset by Mitford's competence at judging the bodies of men and dogs, an integral part of her knowledge of sporting pursuits. Mitford neither blushed at canine anatomy nor failed to notice that sex made a difference to a dog. And she distinguished between the parlor and the field as spheres of activity and staked her preference firmly for the field:

> [W]hat a superb dog was Hector! – a model of grace and symmetry, necked and crested like an Arabian, and bearing himself with a stateliness and

gallantry that shewed some 'conscience of his worth.' He was the largest dog I ever saw; but so finely proportioned, that the most determined fault-finder could call him neither too long nor too heavy. There was not an inch too much of him. His colour was the purest white, entirely unspotted, except that his head was very regularly and richly marked with black. Hector was certainly a perfect beauty. But the little bitches, on which his master piqued himself still more, were not in my poor judgment so admirable. They were pretty little round, graceful things, sleek and glossy, and for the most part milk-white, with the smallest heads, and the most dove-like eyes that were ever seen . . . But, to my thinking, these pretty creatures were fitter for the parlour than the field. They were strong, certainly, excellently loined, cat-footed, and chested like a war-horse; but there was a want of length about them – a want of room, as the coursers say; something a little, a very little, inclining to the clumsy; a dumpiness, a pointer look.[43]

Did Mitford know how audacious her description of her farmer friend's greyhounds would seem to urban male readers? For a woman to sit in judgment on bodily beauty was disturbing. For her to dismiss distinctly feminine characteristics as inferior to masculine ones, as if she could choose her side of the anatomical line, and then follow, as enthusiastically as any sportsman, coursing dogs seeking to kill hares, was positively offensive. By 1824, educated women were meant to have distanced themselves from the gross physical pleasures of the field, leaving them to be enjoyed by gentlemen farmers. In previous centuries the presence of women had been assumed if not actively encouraged (see Chapter 6), just as literary and sporting cultures had been more or less integrated.

Political governance and the sporting ideal

In the seventeenth and early eighteenth centuries both kings and squires sought solace and exercise in the hunting field, and the women of their families did too. Field sports were the health-giving, honor-securing means of negotiating life in the country, away from business and affairs of state. In 1700 John Dryden opined that it was 'Better to hunt in Fields, for Health unbought, / Than fee the Doctor for a nauseous Draught.'[44] The Rev. Gerald Fitzgerald echoed Milton's *Paradise Lost* and Pope's *Essay on Man* in his praise of the health-giving effects of shooting in 1780:

> Seek the fair Fields, and court the blushing Morn;
> With sturdy Sinews, brush the frozen Snow,
> While crimson Colours on our Faces glow,

> Since Life is short, prolong it while we can,
> *And vindicate the Ways of* Health *to Man*.⁴⁵

Field sports also encouraged a healthy interest in close observation of nature. Only by venturing forth with dog and gun on a winter's morning, Fitzgerald insisted, would he have seen these sights:

> Heav'n! what Delights my active Mind renew,
> When out-spread Nature opens to my View,
> The Carpet-cover'd Earth of spangled white,
> The vaulted Sky, just ting'd with purple Light;
> The busy Blackbird hops from Spray to Spray,
> The Gull, self-balanc'd, floats his liquid Way;
> The Morning Breeze in milder Air retires,
> And rising Rapture all my Bosom fires.
>
> (p. 11)

Closely watched gulls presented Fitzgerald with one of his most felicitous poetic moments. 'Self-balanc'd' echoes Milton's description of the Earth's emergence at the moment of creation: 'And Earth self-balanc't on her Centre hung' (*Paradise Lost*, Book 7: 242).⁴⁶ Hovering medially in Fitzgerald's line as Earth does in Milton's, the 'self-balanc'd' gull is a learned allusion both pleasantly and naturalistically described, while 'floats his liquid Way' suggests at once the gull's movement and marine environment.

Hunting could be an enactment of political governance as well as a relief from it. Even the rural sports of country gentlemen, fox- and hare-hunting, were a means of securing public favor as well as a recreation from public office. Dryden asserted in the poem to his kinsman, John Driden, that local communities benefited from the exertions of hunting members of Parliament and local gentry as much as the nation did from the safety-valve offered to glory-seeking courtiers by means of hunting and killing royal deer – rather than royal heads of state. The 'fiery Game' (l. 58) of fox-hunting was especially suited to vigorous youth, and it provided both a community spectacle and a ritual, even sacrificial, but also legalistic, blood-letting. The 'wily' fox, whose cleverness medieval fables and Machiavelli had made exemplary, is figured as like a murdering 'Felon' among the flocks, who must be punished in the interests of the 'Common Good':

> With Crowds attended of your ancient Race,
> You seek the Champian-Sports, or Sylvan-Chace:

> With well-breath'd Beagles, you surround the Wood;
> Ev'n then, industrious of the Common Good:
> And often have you brought the wily Fox
> To suffer for the Firstlings of the Flocks;
> Chas'd ev'n amid the Folds; and made to bleed,
> Like Felons, where they did they murd'rous Deed.
>
> (ll. 50–7)

The fox must be hunted hard like a criminal, but the hare offers a more philosophical recreation, suitable for gentlemen of advanced years, and emblematic of how human and animal lives follow the same cyclic patterns, for the hare, 'after all his wand'ring Ways are done, / His Circle fills, and ends where he begun, / Just as the Setting meets the Rising Sun' (ll. 64–6).

Royalists had no monopoly on the imagery or symbolic resonance of hunting. The chase was a trope beloved of Leveller satirists in seventeenth-century newsbooks. In 1649, Marchamont Nedham turned history into the chase, cautioning that anyone who spoke out against the Cromwellian regime was in danger: 'Now must the *Beagles* go a *hunting* again, and I must be the *Hare*.'[47] Another pamphlet of the same year, attributed to the Leveller John Lilburne, employs the same trope: *The hunting of the foxes from New-market and Triploe-heaths to White-hall, by five small beagles, late of the armie, or The grandie-deceivers unmasked.*[48] In 1655 Andrew Marvell defended Cromwell as a huntsman chasing the beast of misrule, portraying his own muse as a hunt follower, 'hollowing' 'far behind / Angelique *Cromwell* who outwings the wind'.[49] As late as 1933, John Buchan would imagine Adam Melfort, the hero of *A Prince of the Captivity*, announcing during the First World War, ' "I know the Germans pretty well, and they like to hunt one hare at a time . . . I will give them a run for their money." '[50]

Coopers Hill as touchstone: hunting

The description of a royal stag chase in John Denham's *Coopers Hill* (first published 1642, revised 1655) was much imitated by later writers, and the meaning of the hunt within the political allegory of the poem has been much debated.[51] The sacrifice of the stag has often been assumed to refer to the execution of the King's adviser, the Earl of Strafford, in 1641. After the beheading of Charles I in 1649, the killing of the stag in later versions of the poem became associated with regicide. These interpretations emphasize the allegorical dimension of hunting, rather than the historicity of the practice. In 1641 Denham's narrator describes Charles I as going stag-hunting at Windsor for relief from the cares of state:

> Here have I seene our Charles (when great affaires
> Give leave to slacken & unbend his Cares)
>
> (ll. 243–4)[52]

But the dynamics of state power could not be escaped so easily, for those who made up the field in the King's train were hungry for praise and advancement. In 1655 the King was represented as:

> Attended to the Chase by all the flower
> Of youth, whose hopes a Nobler prey devour:
> Pleasure with Praise, & danger, they would buy,
> And wish a foe that would not only fly.
>
> (ll. 243–6)[53]

It would prove all too easy – as it had in 1649 – for the ambitious to substitute for a royal stag a 'Nobler', because human 'prey', a king made royal martyr, especially if he were to show fight.

In the most thorough investigation of the poem's initial composition, John M. Wallace has argued convincingly for Denham's first draft having been written at the close of the stag-hunting season on Holyrood Day, 14 September 1641. This is the 'poem's ideal date', according to Wallace.[54] We should read Denham's monarch of the glen not as a figure for royalist political martyrdom, in the shape of the recently executed Strafford, but rather as a personification of unchecked royal prerogative, 'arbitrary power, or tyranny, or "lawless power" as Denham called it in 1655 (line 326)' (p. 519). As a Parliamentary royalist, Denham wished to promote reconciliation between the Parliamentary and Royalist factions, to advocate peace, not civil war. By showing Charles I killing off arbitrary power, Denham was arguing at this early period in the conflict that 'Charles had already made the necessary concessions, and the time was ripe for the people to make theirs' (p. 524). By invoking the stag hunt on the day of the hunting season's close, Denham signaled his wish that the 'close season' begin on 'political stag-hunting' (p. 532). (Wallace notes that the modern stag-hunting season extends from 10 August to 10 October [p. 525].) By 1655, and publication of the revised edition in the wake of Civil War, the King's death and Cromwellian republicanism, it was clear that Denham's hopes for the preservation of a delicate balance through 'judicious concessions by both sides' had proved short-lived (p. 537):

> So fast he flyes, that his reviewing eye
> Has lost the chasers, and his ear the cry;

> Exulting, till he finds, their Nobler sense
> Their disproportion'd speed does recompense.
> Then curses his conspiring feet, whose scent
> Betrays that safety which their swiftness lent.
> . . .
> So when the King a mortal shaft lets fly
> From his unerring hand, then glad to dy.
> Proud of the wound, to it resigns his bloud,
> And stains the Crystal with a Purple floud.
> This a more Innocent, and happy chase,
> Than when of old, but in the self-same place,
> Fair liberty pursu'd, and meant a Prey
> To lawless power, here turn'd, and stood at bay.
>
> ('B' Text, ll. 263–8, 319–26)

'A king chasing a king, and only a king should kill a king', as Nigel Smith points out: by 1655 'a kind of political forgetfulness is taking place (true history is being ignored).'[55] Royalist nostalgia might make certain readers impatient with the awkward fact of Parliamentary regicide. They might well prefer to romanticize the royal martyrdom by invoking instead a Royalist rhetoric from days of yore regarding human kings dispatching kingly stags.

In the context of England's mid-century population crisis and the struggle over natural resources that occurred during this period, Denham's justification of the stag's death might take on another meaning as well. If the monarch of the glen must be seen to be dispatched justly, this justice might also figure as a rather desperate effort to assert a sense of plenitude – of deer and other beasts of chase, of game, of forests, of rich resources as yet unimproved and unspoiled. The poem, especially in its revised versions, might be seen as a national allegory in an ecological as well as political sense. Denham perpetuates in verse the traditions of royal stag-hunting which the upheavals of the Civil War had in fact decimated, along with the deer population. 'They came for venison and venison they would have,' as the Waltham Forest rioters proclaimed.[56] Royalists' parks, as well as royal ones, had been raided and destroyed during the war, and as early in the Restoration as November 1661, Charles II was importing red and fallow deer from Germany to replenish Windsor and Sherwood Forests.[57]

Men and dogs

Natural bounty remained intrinsic to the myth-making of later sporting culture, as did the fundamental unity of human wishes and animal destinies.

Nowhere was this identity of interests more strongly figured than in representations of sporting dogs. Gerald Fitzgerald waxed lyrical about setting forth to shoot on a winter's day in Ireland because, in addition to human companionship in the form of a fellow sportsman, the poet enjoyed the company of his faithful spaniel:

> And thou, dear Spaniel! Friend in other Form!
> Obsequious come, thy Duty to perform,
> Whose fond Affection ever glows the same,
> Lives in each Look, and vibrates thro' thy Frame.
>
> (p. 16)

Of all dogs, spaniels were thought to be the most 'loving and fond', as Blome put it.[58] Taplin in *The Sportsman's Cabinet* praised their 'inexpressible diffidence and solicitation of notice, accompanied by an aspect of affability, humility, and an anticipation of gratitude'.[59] Dogs for setting and netting or shooting were supposed to be trained exclusively by their masters to achieve this particular closeness that could look like cringing. 'The setting-dog has more continual and intimate relations with man, than almost any other of the species,' commented Taplin; man and dog could give each other pleasure, and in their 'mutual gratification' when 'the game is in the net' lay 'the very basis of reciprocal affection' (2: 176). Somervile described the discipline of the setter as crucial for success in netting birds – an older, more stealthy means of taking them than shooting, and one requiring considerable skill:

> My Setter ranges in the new-shorn Fields,
> His Nose in Air erect; from Ridge to Ridge
> Panting he bounds, his quarter'd Ground divides
> In equal Intervals, nor careless leaves
> One Inch untry'd. At length the tainted Gales
> His Nostrils wide inhale; quick Joy elates
> His beating Heart, which aw'd by Discipline
> Severe, he dares not own; but cautious creeps
> Low-cow'ring, Step by Step; at last attains
> His proper Distance; there he stops at once,
> And points with his instructive Nose upon
> The trembling Prey. On Wings of Winds upborn
> The floating Net unfolded flies; then, drops,
> And the poor flutt'ring Captives rise in vain.
>
> (*Field Sports* [1742], pp. 12–13)[60]

Awed by a severe discipline, the setter checks his desire to give tongue when he has scented game and waits for his master to deploy the net. Blome recommends 'cherishing' – offering words and gestures of praise[61] – the same term used in the training of horses by Michaell Baret, among others, who advises, regarding how 'to bring your Horse to a perfect and true Trot', that as soon as 'you feele him begin to take vp his body, andd treade shorter, which so soone as hee doth, immediately let him stand, and cherish him, that hee may the better conceiue wherefore hee was troubled.'[62] However instilled in the setter or spaniel by 'cherishing' and rebuking – in other words, however constructed by humans – this obsequious loyalty in the dog emblematized gentlemanly entitlement. Mr Andrews's spaniel, for instance, in Gainsborough's *Mr and Mrs Robert Andrews*, looks at his master adoringly, 'naturalizing that gentleman's position of power and authority by implying that the dog's master is someone to whom deference and obedience are naturally due'.[63]

If the fowling or shooting man loved his spaniels, setters, retrievers and other gun or net dogs, the huntsman or master of hounds loved his pack. Writing to his fellow Cambridge undergraduate, Horace Walpole, Thomas Gray captured the shambolic horsiness and dogginess of eighteenth-century provincial hunting countries when describing a visit to the establishment of his maternal uncle, Jonathan Rogers, at Burnham in Buckinghamshire:

> I live with my Uncle, a great hunter in imagination; his Dogs take up every chair in the house, so I'm forced to stand at this present writing, & tho' the Gout forbids him galloping after 'em in the field, yet he continues still to regale his Ears & Nose with their comfortable Noise and Stink; he holds me mighty cheap I perceive for walking, when I should ride, & reading, when I should hunt . . .[64]

The 'comfortable Noise and Stink' of a pack of hounds was usually appreciated in kennels. When every chair indoors was occupied by a dog, the self-styled hard man to hounds might well have been mistaken for an animal-lover.

As we will see in Chapters 6 and 7, the pleasures of hunting as hunting, rather than as, say, riding across country, having been given permission to trespass on private land, derived from watching and listening to hounds going about their work. Killing stags, hares or foxes might consummate the ritual, but only if hounds had got the best of the beast of chase. Thus the pleasure of hunting for the huntsman or keen follower of hounds has traditionally been not so much about killing as about identifying with the hunting animals' pursuit of the hunted ones. Triumph at the death of a fox-as-predator is the farmer's special pleasure. As a former Master of Foxhounds this century put it, not until he hunted a pack of hounds again would he expect to take pleasure in the death of a fox:

> In a future season, perhaps, I may again hunt a pack of hounds; then, but not till then, shall I wholeheartedly enjoy the death of a fox; for then the destruction of a brave, and beautiful, and clever creature will, to me, be insignificant beside the pleasure which it will give to the forty or so other creatures which will, once again, be a part of me.[65]

This contradiction is an aspect of the sporting life that has often been overlooked in twentieth-century discussions of hunting.

Even Edward Augustus Kendall, in the context of proposing animal rights, had to acknowledge the naturalness of dogs following their hunting instincts. 'I cannot help anticipating the time,' Kendall wrote in his dedicatory preface to *Keeper's Travels in search of His Master* (1798), 'when men shall acknowledge the RIGHTS; instead of bestowing their COMPASSION upon the creatures, whom, with themselves, GOD made, and made to be happy!'[66] During Keeper's adventures making his way home after separation from his master, canine loyalty and human cruelty are demonstrated. When Keeper comes upon a hare, he cannot help himself:

> He was interrupted during a few minutes, by a hare, that crossed his path; in pursuit of whom he traversed several acres of crisp and frost-whitened wheat. Having driven puss into a thorny thicket, whither he found it difficult to follow her, he gave up the chase, and returned with the haste of a truant to the road of his journey.
>
> Though this frolic had wasted a small portion of his time, and contributed to weary his feet, yet was it, on the whole, very beneficial to him. The violence of the exertion had warmed his frozen limbs, and he returned with renewed vigour to his path. (pp. 32–3)

Keeper would not have been a proper dog if he had not benefited from such an opportunity for a little course.

So much for qualified sportsmen: poachers loved their dogs no less, and depended upon them more, communicating with them so intimately that gentlemen came to envy them. Bampfylde-Moore Carew, 'the noted Devonshire Stroller and Dog-Stealer' of the 1740s, while at Tiverton School, joined forces with his school-fellows John Martin, Thomas Coleman and John Escott, in devoting themselves 'wholly to Hunting; and frequently used, by an Art known to themselves only, to steal Dogs, Hounds, Setters and others, from Gentlemen and other Persons in the Neighbourhood'.[67] Carew eventually joined the Gypsies, continuing to put his peculiar ability to communicate with dogs to profitable effect:

> *Bampfylde*, during the whole Course of his Travels, successfully followed the Trade of Dog-stealing, frequently making bold with the Hounds, Setting-Dogs, &c. of the Gentlemen where he came, which he sold again to other Gentlemen at a Distance therefrom; whereby he got abundance of Money, commonly selling a single Dog for several Guineas. (pp. 11–12)

Terriers and ferrets were essential to the poacher's art of stealthy netting, because useful for bolting rabbits from their burrows, but trotting purposefully at the poacher's side might often also be a leggy dog capable of coursing rabbits and hares – a lurcher, today one of the most popular dogs among non-farming country dwellers, especially women who enjoy long walks. *The Sportsman's Cabinet* describes the lurcher as originally a cross between a collie, or 'shepherd's-dog', and a greyhound, 'which from breeding *in and in* with the latter', has lost all but the shepherd's-dog's temperament.[68] Taplin devotes some of his most high-flown prose to this low-life denizen of the highways, sympathizing with a dog, which, having been '[p]revented by nature from every chance of dependent society with the great', 'calmly resigns himself to the fate so evidently prepared for him' and becomes the ideal poacher's dog because of his great intelligence and responsiveness to subtle commands, coupled with speed (2: 102):

> [A] single glance of the eye is sufficient; and by that alone he frequently understands the signal of his [master's] will, and proves himself all zeal, all ardour, and all obedience ... [I]n nocturnal excursions he progressively becomes a proficient, and will easily and readily pull down a fallow-deer so soon as the signal is given for pursuit; which done, he will explore the way to his master, and conduct him to the game subdued, wherever he may have left it. To the success of poaching they are every way instrumental, and more particularly in the almost incredible destruction of hares; for when the nets are fixed at the gates, and the wires at the meuses, they are dispatched, by a single word of command, to scour the field, paddock, or plantation, which, by their running mute, is effected so silently, that a harvest is soon obtained, in a plentiful country, with very little fear of detection. (2: 103–4)

A modern self-styled poacher, Brian Vesey-FitzGerald, has insisted that the lurcher is not so much the poacher's dog as 'the Gypsy's dog', for only 'Romani-Tinker' or gypsy dog-trainers can effect the kind of obedience just described.[69] James Hawker, wishing he himself were a gypsy, would have agreed: 'Gipsies don't often have Noisy Dogs. Still waters run Deep.'[70]

Read in the light of this sagacity and loyalty of poachers' or gypsies' dogs, Wordsworth's description of the 'rough Terrier of the hills' with whom he consorted during his summer vacation in Book 4 of *The Prelude* (1805) acquires new meaning. Born to vermin-catching, the terrier was adopted by the Wordsworths to perform a 'gentler' service, poet-minding:

> By birth and call of Nature pre-ordained
> To hunt the badger and unearth the fox
> Among the impervious crags; but, having been
> From youth our own adopted, he had passed
> Into a gentler service . . .
>   . . . this Dog was used
> To watch me, an attendant and a friend
> Obsequious to my steps, early and late,
> . . .
> And when, in the public roads at eventide
> I sauntered, like a river murmuring
> And talking to myself, at such a season
> It was his custom to jog on before;
> But, duly, whensoever he had met
> A passenger approaching, would he turn
> To give me timely notice, and straitway,
> Punctual to such admonishment, I hushed
> My voice, composed my gait, and shaped myself
> To give and take a greeting that might save
> My name from piteous rumours, such as wait
> On men suspected to be crazed in brain.
>
> (4: 86–120)

A terrier is not a lurcher, but the dog's watchfulness and quick-witted warning of the approach of passers-by mimic the lurcher's usefulness to the poacher. Ironically, it is Wordsworth, not the dog, who is cautioned and admonished and signaled to. Wordsworth is 'worked' by the terrier, not the terrier by Wordsworth. Wordsworth, unlike the laborer Clare, is not worried about being mistaken for a gypsy poacher or prosecuted as a vagrant in his wanderings about the public roads. He worries only about being thought mad, since composing verse caused him to talk aloud to himself. In this respect, the terrier covers for him in two ways – by warning him of approaching folk, thus giving him time to compose himself, and by giving him the appearance of a proper countryman, even a sportsman, since accompanied by a dog. Might we not be tempted to think that Wordsworth would have preferred to have been thought a poacher, purposeful if pernicious, rather than a madman?

The identification with dogs rather than with birds, hares or foxes is an aspect of the culture of field sports often ignored in discussions of hunting and shooting today. Cruelty to foxes is the anti-hunting battle cry, but the satisfaction of hounds' natural, though encouraged and disciplined, desire to hunt is rarely addressed. (Anyone who has walked dogs in open country and encountered hounds on the line of a fox will know how stimulating such an event is to a dog.) This is in some ways an odd phenomenon for modern observers not to notice. The keeping of pet dogs might be said to represent a trace memory of former times. Dog walking is like a specter of sporting practice haunting metropolitan Britain, especially when dog owners seek to get away from it all and go to the country. For the dog trotting by the owner's side, though more usually regarded as an excuse, or a necessary spur, for taking a walk, is nothing so much as an obsolete hunting machine.

5
The Origins of the Anti-Hunting Campaign

> Fearful of the hare with the manners of a lady,
> Of the sow's loaded side and the boar's brown fang,
>
> Fearful of the bull's tongue snaring and rending,
> And of the sheep's jaw moving without mercy,
>
> Tripped on Eternity's stone threshold.
>
>   Staring into the emptiness,
> Unable to move, he hears the hounds of the grass.
>
>   Ted Hughes, 'A Vegetarian', *Wodwo* (1967)

2nd Pedestrian. ... My friend and myself, being anxious to see other places besides our own, and other men and other manners than those of our own confined circle, travel thus, at a light expense, in search of knowledge and amusement.

Bob Kiddy. Why, they're crack'd.
        James Plumptre, *The Lakers: A Comic Opera, in Three Acts* (1798)

When Wordsworth was composing verse, the music of hounds still sounded in the background, but English writers were much less likely to be sportsmen than had been the case when Shakespeare imagined Hippolyta recalling the 'gallant chiding' of Spartan hounds: 'I never heard / So musical a discord, such sweet thunder' (*A Midsummer-Night's Dream* 4.1.120, 123–4). The urbanization of literary culture opened a great divide between polite and rural pastimes, and between intellectual production and country life. In 1735 Somervile could hope that his poem *The Chace* would appeal to literate sportsmen as well as readers of verse. In 1796, John Aikin, author of the *Essay on the Application of Natural History to Poetry*, remarked that Somervile would be read not by sportsmen alone but by 'readers of English poetry in general'.[1] Because 'the sports of the hunter are noisy, tumultuous, attended with parade, and generally ending in conviviality', according to Aikin, they 'ill accord' with the 'calm, retired, reflective disposition of the lover of nature and votary of philosophy'.² Somervile, he assured us, was exceptional in possessing a taste for both hunting and the contemplation of nature, for both sporting culture and philosophy.

Something of this split can be discerned much earlier in marginalia by a reader of the Rev. Wetenhall Wilkes's anonymously published *Hounslow-Heath, a POEM* (1747). This is one of the rare instances of reader-reception of a sporting poem I have come across, so it is worthy of close examination. Wilkes, hoping to celebrate his parish as a sporting paradise, conventionally required the Muses' aid to make him a versifier worthy of the task: '– the Nine conspire / To swell my Numbers with Poetic Fire' (p. 3).³ An anonymous contemporary reader was moved to write in the margin: 'They shoul'd conspire / To burn thy Numbers in a blazing fire.' This censorious view of Wilkes's efforts seems to have been maintained throughout this reader's engagement with the poem; neither the sports nor the verse could please. Wilkes himself seems to have anticipated such a reaction, since his lines are frequently apologetic and self-conscious. After cataloguing the names of harriers, he added:

> Let us bring *Doxy* and old *Piper* in;
> *Cum multis aliis* – too prolix for Rhime;
> And too encroaching on my Reader's Time.
>
> (p. 11)

The censorious reader, as we might expect, responded with 'very True'. A later catalogue of foxhounds, ending with '*Seamstress* and *Rover* (all I cannot name)', elicited ' 'Tis no matter' (p. 14). This reader's anti-hunting agenda is further revealed when the response to Wilkes's lines 'When once poor *Reynard* sees himself inclos'd, / By Horses, Men, and greedy Dogs oppos'd', is simply 'a fine group of' (p. 15). The category to which the group belongs is unspeakable. The poem's conclusion, a description of the post-chase festivities, beginning 'At length a jocund Bottle crowns the Day', provoked advice to future readers, that they 'need not read further' (p. 20). Whoever this reader was, he or she disapproved of the poem on the grounds of its subject matter as well as its limited claims to literary merit. Huntin' poems were no longer so readily admired as they had once been. Yet the poem was sufficiently successful to warrant a second edition, 'Carefully CORRECTED and ENLARGED', and brought out under the author's own name, the following year.⁴ In the 1740s sporting and literary cultures had begun to part company but had not yet gone their separate ways.

Metropolitan intellectuals with little regard for rural pastimes seem to have had a disproportionate influence on opinion shaping during the eighteenth century, as a consequence of the burgeoning of print culture and the professionalization of the literary marketplace.[5] In the later eighteenth century, the most radical protests against the abuse of animals were generated in comparative isolation from the world of animal-keeping or husbandry. Many of the leaders of the movement against cruelty to animals were townspeople and members of middle-class professions removed from any close association with animals other than pet-keeping.[6] According to Keith Thomas and John Berger, the agitation against cruelty did not begin among 'butchers or colliers or farmers, directly involved in working with animals', nor among grooms, cab-drivers and other servants who spent their days in close proximity to their animal charges.[7] An assortment of well-to-do townsmen, educated country clergymen and members of the professional middle classes seeking to distance themselves from the 'warlike traditions of the aristocracy' emerged as the leaders of this movement,[8] and often directed their efforts at reprimanding and regulating the lower classes, the professional animal keepers.[9] Whether they actually lived all their lives in cities or not, these people valued urbanity and despised as barbaric the livestock handling and rural sports of country people, whether aristocracy, gentry or the laboring poor. They were what we would now call urban-identified. Indeed, urbanity might be said to be measurable in terms of distance from the world of agricultural production and livestock-rearing, of hunting, butchering and meat production.

This is exactly the character of 'A Vegetarian' drawn by Ted Hughes.[10] Fearful of animals he doesn't know or understand, fearing their revenge on carnivorous humanity, Hughes's vegetarian is also haunted by an image of hunting buried deep within the nature of which he is so apprehensive. Even vegetable 'Nature' might prove 'red in tooth and claw'. Even the grass might have its pack of hounds.

The vegetarian ideal

Vegetarianism haunted seventeenth- and eighteenth-century philosophical discourse. Then, as now, it often marked a desire for radical social transformation.[11] Celebrating current social hierarchy, as Ben Jonson did in 'Penshurst', could mean figuring Eden as precisely plenitude for human use, where fish and flesh existed as mankind's food. But Milton's *Paradise Lost*, highly influential among eighteenth-century writers, represented a vegetarian Eden and attacked guns and gunpowder as devilish instruments. Imagining the Garden of Eden before man's fall as a peaceable kingdom in which all species existed together harmoniously, rather than preying upon one another, was a familiar strategy for endeavoring to model a more perfect world than the present corrupt one. And imitating Eden required a more just appreciation than was commonly felt of the sentience of other animals, of their status as beings capable of suffering and thus having some entitlement to compassion and consideration of their welfare.

James Thomson coupled Edenic vegetarianism and hostility to field sports in *The Seasons* (first published 1726–30; revised 1744). The vegetarian recommendations of George Cheyne, a 'corpulent, indeed mountainous' physician, who was, like Thomson, a transplanted Scot, found their way into Thomson's verse, though not into his diet.[12] Reminding readers of Pythagoras's critique of meat-eating, the poet regretted that in the modern world 'the wholesome herb neglected dies' ('Spring', l. 336), while flocks of sheep, who are a 'peaceful people' ('Spring', l. 359), and the 'harmless, honest, guileless' ox are slaughtered for human greed ('Spring', l. 363).[13] In 'Spring', the 'beast of prey, / Blood-stained, deserves to bleed' (ll. 357–8), Thomson allowed, but in 'Autumn' he protested that his muse took no pleasure in field sports, that 'falsely cheerful barbarous game of death' (l. 384). Later writers opposed to hunting would quote Thomson's descriptions in 'Autumn' of the hare chase – beginning 'Poor is the triumph o'er the timid hare!' (l. 401) – and the stag-hunt – ending 'while the growling pack, / Blood-happy, hang at his fair jutting chest, / And mark his beauteous chequered sides with gore' (ll. 455–7).[14] If readers of *The Seasons* were sufficiently 'comfortably insulated from nature's cruelty' as not to have to take 'any preventative action' but only feel sympathy for rustic human victims, as Tim Fulford has claimed, the spectacle of animal suffering more pointedly recommends some social action.[15]

Alexander Pope followed Thomson's lead with a vegetarian manifesto in the third epistle of the *Essay on Man* (1733), but was not optimistic about vegetarianism's effectivity in the present. Before the fall:

> Man walk'd with beast, joint tenant of the shade;
> The same his table, and the same his bed;
> No murder cloath'd him, and no murder fed.
> . . .
> Heav'n's attribute was Universal Care
> And Man's prerogative to rule, but spare.
> Ah! how unlike the man of times to come!
> Of half that live the butcher and the tomb;
> Who, foe to Nature, hears the gen'ral groan,
> Murders their species, and betrays his own.
> But just disease to luxury succeeds,
> And ev'ry death its own avenger breeds;
> The Fury-passions from that blood began,
> And turn'd on Man a fiercer savage, Man.
>
> (*Essay on Man* 3: 152–4, 159–68)

In the beginning, Nature was humanity's teacher, and animals were moral tutors of mankind, even-handedly offering political models of both 'The Ant's republic, and the realm of Bees' (3: 184). Naturally, humankind would not be content with these, as Nature well knew:

> In vain thy Reason finer webs shall draw,
> Entangle Justice in her net of Law,
> And right, too rigid, harden into wrong;
> Still for the strong too weak, the weak too strong.
> Yet go! and thus o'er all the creatures sway.
>
> (3: 191–5)

The rational animal thought Nature's webs not fine enough, and set about drawing and making others, entangling Justice like a hare or rabbit in a net. Poachers to a man, humanity would not be content to learn pacifist ecology lessons in the greenwood. Oh, for a vegetarian paradise, rather than blood-soaked earth! But Pope was not optimistic about returning to Eden, any more than he was capable of believing that the peaceable kingdom could have survived into human history.

Metropolitan intellectuals

Pope typifies the metropolitan intellectual mildly contemptuous of country sports in spite of, or rather because of, a long association with sporting peers and country gentlemen. Writing to Martha and Teresa Blount in 1718, Pope described riding out 'a hunting on the Downes' with Lord Bathurst at Cirencester.[16] But he tells us nothing about what sort of sport they had. Either he had no idea and just went along for the ride, or he did not care to dwell on the events; whether in recollection generally or in writing to the Blount sisters specifically, we cannot tell. There is also the possibility, of course, that he and Bathurst just went riding, and that Pope, writing a letter to women he wished to impress, added 'a hunting' because it sounded right. If Pope went 'riding out a hunting', he hunted to ride, rather than rode to hunt; he was no sportsman, and there was no hound music to be heard.

In Pope's *Windsor-Forest* (1713), the very terrain made famous by Denham appears blood-soaked by a yearly round of sport. Partridge-netting and pheasant-shooting give way to hare-hunting, the shooting of woodcock, lapwings and larks, fishing and the pursuit of the hart, and that's how we know all's right with England and the Empire. A precondition for the *Pax Britannica* that is celebrated at the end of the poem would seem to be for the whole colonized world to have been comparably bloodied. Once the British have extended their empire of field sports sufficiently overseas, the colonized will return to the imperial metropolis to gape at the English, who will be forever madly chasing some creature or other: 'And naked Youths and painted Chiefs admire / Our Speech, our Colour, and our strange Attire!' (ll. 405–6). Like Denham's stag, Pope's shot pheasant comes to grief in a blaze of color, his royal 'Purple Crest' supplemented by scarlet, green and gold plumage, as he 'Flutters in Blood, and panting beats the Ground' (ll. 114–18). Pheasant-shooting and hart-hunting still represent the dynamics of state power, but the pheasant is also a shot, suffering bird. Windsor Forest is an apparatus of royal power and pleasure, but it is also a naturalist's paradise, and, dare we say, an ecosystem.[17]

Shooting had its more vehement critics, such as John Aldington, who protested the *Cruelty of Shooting* (1769) in verse:

> At last, his Bag well cram'd, he trudges home
> Elate, and loudly joyous, long harangues
> The listening Inmates, on *Pompey*'s Feats,
> Mighty Feats and Goodness, his lovely gun,
> His own unequal'd Skill, counts out his Spoils,
> And glories in the Murder of the Day.[18]

For Aldington, shooting represented not so much an adventure in natural history as a form of criminal activity. It is hard to see much difference between shooting and hunting with hounds in his account, since his sympathies lie solely with the birds and animals that are the victims equally of human insensitivity and the gun dog's, or spaniel's, bloody 'researches'.[19]

Such lessons were absorbed into eighteenth-century polite culture, equating meat-eating with bloodthirstiness, and bloodthirstiness with blood sports, putting the sporting culture of earlier generations into doubt. Vegetarianism remained a radical ideal rather than becoming a common practice, but a new equivalence between gentility and anti-cruelty concerning animals came by the end of the century to be accepted as the dominant view. The discourse of sensibility and benevolence so important in anti-slavery debates and in legitimating the expansion of empire took a domestic form in British confidence regarding the proper treatment of animals.[20] It became patriotic to be kind.[21] As Harriet Ritvo has argued, animals lost their agency and became property – commodities to be traded, but also objects of experimental breeding for 'improvement', protection and affection.[22]

In 1785 Edward Lovibond pondered what kind of beings his fellow humans were if they indulged in 'Rural Sports': 'But what are ye, who chear the bay of hounds, / Whose levell'd thunder frightens Morn's repose . . . ?' (ll. 5–6).[23] The answer offered by the poem's end is that such people had misinterpreted divine imperatives, mistaking warlike ambition for humankind's proper duty of benevolence:

> Mistaken mortal! 'tis that God's decree
> To spare thy own, nor shed another's blood:
> Heaven breathes Benevolence, to all, to thee;
> Each Being's bliss consummates general Good.
>
> (ll. 101–4)

Poetry, pets and the beginnings of animal rights

In the later decades of the century, a number of influential poets identified their own social alienation with the suffering of hunted animals. In 'The Deserted Village' (1770), Oliver Goldsmith's narrator, exiled from the now depopulated village of Auburn, compared himself with a hunted hare:

> And, as a hare, whom hounds and horns pursue,
> Pants to the place from whence at first she flew,
> I still had hopes, my long vexations past,
> Here to return – and die at home at last.
>
> (ll. 93–6)[24]

The disappearance of traditional common field agriculture, practiced by yeomen farmers maintained by their 'rood of ground' (l. 58), a 'bold peasantry' (l. 55) in Goldsmith's phrase, had meant the disappearance of both agricultural prosperity and village sports:

> ... The man of wealth and pride
> Takes up a space that many poor supplied;
> Space for his lake, his park's extended bounds,
> Space for his horses, equipage and hounds;
> The robe that wraps his limbs in silken sloth
> Has robb'd the neighbouring field of half their growth;
> His seat, where solitary sports are seen,
> Indignant spurns the cottage from the green.
>
> (ll. 275–82)

Large landowners were now converting tillage into parkland, feeding horses and hounds for elite blood sports – chiefly fox hunting: hence the nostalgia attached to the hunted hare – and implicitly starving the populace. Vegetarianism was a practice reserved for intellectuals; Goldsmith assumes that rural laborers will look to eat as much meat as they can get, though that won't be much. But now both the laborers and their sports have vanished. No longer was it possible, the narrator argues, for a poet to write a triumphant georgic poem, celebrating England's greatness through her agriculture and the recreational amenities of her countryside. The ruination of English yeomen's agriculture and of plebeian sports had happened simultaneously, funded by urban and colonial wealth. The poet who once wished to retire to the village of Auburn, like the hunted hare returning to her form, now finds no safe haven to which to return. Poet, peasantry and hare are all victims of the second, or landlords', agricultural revolution, of which modern fox-hunting was already becoming a potent symbol.

In 1785 Cowper made hunting and cruelty to animals synonymous. The passage in *The Task* most directly parallel to Goldsmith's about the hunted hare is Cowper's identification with a wounded deer:

> I was a stricken deer, that left the herd
> Long since; with many an arrow deep infixt
> My panting side was charged when I withdrew
> To seek a tranquil death in distant shades.

(3: 108–11)

Saved by 'one who had himself / Been hurt by th' archers' (3: 112–13), Cowper shifts his rhetoric from anti-blood sports to Christian allegory, only to return to anti-blood sports polemic 174 lines later. Introducing his tame hare, for ten years the 'Innocent partner' of his 'peaceful home' (3: 337), who 'Has never heard the sanguinary yell / Of cruel man, exulting in her woes' (3: 335–6), Cowper launches an attack upon:

> . . . detested sport,
> That owes its pleasures to another's pain,
> That feeds upon the sobs and dying shrieks
> Of harmless nature, dumb, but yet endued
> With eloquence that agonies inspire
> Of silent tears and heart-distending sighs!
> Vain tears alas! and sighs that never find
> A corresponding tone in jovial souls.
> Well – one at least is safe. One shelter'd hare.

(3: 326–34)

And so for Cowper the proper countryman is not a countryman, born and bred, not even a 'jovial' one, but rather a refugee from urban corruption, like himself, seeking solace in a garden, a greenhouse, country walks and the companionship of tame animals.

The Origins of the Anti-Hunting Campaign 121

David Perkins has recently argued that Cowper's attack on hunting in *The Task* is that of the animal-lover and pet-owner.[25] Cowper reported in the *Gentleman's Magazine* that his three pet hares took very differently to living with humans: Puss became a tender companion – a companion-animal, we would now say – while Tiney remained eternally suspicious of people.[26] In order for Cowper to observe them closely, the hares had to be domesticated in the literal sense of being kept indoors. 'Epitaph on a Hare', commemorating Tiney, represents an undoing of the history of blood sports on Cowper's Turkey carpet:

> HERE lies, whom hound did ne'er pursue,
>   Nor swifter greyhound follow,
> Whose foot ne'er tainted morning dew,
>   Nor ear heard huntsman's hallo,
>
> Old Tiney, surliest of his kind,
>   Who, nurs'd with tender care,
> And to domestic bounds confin'd,
>   Was still a wild Jack-hare.
>
> Though duly from my hand he took
>   His pittance ev'ry night,
> He did it with a jealous look,
>   And, when he could, would bite.
>
> His diet was of wheaten bread,
>   And milk, and oats, and straw,
> Thistles, or lettuces instead,
>   With sand to scour his maw.
> . . .
>
> A Turkey carpet was his lawn,
>   Whereon he lov'd to bound,
> To skip and gambol like a fawn,
>   And swing his rump around.
> . . .
>
> I kept him for his humour's sake,
>   For he would oft beguile
> My heart of thoughts that made it ache,
>   And force me to a smile.
>
>   (ll. 1–16, 21–4, 33–6)[27]

Although pet-keeping is presented by Cowper as the humane alternative to field sports, it sounds very much like a form of slavery. Tiney's biting comes as no surprise. The hare as pet has displaced the hare as game or food, kept 'for his humour's sake', rather than watched for his wiles in the field or savored as a delicacy. But the hare is still servant to the man.

Not all nature study need take place indoors in Cowper's world. In the sixth book of *The Task*, 'The Winter Walk at Noon', Cowper's narrator exults in having become so familiar a sight to hares, doves and squirrels that they have ceased to take fright at his approach:

> The tim'rous hare
> Grown so familiar with her frequent guest
> Scarce shuns me; and the stock dove unalarm'd
> Sits cooing in the pine-tree, . . .
> The squirrel, flippant, pert, and full of play.
> He sees me, and at once, swift as a bird
> Ascends the neighb'ring beech; there whisks his brush
> And perks his ears, and stamps and scolds aloud,
> With all the prettiness of feign'd alarm,
> And anger insignificantly fierce.
>
> (6: 305–8, 315–20)

'He sees me': Cowper's narrator likes nothing better than being the object of another's gaze, provided he is the only human present. He also confidently interprets the squirrel's scolding as insignificant in its fierceness; the squirrel's apparent anger does not signify, it poses no threat. The poem almost invites us to think the squirrel is pleased to see the poet and is feigning alarm and anger. Cowper does not go so far as to state this explicitly, but the phrase 'anger insignificantly fierce' suggests there is something exaggerated or camp about the squirrel's performance.

Paradoxically, readers are invited to imitate this solitary socializing with wild creatures. Domestic animals too provide pleasure and instruction, and Cowper gathers his moral forces in Book 6 to denounce cruelty towards all animals:

> The heart is hard in nature, and unfit
> For human fellowship, as being void
> Of sympathy, and therefore dead alike
> To love and friendship both, that is not pleased
> With sight of animals enjoying life,
> Nor feels their happiness augment his own.
> The bounding fawn that darts across the glade

> When none pursues, through mere delight of heart,
> And spirits buoyant with excess of glee;
> The horse, as wanton and almost as fleet,
> That skims the spacious meadow at full speed,
> Then stops and snorts, and throwing high his heels
> Starts to the voluntary race again;
> The very kine that gambol at high noon,
> The total herd receiving first from one
> That leads the dance, a summons to be gay,
> Though wild their strange vagaries, and uncouth
> Their efforts, yet resolved with one consent
> To give such act and utt'rance as they may
> To extasy too big to be suppress'd –
> These, and a thousand images of bliss,
> With which kind nature graces ev'ry scene
> Where cruel man defeats not her design,
> Impart to the benevolent, who wish
> All that are capable of pleasure, pleased,
> A far superior happiness to theirs,
> The comfort of a reasonable joy.
>
> (6: 321–47)

When Cowper looks at animals, he does not see potential sporting quarry or valuable livestock, but sentient fellow beings. Clearly, his country walks are meant to be morally instructive, exemplary of proper feeling as well as aesthetically satisfying; one can see what Jane Austen admired in them. We have already noticed that Cowper views vanishing habitats for wildlife with considerable anxiety (in Chapter 2). His anger at cruelty and exploitation of animals is inseparable from his grief at the felling of trees, part of an ecological vision resistant to agricultural market economics. Cowper advocates landlordly benevolence, but despairs that it is now exceptional.

Seeing animals at leisure, expressing themselves and demonstrating the pleasures of freedom, especially freedom from human constraints, does nothing for Cowper so much as consolidate his sense of rational superiority to other humans as well as animals, his own humane benevolence. However happy animals appear, if we are happy to see them pleasing themselves, we shall be happier still. 'The comfort of a reasonable joy' will be our reward for giving up hunting, shooting, beating, overworking and generally oppressing animals. But not eating them; Cowper follows Milton, Thomson and Pope in stressing the vegetarianism of Eden, but is resigned to fallen humanity's carnivorous diet. 'Eden was a scene of harmless sport' (6: 364), but one never to be returned to. Cowper offers a compromise, which today would be called eating only 'happy meat' – meat-eating without cruelty toward the living while they are alive:

> . . . Feed then, and yield
> Thanks for thy food. Carnivorous through sin
> Feed on the slain, but spare the living brute [.]
>
> (6: 456–8)

Cowper's advocacy ends with inviting the benevolent, however compromised by carnivorous sin, to feel pleased with themselves. He turns away from the animals in the field to make his case for how kindness to animals would improve human happiness. This strain of argument is common among animal rights advocates today, whose hands-off approach to animals can be identified as urban-based and fetishistic. 'Humans should live a life which does not touch animals in any way except as equals, and if they are our equals, we should not touch them anyway,' as Keith Tester sums up the position.[28] Thus animal rights, pursued to its logical conclusion, is not about loving animals; animals rights advocates would put paid to pet-ownership.[29] Today Cowper's position would be regarded as confused by the animal rights lobby, but he did anticipate certain tendencies within it. Not least of these is the divergence between animal rights theory and ecological thinking. Animal rights advocates hold ecologists' emphasis on the conservation of species 'in complete anathema', according to Tester, because from an ecological point of view, the 'well-being or suffering of any individual animal is a distant second to global worries'.[30] Cowper would seem to have anticipated the views of Ronnie Lee, former hunt saboteur and founder of the militant animal rights group the Band of Mercy, who has argued, ' "An animal doesn't care if its species is facing extinction – it cares if it is feeling pain."'[31]

The recognition of ecological interdependence, present in White's writing a decade earlier, had made its way into anti-cruelty treatises for children by 1799. Fond pet-owners were not necessarily ecologically knowledgeable, as Edward Augustus Kendall recognized. In the voice of a yellow-hammer, he deplored human indifference to wild species: 'Some few they prize for their beauty; but, in general, those that they do not eat, they stigmatize as useless. This is so far from being true that, the extinction of a single species of birds, would derange the economy of the world.'[32]

Whether they were vegetarians, anti-cruelty advocates, proto-ecologists or simply fashionable dissidents from the mainstream, by the 1780s men of letters were assumed to be removed from sporting culture. Samuel Johnson, mounted 'with a good firmness' on Henry Thrale's old hunter, might follow hounds for 50 miles in a day, but he nevertheless complained to Hester Lynch Thrale that he found hunting ' "no diversion at all"', regretting that ' "the paucity of human pleasures should persuade us ever to call hunting one of them"'.[33] Nevertheless, Thrale confided, Johnson was 'proud to be amongst the sportsmen'.[34] He was never more pleased, according to her, than when 'Mr. Hamilton called out one day upon Brighthelmstone Downs, Why Johnson rides as well, for aught I see, as the most illiterate fellow in England.'[35] Johnson remarked of Somervile, poet of *The Chace*, that he wrote '*very well for a gentleman*' and had shown that it was 'practicable to be at once a skilful sportsman and a man of letters'.[36] Gentlemen amateurs were no longer likely to be men of letters; the marketplace had become professionalized. And with the departure of country gentlemen, how far from sporting culture had popular perceptions of literary men come by the second half of the eighteenth century!

Anti-hunting sentiments and the rise of pedestrianism

Nothing is more characteristically English than going for a country walk. Today walking in the countryside is the second most popular recreation in Britain, after watching television. 'On a summer Sunday afternoon, the Countryside Commission estimate that eighteen million people, two fifths of the population, like to get away from it all and go to the country.'[37] And when they get there, they walk, even if only from car park to viewing point and back again. When the one-armed veteran of the Siege of Gibraltar, Captain Joseph Budworth, ascended Harrison's Stickle in Great Langdale in 1797, local folk remarked that ' "they never remembered foine folk aiming at et afore" '.[38] The taste for getting away from it all in unspoiled or picturesque scenery, whether tackled energetically by means of mountaineering or hill-walking, or casually in a Sunday stroll, has remained popular with working-class ramblers as well as 'fine folk' ever since the late eighteenth century.[39] Yet how did 'Shanks's pony' or 'Shanks's mare', once the last resort of the traveling poor, become a fashionable form of recreation?

Between 1560 and 1640, vagrancy had been the most intractable of social problems, and because the vagabond's status itself was criminalized, 'the social crime par excellence'.[40] Yet by 1675, in William Wycherley's play *The Country Wife*, we hear the eponymous heroine, Margery Pinchwife, inquiring of her London sister-in-law, Alithea, 'Pray, Sister, where are the best fields and woods to walk in, in London?' It would seem that, lacking other diversions, taking the air on foot had become a genteel recreation. Alithea names Mulberry Garden, St. James's Park and 'for close walks, the New Exchange', but soon expostulates with urban disdain, 'A-walking! ha! ha! Lord, a country-gentlewoman's leisure is the drudgery of a footpost; and she requires as much airing as her husband's horses.'[41] Millamant, in William Congreve's *The Way of the World* (1700), takes an even stronger stand in trying to evade being courted during an evening walk with Sir Wilfull Witwoud, a Shropshire squire:

'I nauseate walking; 'tis a country diversion; I loathe the country and everything that relates to it.'[42]

By the 1780s, however, opportunities to play the vagrant in beautiful scenery attracted pedestrian tourists to remote places. In the 1780s, French visitors were struck by the size of Englishwomen's feet, enlarged, it was thought, by this peculiarly English predilection for walking. Such eccentricity was as amazing to Continental observers as women riding boldly to hounds, another peculiarly English phenomenon. '"It gives me no pleasure to see it,"' de la Rochefoucauld declared, '"but they jump like men and are always the first over."'[43]

The earliest pedestrian tourists were 'the moneyed intellectuals, the professors, the retired clergymen, the not-quite-so-retired business men', but they were quickly followed by working-class ramblers drawn from the northern industrial towns.[44] Middle-class poets and intellectuals taking to the roads and footpaths in the later decades of the eighteenth century idealized the gypsy life and what looked to them like the freedom of the poor. 'The poet and the vagrant together constitute a society based on the twin principles of freedom of speech and freedom of movement,' Celeste Langan remarks, noting a link between walking and liberalism.[45] Thinking of vagrancy in this way requires 'a certain idealization of the vagrant: a reduction and an abstraction'.[46] Working-class ramblers, escaping from the confinement of factories and crowded towns, were no less susceptible to the pleasures of the outdoor life. They too might feel 'a love for the wind and the rain, the snow and the frost, the hill and the vale, the widest open spaces and the choicest pastoral and arboreal retreats', as G.H.B. Ward, socialist founder of the Clarion Ramblers of Sheffield, phrased it.[47]

As enclosure and improvement brought better roads and increased commercial traffic, secure from banditry, throughout the country, while de-particularizing agricultural landscapes, both means and motive arose for undertaking the novelty of pedestrian travel in remoter parts of the country. What Philip S. Bagwell calls the 'turnpike era' saw the reduction of journey times between the principal centers of population. If four days between London and Manchester could be bragged of in 1754 – ' "However incredible it may appear this coach will actually arrive in London four days after leaving Manchester" ' – by 1760, three days ' "or thereabouts" ', was advertised by the Handforth, Howe, Glanville and Richardson coach, and by 1784, competition between rival firms had reduced the journey to two days.[48]

This materialist explanation of the vogue for pedestrian travel needs to be supplemented by attending to 'what walking *signified*'.[49] According to Robin Jarvis, the concentration of early pedestrian travelers among the ranks of free-thinking undergraduates, members of the lower clergy – 'a traditionally oppressed and disgruntled social group' – and others with politically levelling views suggests that there was 'an element of deliberate social nonconformism, of oppositionality, in the self-levelling expeditions of most early pedestrians'.[50] Joseph Hucks, friend and fellow undergraduate of Samuel Taylor Coleridge, expressed his sense of taking to the road in 1794 in the following high-minded way:

> [W]e carry our clothes, &c. in a wallet or knapsack, from which we have not hitherto experienced the slightest inconvenience: as for all ideas of appearance and gentility, they are entirely out of the question – our object is to see, not to be seen; and if I thought I had one acquaintance who would be ashamed of me and my knapsack, seated by the fire side of an honest Welsh peasant, in a country village, I should not only make myself perfectly easy on my own account, but should be induced to pity and despise him for his weakness.[51]

Such nonconformism also took the form of voluntary dismounts from gentlemanly steeds. Choosing pedestrian travel represented a democratizing gesture with regard to those who could afford no other form of transport, and a refusal of even the appearance of sporting privileges. There was a democratizing of human relations with the animal world as well as a leveling of class privilege. Seeking out picturesque views, sublime experiences and sometimes a knowledge of rural manners and natural history by rambling in the countryside represented a new alternative to hunting and shooting.

This new taste was not necessarily accompanied by vegetarian abstemiousness, but we do hear often of walkers dining on bread and cheese. When Hucks and Coleridge and two other Cambridge men ascended Caer Idris in July 1794, they armed their guide with 'ham, fowl, bread, and cheese, and brandy'.[52] The Wordsworths might carry 'cold pork in their pockets',[53] or eat 'some potted Beef on Horseback, and sweet cake' when traveling.[54] Dinners or suppers at inns were the chief concession to creature comforts walkers allowed themselves. The Rev. Richard Warner was ecstatic at the cheapness of food and accommodation in Wales in 1797, when the bill for two for one night's lodging and food came to only 5s. 2d., with the supper consisting of 'a sole, a trout, and a gwyniad (a delicious fish, somewhat like the trout, and peculiar to Alpine countries)'[55] with 'mutton steaks, vegetables, excellent bread and cheese, and three tankards of London porter' (p. 88). These walkers' comparative wealth, however, was exposed by the 'frugal repast' of 'Scotch pedlars', also traveling in Wales and carrying their own food – 'an oaten cake, and rock-like cheese' diluted with '"acid tiff"' (p. 57).

James Plumptre's comic opera of 1798 on Lake District tourism foregrounds the contrast between pedestrian and sporting cultures. Two young gentlemen pedestrians, dressed as sailors, describe their interest in a walking tour as downwardly mobile and democratic and declare themselves economical and health-conscious in their choice of food and drink. When asked by the protagonist Sir Charles Portinscale to explain their appearance – 'May I request to know your names and conditions? Your words do not accord with your outward appearance' – the First Pedestrian replies:

> We trust, Sir, that we are gentlemen, though thus habited, and taking our tour on foot to gain a knowledge of our country and of mankind.[56]

Miss Beccabunga Veronica, a would-be botanist and keeper of a picturesque-style journal, exclaims:

> How fortunate! Pedestrians! How picturesque too their appearance – Another incident for my tour.[57]

Pedestrians were themselves becoming part of the rural spectacle other tourists had paid to see.[58] But the chief butt of satire in this lighthearted opera is none other than Miss Veronica's nephew, Bob Kiddy, who fancies himself a 'crack' coachman and all-around sporting gent. Inquiring one minute for blood sports – 'Has any body got a bear, or a badger to bait? or fighting cocks? Famous fun that' (p. 35) – and the next for food, Bob Kiddy represents the sordid side of sporting culture. When the landlord replies to the latter request with 'We've a nice grouse in the pantry, Sir,' Bob's response is, 'Does it stink?' (p. 35). Upon being told the grouse is 'As sweet as a nut, Sir,' Bob gives an order: 'Then keep it till it does, for then aunt and coz won't touch it; and I shall have it all to myself' (p. 36). This stink-seeking, bloody-minded fellow, however funny, looks distinctly sinister compared with Plumptre's other characters. Needless to say, the two Pedestrians refuse Bob Kiddy's offer of a bottle of wine apiece and a bit of dinner at his expense, preferring 'a draught of good ale' with their food as offering 'more health and nourishment' (p. 39).

Coopers Hill as touchstone: walking

Influential in its representation of stag-hunting, *Coopers Hill* proved equally important to the tradition of English topographical verse from which the walking poem developed. Although Denham's narrator does not represent himself as ascending the hill, actually walking up it to attain a prospect view, the description of the surrounding landscape from the summit will influence topographical verse for decades to come, and even find its way into the more strenuous walking poems of late eighteenth- and early nineteenth-century writers. The landscape performs for Denham, acting upon him and beckoning him to explore its intricacies. This is a landscape that will not allow itself to be ignored. From Cooper's Hill, Windsor Castle 'swells' above the Thames valley into Denham's 'eye' (ll. 40–1); the landscape has, as it were, sought him out, not he it:

> . . . and doth it self present
> With such an easie and unforc't ascent,
> That no stupendious precipice denies
> Access, no horror turns away our eyes:
> But such a Rise, as doth at once invite
> A pleasure, and a reverence from the sight.
>
> ('B' text, ll. 41–6)

Although there would seem to be no compulsion for the narrator to walk there, Denham is pleased that the castle looks as if its approach would be neither an exhausting nor a terrifyingly precipitous climb. No strenuous pedestrianizing for him! That is a taste that will come later, when other forms of transport have been so regularized that only walking will seem to retain any radical edge. Such a scene invites patriotic reflections in its harmony and majesty; Denham cannot avoid taking pleasure from this prospect of royal habitation and feeling reverence for it.

By writing *Coopers Hill*, Denham founded what Samuel Johnson called '*local poetry*, of which the fundamental subject is some particular landscape to be poetically described, with the addition of such embellishments as may be supplied by historical retrospection or incidental meditation'.[59] From 1642 to 1826 Denham's poem was published at least 24 times.[60] In 1788, the *Gentleman's Magazine* complained that '"readers have been used to see the Muses labouring up . . . many hills since Cooper's and Grongar, and some gentle Bard reclining on almost every mole-hill"'.[61] Poetic records of prospect views, obtainable through walking up hills, and mole's-eye views, achievable by lazier bards reclining on the flat, were desirable commodities.

By 1788 each step in the ascent of Lewesdon Hill was worthy of recording by William Crowe, climbing before breakfast on a Sunday morning in May:

> . . . along the narrow track
> By which the scanty-pastured sheep ascend
> Up to thy furze-clad summit, let me climb;
> My morning exercise; and thence look round.[62]

'Above the noise and stir of yonder fields / Uplifted', the narrator feels his 'mind / Expand itself in wider liberty' (p. 4). The unbinding of the subject from subjection to everyday cares has allegorical overtones: 'Crowe develops a patriotic vision of his local hill as a seat of native liberty, resistant to unnecessary war and despotism.'[63] Immune to worries from across the Channel and 'sequester'd from the noisy world' at home, Crowe's narrator fancies he could go on living on the hilltop quite happily until the voice of conscience calls him back to the need for 'toil' (p. 5). But it is a toil quite different from his pedestrian exertions, which had left him energized and refreshed.

Often later eighteenth-century poets substituted pedestrian rambling for the sporting excursions so common in seventeenth- and early eighteenth-century writing. Thomas Gray reported in 1736 that he was a great disappointment to his uncle for walking when he should have ridden, and reading when he should have hunted. His comfort while at his uncle's house in Buckinghamshire was that he had a favorite walking and climbing place, now known as Burnham Beeches:

> . . . I have at the distance of half a mile thro' a green Lane, a Forest (the vulgar call it a Common) all my own; at least as good as so, for I spy no human thing in it but myself; it is a little Chaos of Mountains & Precipices; Mountains it is true, that don't ascend much above the Clouds, nor are the Declivities quite so amazing, as Dover-Cliff; but just such hills as people, who love their Necks as well as I do, may venture to climb, & Crags, that give the eye as much pleasure, as if they were more dangerous: both Vale & Hill is cover'd over with most venerable Beeches, & other very reverend vegetables . . . At the foot of one of these squats me I; il Penseroso, and there grow to the Trunk for a whole morning . . .[64]

Some years later in 1750, Gray described himself in 'Elegy Written in a Country Churchyard' remembered posthumously as a crazed youth muttering ' "wayward fancies" ' (l. 106) when not escaping from the noon heat under his favorite beech tree: ' "His listless length at noontide would he stretch, / And pore upon the brook that babbles by" ' (ll. 103–4).

In 1744 John Armstrong, M.D., recommended walking in *The Art of Preserving Health*:

> My liberal walks, save when the skies in rain
> Or fogs relent, no season should confine . . .
> Go, climb the mountain; from the ethereal source
> Imbibe the recent gale.
>
> (Book 3: 54–5, 57–8)[65]

But his recommendation of walking led syntactically to his advocacy of hunting:

> ... The cheerful morn
> Beams o'er the hills; go, mount the exulting steed.
> Already, see, the deep-mouthed beagles catch
> The tainted mazes; ...
>
> (3: 58–61)

A difference is palpable in 1785, when Cowper disembarked from 'The Sofa', in *The Task*, for winter walks in the morning and at noon, congratulating himself on still possessing into middle age:

> Th' elastic spring of an unwearied foot
> That mounts the stile with ease, or leaps the fence,
> The play of lungs inhaling and again
> Respiring freely the fresh air, that makes
> Swift pace or steep ascent no toil to me.
>
> (1: 135–9)

Cowper makes of his walking an aerobic pastime not unlike hunting, leaping fences for the pleasure of it. But in Book 3 he explicitly distinguishes his country walks from hunting, shooting and fishing, railing against 'detested sport', 'the savage din of the swift pack / And clamours of the field' (3: 324–5):

> ... detested sport,
> That owes its pleasures to another's pain,
> That feeds upon the sobs and dying shrieks
> Of harmless nature, dumb, but yet endued
> With eloquence that agonies inspire
>
> (3: 325–30)

Cowper supposes that the death of the hare, and not a contest of intelligence and athleticism between hare and hounds, ending as it may, is the object of hare-hunting. He ignores the traditions of woodcraft and hound work. The music of hounds sounds to his ear like a 'savage din'. By denigrating hunting as a literally bloodthirsty sport, he is able to glorify walking as the only suitable rural pastime for a 'mind / Cultured and capable of sober thought' (3: 323–4).

James Beattie had earlier heroized his Minstrel through strenuous upland walking that was a direct repudiation of hunting:

> Lo! where the stripling, wrapt in wonder, roves
> Beneath the precipice o'er hung with pine;

> And sees, on high, amidst th' encircling groves,
> From cliff to cliff the foaming torrents shine:
> While waters, woods, and winds, in concert join,
> And Echo swells the chorus to the skies.
> Would Edwin this majestic scene resign
> For aught the huntsman's puny craft supplies?
> Ah! no: he better knows great Nature's charms to prize.[66]

The proper prizing of nature's charms through hill-walking and climbing was by the early 1770s already being framed within anti-hunting polemic.

Sporting culture and the rise of pedestrianism

Ironically, although many self-styled pedestrians repudiated field sports, pedestrian feats in the late eighteenth century were often sponsored and celebrated by sportsmen. Before walking became overwhelmingly popular, it was written up as a demonstration of athletic prowess or a bid for celebrity, both of which were wagerable activities.[67] Foster Powell, an attorney's clerk, pioneered the movement in 1773 by walking from London to York and back (402 miles) in five days and $15^{1}/_{2}$ hours, averaging 72 miles a day.[68] In 1801 Captain Barclay won a bet of 5,000 guineas by walking 90 miles in $21^{1}/_{2}$ hours, and in 1809 succeeded in walking 1,000 miles in 1,000 hours at Newmarket.[69] In the late eighteenth century walking was a part of sporting culture before it became what we now know as walking or rambling or climbing culture. When Coleridge and Hucks met up with fellow undergraduates Brookes and Berdmore, Coleridge mocked them as 'rival *pedestrians*, perfect *Powells*' when he caught them taking a post-chaise because one was '*clapped*'.[70] Pedestrian achievements appeared alongside reports of hunting and shooting in *The Sporting Magazine*, the repository of 'Transactions of The Turf, The Chace, And every other Diversion Interesting to The Man of Pleasure and Enterprize'. Pedestrian enterprise offered opportunities for betting, an important component of sporting culture in this period:

> SEPTEMBER 7th, 1780, Capt. Hoare undertook for a considerable wager, to ride three horses 30 miles, and drink three bottles of claret, in three hours, all which he performed with ease within the limited time.[71]

> 1787, January 18th, one of the greatest efforts in walking was performed by a sawyer, of Oxford, in Port Meadow, near that city. He walked fifty miles in nine hours and a half. At eight in the morning, he started, walked till one, when he dined, and at half after five he won his wager. He was allowed ten hours to do it in, but went over his ground with ease, in nine hours and a half, and was so little fatigued with his expedition, that he

refused a carriage, and walked into town, two miles from the field, amid the acclamations of numbers who accompanied him.[72]

May 21 [1792], a very young boy, for a trifling wager, ran twice round the city walls of Chester, (three miles and a half) in 23 minutes; numerous spectators were present, who deemed it a very extraordinary pedestrian exploit, particularly as the youth was only twelve years of age.[73]

September 5 [1792], John Hoole, a hair-dresser, of Twickenham, for a wager ran from the Three Tuns at that place to Hyde Park Corner, (ten miles) in one hour and 18 minutes. Fifteen pounds to ten was betted that he did not do it in an hour and a half, as he is very short in stature, and remarkably bandy-legged.[74]

There were a number of working-class or artisanal participants in these walking ventures, but gentlemen participated in them as well. Captain Barclay makes an appearance in Nimrod's novel *The Life of a Sportsman* (1832), discussing his 'pedestrian feats', his 'eighty-three' days' hunting,[75] and boxing, as mutually beneficial manly sports.[76] Walking was a wagerable sport, a form of gentleman's sporting venture, before it became an antidote to field sports. Following foot packs continued to offer opportunities for strenuous pedestrianism.

Romantic walking as anti-hunting

That Wordsworth disavowed in 'Hart-Leap Well' (1800) his early attraction to hunting has been argued by David Perkins.[77] Wordsworth's own experience of the wild seems to have begun in the traditional hunting and gathering way. Like generations of shepherds and cottagers before him, and like the modern rock-climber Jim Birkett, who first learned rope technique in order to get at birds' nests,[78] Wordsworth began climbing as a boy to set 'springes' for woodcocks and to steal ravens' eggs from nests: 'In thought and wish / That time, my shoulder all with springes hung, / I was a fell destroyer' (*The Prelude* [1805], 1: 316–18). Not surprisingly, the sounds of hunting were never far from his memories of childhood in the Cumbrian Fells. Yet 'Hart-Leap Well' does attack as bloody and cruel the hunting to death of a stag; Sir Walter's obsession with killing the stag that has outrun all his hounds and horses emblematizes human tyranny over the animal world. The attack on blood sports is more briefly suggested in these lines from 'Home at Grasmere' (also chiefly written in 1800), in which Wordsworth describes first visiting the spot that inspired 'Hart-Leap Well' with his sister Dorothy (named 'Emma' in the poem) and having there a vision of a bloodless future:

> . . . And when the trance
> Came to us, as we stood by Hart-Leap Well,
> The intimation of the milder day
> Which is to come, the fairer world than this,
> And raised us up, dejected as we were
> Among the records of that doleful place
> By sorrow for the hunted beast who here
> Had yielded up his breath, the awful trance –
> The vision of humanity, and of God
> The Mourner, God the Sufferer when the heart
> Of his poor Creatures suffers wrongfully –
> Both in the sadness and the joy we found
> A promise and an earnest that we twain,
> A pair seceding from the common world,
> Might in that hallowed spot to which our steps
> Were tending, in that individual nook,
> Might even thus early for ourselves secure,
> And in the midst of these unhappy times,
> A portion of the blessedness which love
> And knowledge will, we trust, hereafter give
> To all the Vales of earth and all mankind.
>
> (ll. 236–56)

All the ingredients for a modern rejection of blood sports are present here. The hunted hart is persecuted, not engaged with in a sporting or ritualistically reverent way. The hart suffers 'wrongfully'. According to this logic, no hunting could ever be justified. It would be only animal killing, and that would seem to be equated with homicide in Wordsworth's scheme. But notice too how this deploring of blood sports also cements the bond between the poet and his sister. Together they share a 'trance', a kind of out-of-body experience, and together receive an 'intimation' of a 'milder day' in which blood will no longer be shed. They are the sole witnesses to a preview of the general happiness to come. Their intimacy is secured and authorized through this shared vision of bloodlessness. The polemic against blood sports ends up being about them, their special status and insight, their shared bond and private happiness, rather than the emancipation of future harts. There is a contradiction here between a pan-animal vision of humans giving up persecuting their fellow creatures, and the very private view of specially sensitive humans wishing to bring about this pan-animalism. The pan-animal vision of humanity will be more explicitly advocated, under the rubric of Pantisocracy, by the Wordsworths' close friend Coleridge. For him, the extension of equality and fellowship much debated in the politically volatile 1790s should not have ended with humans but been extended to all animal species. Pantisocracy is radical democracy. But there is a contradiction in embracing it. Giving up the sporting life in the 1790s, or in 1800 when Wordsworth wrote 'Home at Grasmere', meant 'seceding from the common world'.

This Wordsworth was more than prepared to do. Congratulating himself on his retirement to Grasmere, a place he had only ever hoped (without much hope) to inhabit, Wordsworth's narrator in 'Home at Grasmere' struggles to reconcile himself to his fortunate fate (as a poet who can afford to live at Dove Cottage on the income bequeathed to him by his friend Raisley Calvert). Echoing Cowper's line about Tiney, the hare saved from hunting, 'Well – one at least is safe. One sheltered hare' (*The Task* 3: 334), Wordsworth's narrator refers to himself: 'But I am safe, yes, one at least is safe' (l. 74). Wordsworth has allusively identified this image of himself with Cowper's tame hare. The poet's pleasant retirement from the necessity of earning a living is metaphorically equated with the happy fate of a game animal rescued from hunting.

'Resolution and Independence' (composed 1802; published 1807) is not only one of Wordsworth's most successful pedestrian poems – establishing that walking on the moors is a suitable occupation for a grown man, and that he need not kill something in order to justify liking a long wander outdoors – but also one in which the rambling poet on the moors glories in an unhunted landscape. Wordsworth's narrator could not be further from Melesina Bowen's local Guide in *Ystradffin*, reveling in the music of hounds. He revels instead in watching a hare disporting herself, far from any hounds. The hare is an image of unfettered freedom parallel to the poet's freedom from financial exigency and everyday care. If we acknowledge the venerable equation, in Wordsworth's day, of hares with gentlemanly sport, if we remember that putting up a hare as the poet does was tantamount to hunting her with harriers, coursing her with greyhounds or shooting her, the hare's 'mirth' can be read as a direct repudiation of blood sports. In addition to Cowper's hares in verse, Wordsworth may have had in mind Burns's 'On Seeing a Wounded Hare Limp By Me Which a Fellow Had Just Shot At' (written 1789, published 1793), which concludes: 'I'll miss thee sporting o'er the dewy lawn, / And curse the ruffian's aim, and mourn thy hapless fate.'[79] The poet's moor-walking undoes metaphorically the history of hunting and shooting:

> All things that love the sun are out of doors;
> The sky rejoices in the morning's birth;
> The grass is bright with raindrops; on the moors
> The Hare is running races in her mirth;

> And with her feet she from the plashy earth
> Raises a mist; which, glittering in the sun,
> Runs with her all the way, wherever she doth run.
>
> I was a Traveller then upon the moor;
> I saw the Hare that raced about with joy;
> I heard the woods, and distant waters, roar;
> Or heard them not, as happy as a Boy:
> The pleasant season did my heart employ:
> My old remembrances went from me wholly;
> And all the ways of men, so vain and melancholy.
>
> (ll. 8–21)

Walking enables the pedestrian to merge with fellow creatures, apart from human society with all its woes. But the effect is short-lived in this case, for the narrator is soon beset by anxiety: '[F]ears, and fancies, thick upon me came; / Dim sadness, and blind thoughts I knew not nor could name' (ll. 27–8). Although the narrator then claims, 'I heard the Sky-lark singing in the sky; / And I bethought me of the playful Hare; / Even such a happy Child of earth am I; / Even as these blissful Creatures do I fare' (ll. 29–32), nevertheless he could not keep dread about future unhappiness at bay: 'But there may come another day to me, / Solitude, pain of heart, distress, and poverty' (ll. 34–5). It will take the sturdy rectitude of the old leech-gatherer to reassure this anxious Wordsworthian figure that he too is capable of 'resolution and independence'.

If read in this context, 'Simon Lee, the Old Huntsman, with an incident in which he was concerned', one of the *Lyrical Ballads* published in 1798, appears to be an even clearer instance of Wordsworth striking a blow at sporting culture.[80] This was the poem that stuck in Clare's mind at Northampton Asylum when he wrote his poem in praise of Wordsworth, beginning 'WORDSWORTH I love; his books are like the fields, / Not filled with flowers, but works of human kind' (ll. 1–2).[81] The single character mentioned from all Wordsworth's 'tenants of the earth' (l. 8) is 'The aged huntsman grubbing up the root –' (l. 7), a clear reference to Simon Lee. Wordsworth's poem describes Simon Lee as a retired pedestrian huntsman, who was renowned for 'four counties round' (l. 19) for his prowess with the horn and on foot, outrunning even a mounted field, though he sometimes fainted from his exertions. The old man still rejoices to hear 'the chiming hounds': 'He dearly loves their voices!' (ll. 46–8), and he still proudly wears a hunt servant's livery, but he and his wife Ruth are 'poorest of the poor' (l. 60). Being a huntsman meant 'When he was young he little knew / Of husbandry or tillage' (ll. 37–8):

> A long blue livery-coat has he,
> That's fair behind, and fair before;
> Yet, meet him where you will, you see
> At once that he is poor.
>
> (ll. 9–12)

His 25 years as 'a running huntsman merry' (l. 14) have left him blind in one eye and with ankles 'swoln and thick' (l. 35), despite his lean frame, so that his wife, though 'not over stout of limb, / Is stouter of the two' (ll. 51–2). Having long ago enclosed a 'scrap' of land from the 'heath', they can no longer till it (ll. 59–64), and though they persist in their labor, 'Alas! 'tis very little, all / Which they can do between them' (ll. 55–6). Lest we blame the squire's family which employed him for his destitute old age, the poem relates that they are all dead: 'Men, dogs, and horses, all are dead; / He is the sole survivor' (ll. 53–4).

Although the poem is set in Cardiganshire, its date of composition suggests a West Country inspiration, and Wordsworth confirmed this when he told Isabella Fenwick in 1843:

> This old man had been huntsman to the Squires of Alfoxden, which, at the time we occupied it, belonged to a minor. The old man's cottage stood upon the Common, a little way from the entrance to Alfoxden Park. But [in 1841] it had disappeared . . . I have, after an interval of forty-five years, the image of the old man as fresh before my eyes as if I had seen him yesterday. The expression when the hounds were out, 'I dearly love their voice,' was word for word from his own lips.[82]

The huntsman Wordsworth had met at Alfoxden was in fact named Christopher Trickey, servant to the St. Albyn family, who hunted with the Quantock staghounds, and was renowned 'for his fleetness of foot and his sounding of the horn.'[83] Since Alfoxden belonged to a minor, the Wordsworths, tenants of the grandest house they would ever occupy, stood as proxy to the squire's family, an awkward position for William, as David Simpson observes.[84] At '23£ *a year, taxes included*!!' the rent for Alfoxden was nearly three times what the Wordsworths would soon be paying for Dove Cottage in Grasmere.[85] Dorothy was awed by the size and gentility of the place but loved its beauty:

> Here we are in a large mansion, in a large park, with seventy head of deer around us . . . The House is a large mansion, with furniture enough for a dozen families like ours . . . This hill is beautiful, scattered irregularly and abundantly with trees, and topped with fern, which spreads a considerable

way down it ... Wherever we turn we have woods, smooth downs, and valleys with small brooks running down them through green meadows ... Walks extend for miles, over the hill-tops; the great beauty of which is their wild simplicity ...[86]

She could not help repeating 'a large mansion': Alfoxden could accommodate a dozen William-and-Dorothy households. Its secluded grounds, complete with aristocratic deer park, opening out seamlessly into long walks across country, filled her with ecstasy.

The incident in this poem – no tale in itself, the narrator insists, but one which could be made into a tale – consists of the narrator watching Simon Lee struggling to chop the root of a dead tree stump:

> 'You're overtasked, good Simon Lee,
> Give me your tool' to him I said;
> And at the word right gladly he
> Received my proffered aid.
> I struck, and with a single blow
> The tangled root I severed,
> At which the poor old man so long
> And vainly had endeavoured.
>
> The tears into his eyes were brought,
> And thanks and praises seemed to run
> So fast out of his heart, I thought
> They never would have done.
> – I've heard of hearts unkind, kind deeds
> With coldness still returning.
> Alas! the gratitude of men
> Has oftner left me mourning.
>
> (ll. 89–104)

The image of the dead tree root resonates with nothing in the poem so much as the dead family at the hall. Simon Lee's years of hunting hounds meant he had no practical experience of farming. Might there not have been some expectation that, his employers being dead, those other hunt followers who benefited from his expertise ought to have provided for him in old age? The Wordsworthian narrator as tenant, proxy to the lords of the manor, might feel it uncomfortably incumbent upon him to rectify this lack. The effortlessness with which the blow is delivered, as Heather Glen has noticed, makes it 'an apt and powerful image for the unwitting ease of that paternalistic "pity"' which diminishes Simon's struggles; it is an image which, 'in its incipient brutality, and in the irrational guilt it seems to inspire', comes closer than anything else in Wordsworth's writing of this period to Blake's 'Pity would be no more, / If we did not make somebody Poor'.[87]

If there is an allegorical point to the narrator's chopping the root which Simon Lee has been unable to sever, it might well be that the poem kills any expectation of benevolence from the hunting fraternity. Without saying so outright, Wordsworth manages to insinuate that the sporting community is responsible for Simon Lee's hard fate. He points the accusing finger at Old Corruption, to use Thomas Paine's phrase, in the form of those who have forgotten the old huntsman once he ceased to show sport. Representing Wordsworth as striking through 'the tangled root that separates man from man in the social as well as the physical order',[88] Simpson finds an echo of Paine's imperative to 'Lay then the axe to the root, and teach governments humanity'.[89] Paine wrote of hereditary succession, 'The moving power in this species of Government, is of necessity, Corruption'.[90] It is Old Corruption, with its regime of violence, whose root is to be axed. The tangled root with which Simon Lee has unsuccessfully struggled is precisely hereditary succession, landlordly privilege, the squire's sporting prerogatives that should be accompanied by responsibilities. Despite Wordsworth's attempts to 'affirm the integrity of all men', his social vision remains 'deeply pessimistic as to human possibility'.[91]

What Wordsworth exposes the absence of in 'Simon Lee' is the subject of Morland's *Benevolent Sportsman* of 1792 (Col. Pl. 1). There is social distance, to be sure, between the man of property, the qualified sportsman mounted on horseback, and the gypsies whose poverty his charity aims to relieve. But this representative of the propertied classes is nevertheless within the frame, not outside it, and dressed and mounted in such a way as to minimize, not maximize, his distance from the poor. Morland's picture might be said to invoke Gilpin in its choice of objects of benevolence; gypsies, outside the legitimate circuits of industry and commerce, pose a vague threat to social hierarchy. In what Stephen Deuchar calls the 'politically and socially nervous' 1790s, admiration for 'the old seventeenth-century-based sporting ideal' – 'country sportsmen's robust physical health, warlike capabilities, hospitality, national loyalty and personal generosity' – was rekindled in the guise of patriotism.[92] Earlier in the century, there had been pacifist appropriations of field sports to advertise the virtues of aristocratic rule. Celebrating the peace treaty of Utrecht in 1713, Pope had hazarded in *Windsor-Forest* that killing birds and animals might satisfactorily take Englishmen's minds off war. But Morland's sportsman, though he leaves his curb rein slack, rides in a double bridle. He is a man of peace, but patriotically equipped for hunting, or war. Morland's nicely ambiguous coding defends the sporting squire as in no need of Paineite reform. Aligned with gypsies, he is himself a picturesque object – downwardly mobile, if not democratic. His personal generosity commands respect and induces patriotic loyalty. The sportsman's benevolence is represented precisely so as to fend off the kind of criticism Wordsworth would make of sporting culture six years later.

Might not Wordsworth's principled opposition to field sports have led him to a moment of levity in the ascent of Snowdon in Book 13 of the 1805 *Prelude*? Before dawn, on a summer's night in thick fog, the climbing party are escorted by the traditional guide, a shepherd, who, as a proper countryman, takes his dog along. The party engage in 'ordinary travellers' chat' (l. 17) with the guide, then fall silent:

> Thus did we breast the ascent, and by myself
> Was nothing either seen or heard the while
> Which took me from my musings, save that once
> The Shepherd's Cur did to his own great joy
> Unearth a hedgehog in the mountain crags
> Round which he made a barking turbulent.
> This small adventure, for even such it seemed
> In that wild place and at the dead of night,
> Being over and forgotten, on we wound
> In silence as before. With forehead bent
> Earthward, as if in opposition set
> Against an enemy, I panted up
> With eager pace, and no less eager thoughts.
>
> (13: 20–32)

Not hound music, then, such as delights the guide at Ystradffin, but a turbulent barking at a hedgehog, a burlesque adventure, enlivens Wordsworth's climb to the top of Snowdon. Wordsworth did not always write in a solemn anti-hunting vein. Antic anti-hunting might also appeal to a metropolitan audience, especially when combined with the strenuous ascent of a mountain which epitomized, even in thick fog, all that was grand about picturesque scenery.

Coleridge, for whom writing poetry was always so fraught with difficulty as to be frequently impossible, seems to have invested more psychic energy than Wordsworth in the physical activities of walking and climbing for their own sake. He wrote less about walking than his friend Wordsworth did, but what he wrote is both startlingly original and paradigmatic of subsequent walking writing. Although he has not been embraced by ramblers as Wordsworth has been, Coleridge perfected the walking poem in 'This Lime-Tree Bower My Prison' (composed 1797, first published 1800) – a verse guide to a walk in the

Quantocks and a fully hob-nailed alternative to hunting and shooting (see Chapter 10).

Pantisocracy substituted other, less bloody pleasures for the need to go out and kill something in the country. The origins of this never-to-be-realized ideal, supposed to be fulfilled in the founding of a private-propertyless community on the banks of the Susquehanna River in Pennsylvania, began with a pedestrian tour.[93] In the summer of 1794, Coleridge set out from Jesus College, Cambridge, with Joseph Hucks to walk to Wales. They went via Oxford to meet Robert Southey, the 20-year-old Balliol poet from Bristol, 'already renowned for his extreme republican views. This meeting delayed their planned three-day stop for three weeks, and saw the birth of the famous "Pantisocratic" scheme.'[94] As Richard Holmes puts it, 'The Pantisocrats would befriend the natural world, and live harmoniously as part of it.'[95]

In 'To a Young Ass, Its Mother Being Tethered Near It' (first published in *The Morning Chronicle*, 30 December 1794), Coleridge extended his vision of universal brotherhood to include the 'POOR little Foal of an oppressèd race!' (l. 1) he often saw on Jesus Piece, with its mother chained to a log nearby:

> Innocent foal! thou poor despis'd forlorn!
> I hail thee *Brother* – spite of the fool's scorn!
> And fain would take thee with me, in the Dell
> Of Peace and mild Equality to dwell, . . .
> Yea! and more musically sweet to me
> Thy dissonant harsh bray of joy would be,
> Than warbled melodies that soothe to rest
> The aching of pale Fashion's vacant breast!
>
> (ll. 25–8, 33–6)

In two manuscript versions, it is 'high-soul'd Pantisocracy' that dwells in the 'Dell'.[96] Naturally, Coleridge would be conscience-stricken by the need to rid his cottage larder at Nether Stowey of mice.[97] And in writing *The Rime of the Ancient Mariner* (first published 1798, glosses added 1817), he would compose one of the greennest of green poems in the animal rights' vein.[98] Famously anti-crossbow shooting (the poem cannot justify the mariner's shooting of the albatross), the *Ancient Mariner* advocates being kind to all creatures 'great and small':

> Farewell, farewell! but this I tell
> To thee, thou Wedding-Guest!

> He prayeth well, who loveth well
> Both man and bird and beast.
>
> He prayeth best, who loveth best
> All things both great and small;
> For the dear God who loveth us,
> He made and loveth all.
>
> (ll. 610–17)

The poem makes a statement that could be read, in the light of subsequent ecological thinking, as promoting biodiversity, from albatrosses to watersnakes. Coleridge is particularly good at singling out less than obviously cuddly species for his embrace. Thus he could be said to move from the animal-loving and pet-owning kind of argument common to Cowper and many contemporaries to a more ethical and principled kind of argument based on empathy, not identificatory and possessive sympathy, with fellow creatures. Coleridge should be recognized as having contributed to green thinking, and to the tradition of walking poems that are anti-hunting and shooting, in addition to his other accomplishments.

Part II
Hunting a Country

6
Sportswomen

> ... I pass many hours on Horseback, and I'll assure you ride stag hunting, which I know you stare to hear of. I have arriv'd to vast courrage and skill that way, and am as well pleas'd with it as with the Acquisition of a new sense.
>
> Lady Mary Wortley Montagu, Letter to Lady Mar (August 1725)
>
> January 2, died at Barnstaple, in Devonshire, Barbara Snelgrove, but more generally known by the appelation of Granny Bab; in her 96th year, who till within a few days of her death was able to walk to and from the seat of Lord Fortescue, near 12 miles from Barnstaple. She had been, and continued till she was upwards of 94, the most noted poacher in that part of the country, and frequently boasted of selling to gentlemen fish taken out of their own ponds.
>
> *The Sporting Magazine* (January 1795)

Do women have a relationship with the natural world different from men's? Although sporting culture was highly masculinized, during the sixteenth, seventeenth and eighteenth centuries women at both ends of the social scale hunted, hawked, coursed, netted, shot and fished. Katharine Howard, wife of Henry, Lord Berkeley, and granddaughter of the Duke of Norfolk, was as keen on field sports as her husband, whose 'hounds were held inferior to no mans'.[1] During Queen Mary's reign and the first 13 years of Elizabeth I's, Lady Berkeley followed her husband's hounds as they progressed across country, 'hunting the hare fox and deere, red and fallow' (2: 363), giving 'her self to like delights as the Country usually affordeth' and 'delighting her crosbowe' (2: 285). The crossbow was commonly used for sporting.[2] More unusual was Lady Berkeley's exceptional skill with the difficult, warlike long bow (2: 285). She also 'kept commonly a cast or two of merlins', which sometimes she mewed in her own chamber, 'which falconry cost her husband each yeare one or two gownes and kirtles spoiled by their mutings' (2: 285). Among northern gentry families in the eighteenth century, hunting and shooting were discussed primarily in terms of male competition and camaraderie, but women also participated.[3] Poaching, 'a national pastime in Tudor and early Stuart England', involved both peers and smallholders, servants, adolescents and women, as well as 'masterless men'.[4] Poaching remained a national pastime well into the nineteenth century. For every Barbara Snelgrove, an accomplished fish poacher, who was reported in the press, there were very likely dozens in every part of the country whose activities were not recorded. Granny Bab was, after all, 'the most noted poacher' in the district of Barnstaple, North Devon, regardless of gender.[5]

Science and sporting culture: women in the field

The exclusion of women from field sports which began late in the eighteenth century coincided with their exclusion from science.[6] Neither exclusion was complete, but women's participation in both fields began to be actively discouraged. 'In the early years of the scientific revolution, women of high rank were encouraged to know something about science', and noblewomen 'continued to participate in these informal scientific networks until late in the eighteenth century'.[7] The sporting Lady Berkeley in 'her elder years gave her self to the study of natural philosophy and Astronomy', acquiring a globe, 'a quadrate, Compass, Rule, and other instruments, wherein shee much delighted her self till her death'.[8] It is significant that after the death of Margaret Cavendish in 1674, no woman in England wrote so boldly on natural philosophy.[9] The very category of gender here – men as opposed to women, men commanding the field of nature and women watching from the sidelines – stands exposed in sharp relief as a power differential. Such gender dynamics give the whole question of women's sporting and scientific activities as an object of study a new coherence and urgency.

There were, for example, a remarkably high proportion of women in the field of entomology in the mid-eighteenth century, including Mary Somerset, Duchess of Beaufort, who first recognized that every species of butterfly and moth has its own special food plant. In her gardens at Badminton and Chelsea, guided by the leading botanists of the day, she established 'almost unrivalled collections of rare exotic plants, of which she left a large herbarium'.[10] Later generations of women in the Somerset family have distinguished themselves by writing books about the hunting field.[11] Another leading woman in botanical circles, and the 'paradigmatic aristocratic woman collector of the eighteenth century', was Margaret Harley Bentinck, Duchess of Portland.[12] Employing such naturalists as James Bolton to collect lichens, the

Rev. John Lightfoot to catalogue her collection, Daniel Solander to be curator of her museum, with Richard Pulteney and Thomas Yeats to assist, G.D. Ehret to paint her plants and William Lewin her British birds and eggs, Margaret Bentinck both patronized and promoted the study of natural history from her estate at Bulstrode in Buckinghamshire.[13]

Hunting, like science, offered women opportunities for studying the natural world and exercising agency beyond the boundaries of domesticity. Ann Shteir emphasizes that Margaret Harley's marriage to William Bentinck in 1734 'brought her into a family with distinguished botanical ancestry', in that the first Earl of Portland had been superintendent of William III's gardens at Hampton Court.[14] But there is every reason to connect Margaret's interests in natural history with her own family as well as her husband's, for Margaret Harley was the only daughter and heiress of Edward and Henrietta Harley, the Earl and Countess of Oxford. Lady Henrietta Cavendish Holles, who married in 1713 Edward Harley, son of the Earl of Oxford and friend of Pope, seems to have inherited more than a fortune from her Cavendish forebears, including William, the first Duke, and Margaret, the Duchess of Newcastle.[15] Henrietta Harley had her own pack of harriers and was painted by John Wootton hunting them (Plate 5: *Lady Henrietta Cavendish Holles, Countess of Oxford hunting at Wimpole Park* [1716]).[16] She was a devotee of the field in one sense, her daughter Margaret in another.

Women, hunting and poetry in the seventeenth century

When in 1611 Aemilia Lanyer initiated the genre of the country-house poem with 'The Description of Cooke-ham' in *Salve Deus Rex Judaeorum*, she celebrated her patroness as a huntress and the estate as surrounded by fine hunting country. From her favorite seat beneath Cookham's tallest oak, Margaret, Countess Dowager of Cumberland, could view a prospect encompassing 'thirteen shires'. Lanyer opines that there was probably not a finer view to be had in Europe:

> Where beeing seated, you might plainely see,
> Hills, vales, and woods, as if on bended knee
> They had appeard, your honour to salute,
> Or to preferre some strange unlook'd for sute:
> All interlac'd with brookes and christall springs,
> A Prospect fit to please the eyes of Kings:
> And thirteen shires appear'd all in your sight,
> Europe could not affoard much more delight.
>
> (ll. 67–74)[17]

What does Margaret watch from her commanding heights in the Thames valley but episodes of venery in the surrounding shires? Aristocratic stewardship and national superiority are conflated. The very landscape pays homage to its mistress, or begs for favors from her; she commands all things within the precincts of the estate. At the same time, this prospect – a crown manor, and fit for kings – exceeds all others in the delight it promises.[18] This is a green world not likely to be surpassed, but the tranquility it affords is in direct proportion to the suffering exacted from Margaret and her daughter, Anne Clifford, Countess of Dorset, because of their estrangement from other estates; they have been battling James I and court bureaucracy on behalf of Anne's claim to her late father's Cumberland lands and titles.

The image of Margaret Clifford presented is a formidable one, not least when she appears as both a naturalist, comforted by the sight of wild creatures disporting themselves, and as a huntress, exercising aristocratic prerogative over the natural world:

> The little creatures in the Burrough by
> Would come abroad to sport them in your eye;
> Yet fearefull of the Bowe in your faire Hand,
> Would runne away when you did make a stand.
>
> (ll. 49–52)

A 'stand' is both a bodily stance for taking aim in archery, and a platform erected in a park from which to shoot. The mistress of an estate has proprietary rights over its animal inhabitants as presenting sporting opportunities. Yet who can miss the allegorical implication that the Countess's enemies should beware when she elects to 'make a stand', as in the defense of her daughter's interests? The 'little creatures in the Burrough' know this much at least, which would seem to be more than the King and his advisers know.

Perhaps the most extraordinary confluence of hunting and poetry in writing by a woman occurred in Margaret Cavendish's 'The Hunting of the Hare' and 'The Hunting of the Stag' from *Poems, and Fancies* (1653).[19] These poems succeed in conjuring the hunting field in brilliant detail, invoking all five senses, only to end by criticizing hunting as an enactment of human tyranny over the animal world. These poems could only be the product of a close acquaintance with field sports, though a vexed one.

What, then, was Margaret Cavendish's experience of hunting? The Duke of Newcastle was a keen sportsman and horseman, and the Duchess always wrote of his sporting interests sympathetically. She reported after their return from royalist exile in 1660 her husband's sadness at the devastation of his estates, particularly Clipston-Park, which had been full of Fish and Otters; was well stock'd with Deer, full of Hares, and had great store of Partriges, Poots, Pheasants, &c, besides all sorts of Water-fowl; so that this Park afforded all manner of sports, for Hunting, Hawking, Coursing, Fishing, &c. for which my Lord esteemed it very much.

The Duke immediately set about having this park repaled, and 'got from several Friends Deer to stock it'.[20] In 'A True Relation of my Birth, Breeding and Life', the former Margaret Lucas described her brothers as often 'fencing, wrestling, shooting, and such like', but seldom hawking or hunting, and herself as 'tender-natured, for it troubles my conscience to kill a fly, and the groans of a dying beast strike my soul'.[21] In *Philosophical and Physical Opinions* (1655), Cavendish remarked that although it was 'a usual custome, for Ladies and women of quality, after the hunting a Deer, to stand by until they are ript up, that they might wash their hands in the blood, supposing it will make them white, yet I never did'.[22] In *Orations of Divers Sorts* (1662), she advocated women's participation in field sports:

> [T]o shew Men we are not so Weak and Foolish, as the former Oratoress doth Express us to be, let us Hawk, Hunt, Race, and do the like Exercises as Men have . . . we should Imitate Men, so will our Bodies and Minds appear more Masculine, and our Power will Increase by our Actions.[23]

Exemplifying Cavendish's oratorical virtuosity, the next speaker declaims against such 'Hermaphroditical' ventures.[24] We can fix Cavendish in neither one rhetorical posture nor the other.

Cavendish's materialist philosophy might offer a firmer ground for understanding the seeming contradictoriness of her views on hunting. Influenced by the vitalist debates of the mid-seventeenth century, in *Philosophical Letters* (1664), Cavendish attributed 'Life and Knowledge, which I name Rational and Sensitive Matter' to 'every Creature'.[25] Sense and reason were present in all Creatures, not only Man and Animals, Cavendish argued, for 'Vegetables will as wisely nourish Man, as Men can nourish Vegetables; Also some Vegetables are as malicious and mischievous to Man, as Man is to one another, witness Hemlock, Nightshade, and many more'.[26] The effect of Cavendish's argument is a democratizing of relations between humans and other species, even between humans and other forms of matter – vegetables and minerals. There is certainly a basis here for an ecological, non-instrumental relation to the natural world. 'Like blood for [William] Harvey in 1649 and human tissue for [Francis] Glisson in 1650, matter for Cavendish in 1663 [and after] possesses attributes of motional self-determination hitherto reserved for thinking, soulful human beings,' as John Rogers observes.[27] Rogers argues that much of Cavendish's scientific writing was 'devoted to thwarting the Hobbesian conjecture of the ultimate priority of force',[28] and he grounds Cavendish's liberal feminism in this commitment.[29] Attempting to thwart the rule of force also involves Cavendish in questioning human domination over other species, as we shall see.

In 'The Hunting of the Hare', Cavendish demonstrates an empathetic apprehension of the hare as a separate but fellow creature, closely observed for his own being and behavior. This description we would now call 'ecologically aware'; it is a field naturalist's description:

> BEtwixt two *Ridges* of *Plowd-land*, lay *Wat*,
> *Pressing* his *Body* close to *Earth* lay squat.
> His *Nose* upon his two *Fore-feet* close lies,
> Glaring obliquely with his *great gray Eyes*.
> His *Head* he alwaies sets against the *Wind*;
> If turne his *Taile*, his *Haires* blow up behind:
> Which *he* too cold will grow, but *he* is wise,
> And keepes his *Coat* still downe, so warm *he* lies.
> Thus resting all the *day*, till *Sun* doth set,
> Then riseth up, his *Reliefe* for to get.
> Walking about untill the *Sun* doth rise,
> Then back returnes, downe in his *Forme he* lyes.
>
> (ll. 1–12)

The habits of the hare – squatting between the ridges of a ploughed field to be invisible, turning head to the wind to keep warm, resting inconspicuously by day, and finding relative freedom to move and feed by night – are detailed in this passage. Watching hares being hunted taught observers about hare behavior and concentrated their minds on the animal's point of view in the struggle for survival. During the hunt, Wat takes advantage of hounds having lost the line to clean himself:

> On his two *hinder legs* for ease did sit,
> His *Fore-feet* rub'd his *Face* from *Dust*, and *Sweat*.
> Licking his *Feet*, he wip'd his *Eares* so cleane,
> That none could tell that *Wat* had hunted been.
>
> (ll. 39–42)

The hare's feeding and sheltering habits, his nocturnal movements and return to his 'form', his pausing to wash during a check, are described in a manner simultaneously empathetic and scientific.

In 'The Hunting of the Stag', the companion-piece to 'The Hunting of the Hare' in *Poems, and Fancies*, Cavendish similarly offers close, scientifically observed details of what is species-specific about stags. The stag's '*Nervous*' legs and 'strong' '*Joynts*', his predilection for the green shoots of a crop of wheat, his lying down to rest after feeding, are all naturalistic observations (ll. 4, 53–5). Cavendish's stag is a moral being as well as a creaturely one, resembling in his vanity and pride Denham's stag in *Coopers Hill*: 'Taking such *Pleasure* in his *Stately Crowne*, / His *Pride* forgets that *Dogs* might pull him downe' (ll. 13–14). We are in the realm of royalist allegory: ''Twas not for want of *Courage* he did run, / But that an *Army* against *One* did come' (ll. 125–6). Yet in Cavendish's poem, unlike Denham's, we never lose sight of the struggling animal in the field. As the treatises on hunting so popular in the sixteenth and seventeenth centuries emphasized, knowledge of animal physiology, habitats and behavior was inseparable from hunting practice. This poem too is worthy of classification in the tradition of natural history associated with Gilbert White, though Cavendish was writing more than a century earlier.

In both poems, but more successfully in 'The Hunting of the Hare', Cavendish capitalized on the traditional link between English poetry and hunting to achieve some of her best verse. The mysteries of scent, the eagerness and music of the hounds whose voices are classified according to their breeding, and the competitive thrusting of the riders, all provide opportunities for Cavendish to demonstrate her poetic skill grounded in her knowledge of the hunting field:

> Then every *Nose* is busily imployed,
> And every *Nostrill* is set open, wide:
> And every *Head* doth seek a severall way,
> To find what *Grasse*, or *Track*, the *Sent* on lay.
> *Thus quick Industry, that is not slack,*
> *Is like to Witchery, brings lost things back.*
> For though the *Wind* had tied the *Sent* up close,
> A *Busie Dog* thrust in his *Snuffling Nose*:
> And drew it out, with it did foremost run,
> Then *Hornes* blew loud, for th' *rest* to follow on.
> The *great slow-Hounds*, their throats did set a *Base*,
> The *Fleet Swift Hounds*, as *Tenours* next in place;
> The little *Beagles* they a *Trebble* sing,
> And through the *Aire* their *Voice* a round did ring;
> Which made a *Consort*, as they ran along;
> If they but *words* could speak, might sing a *Song*,
> The *Hornes* kept time, the *Hunters* shout for *Joy*,
> And valiant seeme, *poore Wat* for to destroy:

> Spurring their *Horses* to a full *Careere*,
> *Swim Rivers deep, leap Ditches* without feare;
> Indanger *Life*, and *Limbes*, so fast will ride,
> Onely to see how patiently *Wat* died.
>
> (ll. 57–78)

Cavendish finds her own most musical voice in the music of hounds, but she doesn't enter into their view of things for long. The poem ends with a protest against blood sports as a despoliation of nature:

> When they do but a *shiftlesse Creature* kill;
> To hunt, there needs no *Valiant Souldiers* skill.
> But *Man* doth think that *Exercise*, and *Toile*,
> To keep their *Health*, is best, which makes most spoile . . .
> And that all *Creatures* for his sake alone,
> Was made for him, to *Tyrannize* upon.
>
> (ll. 85–8, 105–6)

Brilliant verse derived from hunting's pleasures turns to a moral injunction against its cruelty, especially when the beast of chase is a '*shiftlesse Creature*', lacking resources for defense. Cavendish, it would seem, in spite of the vigorous excitements of the chase, cannot help protesting against the death of the hare as an instance of man's tyranny. She empathizes with the hare's fear and identifies with animal suffering.

What is gender but a differential relation to power? Cavendish's materialist philosophy, with its democratizing of sense and reason, its extension of vitality throughout the commonwealth of matter, seems designed to challenge patriarchal relations and social hierarchies, but stops short of doing so.[30] The Duchess often sounds resigned to her gendered subordination. Apologizing to readers for her inadequate command of human anatomy because she had never been able to attend a dissection, Cavendish pleaded her gender as an excuse: she would have liked to have known about 'intrals' in their proper place in the body, but 'found that neither the courage of nature, nor the modesty of my sex would permit me'.[31] Her sex aligns her with the '*shiftlesse*' animal creation, subject to the superiority of masculine force. She does not baulk at lecturing sportsmen on the cruelty of blood sports, but nor has she retired from the field, nor relinquished the privileges of rank, field sports among them.

Characterizing this poem as 'very unremittingly sad and painful', Nigel Smith opines that the modern reader 'must wonder how a seventeenth-century person, used to hunting as part of everyday life, would have responded' to it.[32] We can recover something of seventeenth-century reception by entering into the mentality induced by traditional hunting discourse. English poetry had been attuned to hunting rhythms for some centuries. The literature of hunting, with its resonances of fable and moral allegory, was itself often dialogic in its address to the animal world, and often critical of human actions. Consider the genre of hunting poems, within which the plea from the prey to the hunter or other human audience was traditional. 'The Hare, to the Hunter', which ends Gascoigne's treatment of hare-hunting in *The Noble Arte*, would, out of context, be bound to be called an anti-blood sports poem.[33] Humans pursuing hounds chasing hares could be said to constitute an immersion in the natural world rather than a tyrannizing over it. But such an immersion is not what happens at the death in Cavendish's poem.

'The Hunting of the Hare' might well be read as protesting masculine swagger and swank. The hounds' efforts have been forgotten in a moment of human braggadocio. And the hare's death is merely an instance of trophy-taking or spoilage rather than the consequence of a contest in which either hare or hounds must be given best. Cavendish's poems succeed as poems of the hunting field, while remaining critical of human behavior in the field. The Lady Newcastle of 1653 knew what she knew as a badge of her rank: naturalistic knowledge derived from time spent in the hunting field must be acknowledged in Cavendish's works alongside her criticism of blood sports.

Hunting women in the eighteenth century

Women like Cavendish made themselves unpopular in the hunting field. By 1711, the horsy Englishwoman was already a social type being ridiculed by the polite press, an ironical parallel to the effeminate man.[34] Here is Addison in *The Spectator*:

> I have very frequently the Opportunity of seeing, a Rural *Andromache*, who came up to Town last Winter, and is one of the greatest Fox Hunters in the Country. She talks of Hounds and Horses, and makes nothing of leaping over a Six bar Gate. If a Man tells her a waggish Story, she gives him a Push with her Hand in jest, and calls him an impudent Dog; and if her Servant neglects his Business, threatens to kick him out of the House.[35]

Andromache's mistake seems to have been her rural assumption that being sporting and recklessly brave was fun, and that being physical, especially with men, was appropriate behavior for a woman.

By 1730, this object of anxiety was too threatening to portray directly. Instead, there was a wishful imperative that the British fair sex be immune to the 'horrid joy' of hunting: 'Far be the spirit of the chase from them,' wrote James Thomson hopefully in *The Seasons* ('Autumn' [1730; revised

1744], ll. 571-3). 'Autumn proceeds by tracing man's change from primitivism to patriotism,' as Ralph Cohen has observed.[36] Thomson's is an apotropaic gesture, a warding off of what was most strongly feared. At the same moment in 'Autumn' that this Scottish poet was endeavoring to construct a national discourse of identity, addressing his readers as 'Britons' (l. 40), he satirized as cruel and debauched the ethnically English preoccupation with all forms of blood sports. He then immediately set about defining an ideal British femininity – as decidedly *not* of the English riding, hunting, swearing, drinking variety. Dismissing shooting, hare-hunting and stag-hunting as barbaric – 'These are not subjects for the peaceful muse' (l. 379) – Thomson recommended the abolition of blood sports to rescue women from hunting men. It would be more delightful to the peaceful muse to perceive as 'social':

> The whole mixed animal creation round
> Alive and happy. 'Tis not joy to her,
> This falsely cheerful barbarous game of death,
> This rage of pleasure which the restless youth
> Awakes, impatient, with the gleaming morn.
>
> (ll. 381-6)

Unlike animal predators, 'the steady tyrant, man' (l. 390) hunts not from necessity, but for sport:

> Who, with the thoughtless insolence of power
> Inflamed beyond the most infuriate wrath
> Of the worst monster that e'er roamed the waste,
> For sport alone pursues the cruel chase
> Amid the beamings of the gentle days.
>
> (ll. 391-5)

Especially since Britain lacks lions, wolves and boar – worthy, because aggressive, beasts of chase – hunting should be deplored (ll. 458-69). But if Britons must hunt, let them 'pour' their 'sportive fury' on the fox, 'the nightly robber of the fold' (ll. 471-2), that predator upon sheep, poultry and other forms of property.

Thomson commanded his fellow Britons to ride hard to 'the thunder of the chase' (l. 474), but with gathering irony:

> Throw the broad ditch behind you; o'er the hedge
> High bound resistless; nor the deep morass

>   Refuse, but through the shaking wilderness
>   Pick your nice way; into the perilous flood
>   Bear fearless, of the raging instinct full;
>   And, as you ride the torrent, to the banks
>   Your triumph sound sonorous, running round
>   From rock to rock, in circling echo tost;
>   Then scale the mountains to their woody tops;
>   Rush down the dangerous steep; and o'er the lawn,
>   In fancy swallowing up the space between,
>   Pour all your speed into the rapid game.
>   For happy he who tops the wheeling chase;
>   Has every maze evolved, and every guile
>   Disclosed; who knows the merits of the pack;
>   Who saw the villain seized, and dying hard
>   Without complaint, though by an hundred mouths
>   Relentless torn: O glorious he beyond
>   His daring peers, when the retreating horn
>   Calls them to ghostly halls of grey renown,
>   With woodland honours graced – the fox's fur
>   Depending decent from the roof, and spread
>   Round the drear walls, with antic figures fierce,
>   The stag's large front: he then is loudest heard
>   When the night staggers with severer toils,
>   With feats Thessalian Centaurs never knew,
>   And their repeated wonders shake the dome.
>   But first the fuelled chimney blazes wide;
>   The tankards foam; and the strong table groans
>   Beneath the smoking sirloin, stretched immense
>   From side to side, in which with desperate knife
>   They deep incision make, and talk the while
>   Of England's glory, ne'er to be defaced
>   While hence they borrow vigour; . . .
>
>   (ll. 475–508)

It is easy to see why Thomson's 'Argument' for 'Autumn' described this passage as 'A ludicrous account of foxhunting'. Fox-hunters are represented as a fraternity of braggarts, not so much pursuing a fox as opportunities for advertising their own prowess, trumpeting their triumph in crossing a river, priding themselves on so anticipating every trick of the fox as to 'top' the chase and be at the forefront of the mounted field. Thomson relishes sending up the self-congratulatory complacency with which fox-hunting men describe their sport. Fancying himself an heroic thruster, each and every foxhunter sees himself as that 'happy he' who knows his country, knows his hounds (the 'merits of the pack'), and is thus always well up with hounds and in at the death. The fox, always a 'villain', inevitably dies 'hard' but 'Without complaint', in spite of being torn apart by hounds. And the evening after a day's hunting is entirely spent in re-telling and embellishing the day's exploits while getting grotesquely drunk. 'The fur hangs and the sirloin is "stretch'd immense"; England's "Glory" is connected with the grossness of gluttony,' as Cohen puts it.[37] The chase itself occupies 17 lines, the après-chase requires 77 lines before climaxing in a drunken stupor, with the fox-hunting brethren 'drenched in potent sleep till morn' (l. 564). Finally the parson – 'some doctor of tremendous paunch, / Awful and deep, a black abyss of drink' (ll. 565–6) – having drunk everyone else under the table, retires to bed, lamenting how weak the present generation of fox-hunters has become (ll. 568–9).

The next verse paragraph begins, without any break:

> But if the rougher sex by this fierce sport
> Is hurried wild, let not such horrid joy
> E'er stain the bosom of the British fair.
> Far be the spirit of the chase from them!
> Uncomely courage, unbeseeming skill,
> To spring the fence, to reign the prancing steed,
> The cap, the whip, the masculine attire
> In which they roughen to the sense and all
> The winning softness of their sex is lost.
> ...
> O may their eyes no miserable sight,
> Save weeping lovers, see! a nobler game,
> Through love's enchanting wiles pursued, yet fled,
> In chase ambiguous. May their tender limbs
> Float in the loose simplicity of dress!
> ...
> To give society its highest taste;
> Well-ordered home man's best delight to make;
> And, by submissive wisdom, modest skill,
> With every gentle care-eluding art,
> To raise the virtues, animate the bliss,
> Even charm the pains to something more than joy,
> And sweeten all the toils of human life:
> This be the female dignity and praise.
>
> (ll. 570–8, 586–90, 602–9)

From the mode of satire to the mode of moral precept, just like that. Why do femininity and fox-hunting come together in 'Autumn' at all? Because horsewomanship and the excitements and dangers of the hunting field might be erotically stimulating – venery was venery, after all – and hunting did give women agency in a space of movement and pleasure antithetical to domesticity. The equation of the female and the feminine with the domestic was what was most threatened by female horsemastership and hunting knowledge. That was a risk which certain gentlemen thought they could not afford, or in any case would not brook.

Thomson had his own reasons for scorning hunting: he was an urban person, renowned for being fond of food and drink; the last work he published was *The Castle of Indolence*.[38] From Thomson's point of view, the hunting field henceforth should be exclusively the preserve of men, the proving ground of their identity, apart from women. And fox-hunting should be the sole legacy of a tradition of bloody field sports he found both risible and abhorrent for a polite, benevolent and thus deservedly imperial nation.

What the women said

What did women themselves have to say about hunting? Compared with the rich visual record, the eighteenth-century written evidence is frustratingly sketchy. In a letter to her husband Edward Harley dated 3 May 1718, Lady Henrietta seemed entirely preoccupied with assuring that proper arrangements would be made for the arrival of her horses at Bramton Castle, in Herefordshire, the Harley family seat:

> I hope you wil go to Bramton & give Directions there for what you judge necessary.
> I wish you would order the Stable wch is thatch'd to be made up for my own Horses. & then that that near the House wil be for Strangers.
> I have bought Six Coach-Geldings & a pad. Old Hey & Oats ought to be provided. I would have that wch is the Carters Stable clean'd & got ready. & the Coach Horse Stable & the others to be put there in order, the middle of those Stables to be for – my own Saddle Horses.[39]

This is the letter of a keen horsewoman. Of her 18-month-old daughter, Margaret, Lady Henrietta had written to her father-in-law, Robert, Earl of Oxford in 1716: 'Peggy is well and does indeed love Horses as well as her Father or Mother; she has cut a great Tooth.'[40] But we haven't any other evidence of Lady Henrietta Harley's pleasure or agency in the chase from her own hand so far as I have been able to determine; she seems not to have written about it. However, her distant cousin and close girlhood friend,

Lady Mary Pierrepont (after marriage, Lady Mary Wortley Montagu) did write about riding and hunting. The two were near neighbors in Nottinghamshire, Lady Mary at Thoresby, and Lady Henrietta at Welbeck Abbey.[41] Much of Mary's early experience of hunting was likely to have been with Henrietta's harriers. It was, however, stag-hunting with the Prince of Wales, soon to be George II, that converted Lady Mary to a keen participant in field sports, suggesting how important the differences between kinds of hunting could be. In August 1725, Lady Mary wrote to her sister, Lady Mar:

> I pass many hours on Horseback, and I'll assure you ride stag hunting, which I know you stare to hear of. I have arriv'd to vast courrage and skill that way, and am as well pleas'd with it as with the Acquisition of a new sense. His Royal Highness hunts in Richmond Park, and I make one of the Beau monde in his Train.[42]

The acquisition of 'vast' courage and skill in stag-hunting was as pleasing to Montagu as an entirely new form of perception. Another member of the circle, Lady Isabella Finch, reported stag-hunting at Windsor in the 1730s had left her 'Ears . . . full of dust and dirt' after 'a noble Chace near 50 miles in all tho not indeed end ways and we thought twas pretty fair to ride tired horses 12 miles home after ye Sport was over. Princess royal rides so well yt tis a pleasure to follow her.'[43] Here is agency, here is pleasure and here, undoubtedly, is proximity to the 'Beau monde', the space of celebrity as well as sociability.

Henrietta Harley would have experienced something of these pleasures. Lacking the *frisson* of proximity to the royals, she would have had instead the more private pleasure of having her own pack of harriers for hunting in her own country. Wootton's paintings of her vividly convey the thrill of speed and sense of athletic movement over a hilly landscape. Pursuit, not killing, was the object, and the hare is represented as giving the pack a good run. Accompanied by various male attendants and friends, the female master is showing good sport, making it a pleasure to follow her. But hare-hunting, though it had its moments of speed and dash, was not a fast, extended chase, covering miles of country, as stag-hunting was and as fox-hunting was rapidly becoming. It was the hare's 'doubling wiles' that delighted hare-hunters.[44] Eventually, a hare would circle back over the same ground, the field retracing their steps over the same terrain and obstacles, which were less hazardous to negotiate the second time around: '[W]hen Puss is started, she seldom fails to run a Ring, the *first* is generally the *worst* (for Horse or Foot) that may happen in the whole Hunt. For the Fences once leaped, or the Gates once opened, makes a clear passage oftentimes, for every Turn she takes afterward.'[45] The whole field was likely to remain within view as well as hearing distance of hounds for the entirety of the chase. Indeed:

> This slow kind of hunting was admirably adapted to age and the feminine gender; it could be enjoyed by ladies of the greatest timidity as well as gentlemen labouring under infirmity; to both of whom it was a consolation, that if they were occasionally a little way behind, there was barely a possibility of their being thrown out. A pack of this description was perfectly accommodating to the neighbouring rustics, the major part of those not being possessed of horses found it a matter of no great difficulty to be well up with them on foot ... Hare-hunting, however, though not carrying with it all those brilliant rays of diversity so very attracting to those who delight in a recital of the toils, dangers, and difficulties of the day; yet, to the contemplative naturalist, much more of the true spirit of hunting, and the instinct of animals, is to be observed and enjoyed, than in either of the other two [stag- and fox-hunting].[46]

Hare-hunting, whether with beagles or larger harriers, was not only suitable for ladies, but could easily include 'rustics' as foot followers. In both of Wootton's paintings of Lady Henrietta Harley hunting hares, there are indeed rustically garbed men on foot who are following and cheering on the hounds. They are not among the liveried hunt servants, but they could be helping to hunt hounds by whipping in, or they could simply be country folk who have joined in; their status is ambiguous. But their presence contributes to the sense given by these pictures that the whole countryside has been caught up in the excitement of the chase.

By 1860 this sort of social informality, and the interest in the naturalistic aspects of hunting which a slower pace had made possible, was being dismissed as so old-fashioned as hardly to be worthy of being called hunting at all. In the self-proclaimed first treatise on riding aimed entirely at ladies rather than gentlemen, Mrs. J. Stirling Clarke remarked:

> In speaking of hunting, I do not refer to a mere gallop with the harriers on the Brighton Downs, or similar places; for, if they are properly mounted for the purpose, such exercise may be considered legitimate ladies' hunting. They have not far to go from home; they can join them when they like, and leave when they like; and last, but not least, there are usually several of their own sex present to assist them in case of accident. But my counsel and caution refer to the more exciting and hazardous sport of foxhunting, which calls for consummate skill in riding, – a knowledge of the sport that can *alone* be acquired by experience and practice, – the most determined courage, and great power of enduring fatigue – a spirit that holds in contempt both wind and storm, – and a constitution that sets at defiance coughs, colds, and rheumatism. In almost every part of the country two or

three ladies are to be found who greatly distinguish themselves in the hunting field . . .[47]

Hare-hunting, once an exciting diversion that was also a common occurrence in the country, has been reduced to a 'mere gallop with the harriers'. By 1860 hunting was all about riding and very little about hunting; we hear nothing of hares or foxes and their habits, nor about hounds, their music and their cleverness.

This subordination of hunting as hunting to hunting as bold riding had begun as early as the eighteenth century in certain forms of chase. In the eighteenth century, women who fancied themselves as riders rode to staghounds if they could. Neither Montagu nor Finch mentions anything about hounds or hunting; they were only aware of their own experience of riding. Henrietta Harley, hunting her own pack, would have had a different sense of the activity — a naturalist's and sportswoman's sense of hares, hounds, hunting techniques and a detailed knowledge of the country.

By 1782, according to an anonymous poem, *The ROYAL CHASE; A POEM. Wherein Are Described Some humorous Incidents of a Hunt at Windsor*, stag-hunting had become a venue for male thrusters to show off before an audience of women, also vying amongst themselves for pre-eminence on the coachbox, while engaging in celebrity-spotting. Everyone turned out to catch the eye of the notoriously sporting Prince of Wales, who would become George IV. In the autumn of 1781 the Prince described hunting as that 'almost divine amusement', reporting to his brother Frederick that he had 'plenty of excellent & beautiful horses' and that he rode 'ym. well up to ye [sic]', hunting four times a week — twice with the 'King & his hounds' and twice with their uncle the Duke of Cumberland and his.[48] Equestrian prowess was the key to royal favor:

> AND now more pleasing scenes engag'd the eye,
> The stag in view, the huntsmen in full cry:
> O'er hedge and ditch contending nobles flew,
> And glorying in past toils they sought for new;
> Foremost of which appear'd the daring youth,
> Who mock'd such dangers, as, to speak the truth,
> Prov'd him by far more obstinate than wise,
> And proud of every joy the chase supplies.[49]

The 'daring youth' is none other than the Prince, the poem's addressee. To attract his attention young women drove daring equipages to the meet. The stag seems almost incidental to this 'royal chase', with such a human quarry in view. According to the anonymous poet, the stag-hunt had lost its allegorical power to signify great matters of state. Stag-hunting had become entirely a space of sociability and celebrity.

Here, as in Montagu's and Finch's reports, the presence of royalty added a note of ceremony and social exclusivity which would have been absent from hare- and fox-hunting, and seems to have proved especially attractive to women. Later eighteenth-century fox-hunting had its notable female thrusters, such as the Marchioness of Salisbury, who hunted a pack of dwarf foxhounds at Hatfield:

> THE Hertfordshire hunt had, what they call, a most gallant day last week; having run a burst of more than an hour, they crossed upon a fresh fox at Branfield, and clattered him two hours and a half more, when he was run to *earth* near *Baldock*. The two chaces, which admitted of no interruption from hard running were full forty miles in extent! Out of a field of fourscore, only *nine* were in at the earth, at the head of whom was Lady Salisbury. After giving honest Daniel, the old huntsman, the go by, she pressed Mr. Hale *neck* and *neck*, – soon blowed the Whipper-in, and contrived, indeed, through the whole chace, to be always nearest the *brush*![50]

Despite such publicized runs as these, the royal staghounds continued to attract the most notably accomplished horsewomen, whose exploits were covered in the new sporting press (*The Sporting Magazine* began in 1792). What Laetitia Lade (*née* Darby, alias Smith) actually thought about stag-hunting we do not know, but her appearance in the hunting field was always news. Celebrity status attached itself to this female rider of vast courage and skill whose origins were obscure. She was reputed to be a gypsy, and to have been a servant in a brothel (perhaps a cook) at Broad Street, St. Giles, and the mistress of a highwayman, John Rann, before meeting the Prince of Wales at a masquerade in 1781 and marrying his friend and equestrian adviser, Sir John Lade, in 1787.[51] The Prince of Wales both commissioned a portrait of her mastering a rearing horse from Stubbs in 1793 (Col. Pl. 3) and coined the phrase 'to swear like Lady Lade'.[52] As Stella Walker remarks, 'Immorality and foul language may have been part of her image, but her skill in horsemanship was renowned, a fact indicated by Stubbs in the prancing impetuosity of her horse and the telltale restraining rein for a "puller" running from bridle to saddle.'[53] In 1794 *The Sporting Magazine* published an engraving entitled 'The Accomplished Sportswoman' (Fig. 6) clearly alluding to Lady Lade as 'a lady of rank and fortune, leaping over a five-barred gate in the neighbourhood of Windsor, and who may be frequently seen with his majesty's hounds during the hunting season'.[54] In 1796 she was celebrated as a rider of thoroughbred horses, the only mounts that could keep pace with the royal staghounds, and pronounced 'the first horsewoman in the kingdom, being constantly one of

Figure 6 'The Accomplished Sportswoman', opposite page 154 from *The Sporting Magazine*, Volume 4, June 1794. Reference shelfmark, Vet.A5 e.875. By courtesy of the Bodleian Library, University of Oxford.

the only five or six that are invariably with the hounds'.[55] She lived until 1825, inspiring in 1849 the 'Amazon', Letitia, Lady Lade, in George W.M. Reynolds's *The Mysteries of the Court of London*, a character who is introduced in vol. 1, p. 56, and appears frequently thereafter.

By the end of the century, especially in the hard-riding, hard-drinking and sexually sporting circles within which the Prince of Wales moved, there were avenues to upward social mobility for women in the hunting field. Horsemastery could lead to concubinage and sometimes marriage. Combined with the aerobic euphoria of the chase, an opportunity to demonstrate a mastery of horseflesh few men could equal, and proximity to the *beau monde*, the prospect of social advancement must have been a powerful elixir. Sporting females with doubtful origins as horse trainers, circus performers and actresses, who married into the wealthy gentry and aristocracy, became stock figures in nineteenth-century writing. The 'beautiful and tolerably virtuous' Lucy Glitters, creation of Robert Smith Surtees, takes the followers of Sir Harry Scattercash's hounds by storm (Fig. 7), having most recently displayed her equestrian skills in 'flag-exercises' at 'Astley's Royal Amphitheatre' in London.[56] Mounting her old friend Lady Scattercash's Arab palfrey, White Surrey, Lucy shows how the thing should be done:

> Taking the horse gently by the mouth, she gave him the slightest possible touch of the whip, and moved him about at will, instead of fretting and fighting him as the clumsy, heavy-handed [Mr. Orlando] Bugles had done. She looked beautiful on horseback, and for a time riveted the attention of our sportsmen.[57]

Where once upon a time the wives and daughters of country gentlemen could hunt without comment, in the nineteenth century, there was constant debate about whether respectable women did or should hunt.

The side-saddle as a gender machine

When fox-hunting, with its potential for a whole day's hard galloping and jumping, became more fashionable than stag- or hare-hunting, women were at a particular disadvantage. Their presence in the field became increasingly exceptional for technical as well as social reasons. A side-saddle may not have moving parts (it is the woman using it who moves, grips, balances), but it functions as a social machine that distinguishes 'women' from 'men'.[58] The side-saddle seat supposedly protected a woman from damaging herself on the cross-saddle – after all, ruptures were a common complaint of cavalrymen forced to ride with over-long stirrups on rough-gaited horses and to sit the trot, rather than rising to it, well into the nineteenth century.[59] Women, it

164

Figure 7 John Leech (1817–64), 'Lucy Glitters showing the way', opposite page 379 from R.S. Surtees, *Mr Sponge's Sporting Tour* (London: Bradbury & Evans, 1853). Reference shelfmark, C.70 d.8. By courtesy of the British Library, London.

was feared, might also be insufficiently muscular in their thighs to maintain a strong seat, especially on an active horse: 'As they have round and weak thighs, women can never develop a firm seat on the cross-saddle,' declared James Fillis in 1890.[60] Jules-Théodore Pellier observed in 1897 that 'the side-seat is more elegant, gives the woman-rider a reassuringly firm position, and is the most satisfying from the viewpoint of decency'.[61]

But the side-saddle seat also supposedly showed to advantage, more elegantly than did sitting astride on a cross-saddle, that desirably feminine figure – wasp-waisted from stays, but perhaps rather broad in the hips and big-bottomed, for which no restraining devices could be devised. In fact, this last appeal to marking gender difference in a heterosexual economy of competitive femininity seems a crucial rationale for a device that took several centuries to perfect in order to make riding side-saddle potentially as effective a means of controlling a horse at speed as riding astride.[62]

Catherine de Medici (1519–89), Italian wife of Henry II of France, an accomplished and bold horsewoman and follower of the hunt while riding astride, is most often credited with the invention of the side-saddle in the sixteenth century, and the justification most often introduced is one of vanity, following the French historian Brantôme, who claims to have been told it by the queen's own ladies in waiting. Lida Fleitmann Bloodgood, in *The Saddle of Queens*, gives us a Catherine showing off her good legs side-saddle, with skirt trailing behind her:

> Thus was born the idea of a second crutch between which and the original pommel she could wedge her right leg, pulling the skirt up in front to reveal her pretty hose and dainty ankle, while allowing the gown to trail gracefully down behind in a style that persisted until well into the nineteenth century.[63]

In fact, recent research has turned up an etching by Albrecht Dürer from 1497 clearly showing the second pommel or crutch and a woman in the modern, forward-facing side-saddle seat;[64] so Catherine de Medici did not invent so much as popularize the side-saddle. The side-saddle was to undergo a number of other changes in subsequent centuries, particularly the introduction of the 'leaping head' around 1830, attributed to a Frenchman, Jules Charles Pellier, or to an Englishman, Thomas Oldaker.[65] The 'leaping head', an additional pommel, was fitted just above the left leg, enabling a woman to obtain a purchase against it with her left thigh:

> Thus, and only thus, can she grip sufficiently to enable her to gallop on and jump fences at speed without using a hand to hold on to the back of the saddle. It is more remarkable in these circumstances that a few women

succeeded in managing the standing jumps required at the time than that they were subsequently unable to cope with the transition to 'flying leaps'.[66]

By the mid-nineteenth century riding side-saddle would be common practice for many women in North America, as well as England and France,[67] and even Italy.[68]

It is significant that Montagu, who had the rare advantage of being able to compare riding astride with riding side-saddle through her long residence in Italy in later life, gladly abandoned her English equipment as soon as she got the chance. When she first decamped to the Continent, she saw to the shipping of her saddle with as much care as the shipping of her books.[69] Soon she was reporting, however, 'I have got a little horse, and sometimes amble about after the maner of the D[uchess] of Cleveland, which is the only fashion of rideing here'.[70] In January 1748, she received '30 Horse of Ladys and Gentlemen with their servants' at her house in Gottolengo, south of Brescia, again reporting that 'the Ladies all ride like the [by now] late D[uchess] of Cleaveland', and that it was gratifying to see a woman of her own age dancing and 'jump[ing] and Gallop[ing] with the best of them'.[71] Italian women who hunted rode astride.[72] In order to convey these different conventions to her husband and daughter in England, Montagu invoked the eccentric English exception who proved the rule, the second wife of the second Duke of Cleveland, Anne Pulteney Fitzroy (1663–1746), the daughter of Sir William Pulteney of Misterton, Leicestershire, that veritable ocean of turf.[73] It was possible for a Duchess to get away with riding astride in England in the first half of the eighteenth century, though she was clearly an exception.

After her conversion to riding astride, Montagu came close to pronouncing the side-saddle a masculine plot to keep women handicapped on horseback. In 1749 she wrote to her daughter that at 60, she was 'a better horse Woman' than she had ever been in her life, 'having compli'd with the fashion of this Country, which is every way so much better than ours I cannot help being amaz'd at the obstinate Folly by which the English Ladys venture every day their Lives and Limbs'.[74] We have her testimony that, far from being a safety device, the side-saddle was the sign of Englishwomen's 'obstinate Folly' – their vanity and pride in their *puissant* femininity.

The side-saddle, so much more popular in Britain and France than elsewhere, was a social machine for producing gender and for managing anxieties about masculinity and femininity during the very period in which modern gender difference was being formulated and then formalized.[75] That is why the Amazon riding to hounds emerged as a figure of satire in the eighteenth century. She presented an image of female agency that became pathologized in the course of the nineteenth century,[76] before giving way to a twentieth-century hunting field, in which, in a machine-loving age in which horses are relics of a bygone era, women form at least half the field and on weekdays may outnumber men.[77] Now boys covet sports cars, motorcycles, speedboats and airplanes rather than thoroughbred hunters. The bond between women and horses in Britain and the United States is psychologically complex.[78] The fearful figure of the Amazon still vexes some hunting men's imaginations. Roger Scruton's 'hunting harpie', if she 'rides you off at a jump', will cry ' "Get out of the way!" ':

> She will cut across you as she passes, shouting, 'Do *please* mind my heels!' And if she particularly dislikes you . . . she will make a point of being there at every jump, in order to press in ahead of you, or swear at you for doing the same.[79]

7
The Pleasures of the Chase *circa* 1735 to *circa* 1831

> The persevering speed and fortitude of the game, the constantly improving high-mettled excellence of the hounds, the invincible spirit of the horses, and the unrestrained ardour of their riders, have given it a decided superiority over every other sport or amusement ever yet known to the people of this kingdom. Its salutary and permanent effect upon the human frame has been so long self-evident, that it appears to be too firmly established ever to be shaken even by time itself; the superlative pleasure of every variegated scene, the diversities of the country, the rapturous enjoyment of the aggregate, and the ecstacy with which it is embraced by its infinity of devotees, have exalted the estimation and excellence of this sport to a system of perfection never before known: in fact, so very much so, that some of the most opulent, the most eminent, and the most learned characters are principally and personally engaged in it, in almost every county from one extremity of the kingdom to the other.
>
> [William Taplin], *The Sportsman's Cabinet*, Vol. 2 (1804)

> Our hunting fathers told the story
> Of the sadness of the creatures,
> Pitied the limits and the lack
> Set in their finished features
>
> W.H. Auden, 'Our Hunting Fathers' (ll. 1–4)[1]

In 1733, an anonymous country squire published *An Essay on Hunting*, lamenting how urban luxury had corrupted English youth who lost those 'useful Hours that our Fathers employed on Horse-back in the Fields' 'betwixt a Stinking Pair of Sheets'.[2] Patriotism and natural science were both nurtured by the pursuit of the hare, and England would soon succumb to 'Degeneracy' unless the 'Manly Exercise' of hunting were once again to become popular. The anonymous squire need not have worried, since for the next hundred years the hunting field would be constantly reinvented as a proving ground of English manhood. The rise of modern fox-hunting, as distinct from other forms of chase, can be understood as an expression of the 'new patriotic, patrician machismo', as a 'conscious and aggressive effort on the part of the landed élite to assert its status as arbiter and guardian of the national culture,' as Linda Colley puts it.[3]

Modern fox-hunting, emergent throughout the eighteenth century, became by century's end dominant, while hare-hunting and stag-hunting became residual.[4]

By 1800, the map of desirable hunting countries had definitively altered from Michael Drayton's or Celia Fiennes's preference for the West Country or Wiltshire. From then on only the Midland shires had any claim to fashion (see Chapter 1). Although by the 1920s successful artists such as Lionel Edwards had begun to embrace provincial hunting once again as 'more picturesque' than the shires, it was in the famous Pytchley country that Snaffles (C.J. Payne) in 1921 produced his best-known picture, *The Finest View in Europe* – the view from the back of a hunter, with a stiff country framed by the horse's ears.[5] We have seen how 'picturesque' countryside was regarded by Charles James Apperley in 1835; Nimrod found himself inclined to agree with the gentleman who said, ' 'tis a pretty country enough, but how the devil do they ride over it?'[6] The landscape of agricultural improvement, home to the new improved breeds of cattle, sheep, foxhounds and hunters, was the preferred landscape for modern fox-hunting. Neither its unimproved champion predecessor nor its picturesque alternative could compare with Leicestershire. John Ferneley's painting of 1823, *Sir John Palmer on his favourite mare with his shepherd John Green and prize Leicester long wool sheep* (Plate 6), epitomizes this convergence of kinds of improvement.[7]

Described by Taplin in patriotic tones, fox-hunting in the shires was the national sport, a test of bravery and an advertisement for British imperial governance. Lord Seaton, a veteran of Waterloo, asserted that, 'The same men who will ride straight across a country at a gallop, taking their fences generally as they come . . . will be likely to do anything or everything which may be required of them in action.'[8] Modern fox-hunting meant a long day subjected to harsh wintry weather, crossing the country at racing pace if hounds found, hanging about outside covers if they did not. The Rev. William B. Daniel enthused in 1801 about 'cheerly riding to the Cover side, with all the ecstasy of Hope and Expectation'.[9] There was a downside to this aspect of fox-hunting, as Taplin himself was quick to point out. A self-styled man of business as well as a sportsman, Taplin defended stag-hunting in

The Sportsman's Cabinet as more suitable for busy people, insinuating that dedicated fox-hunters were likely to be clergymen and other chaps with too much time on their hands. He sneered that the Rev. Daniel, who was contemptuous of stag-hunting, had '*most* of his time to *kill*, and very little to *employ*', so that to him:

> a long and dreary day through the gloomy coverts of a distant and dirty country, without a single challenge, or one consolatory chop or drag, must prove a scene of the most ecstatic enjoyment; and in the very zenith of sporting exultation, it must be acknowledged, by professed and energetic juveniles, that riding thirty, or forty miles in wet and dirt (alternately replete with alternate hope, suspense, and expectation), to enjoy the supreme happiness of repeated disappointments, terminating in *a blank day*, must be equal, if not superior to, a stag-hunt of even the first description. (2: 13)

The stag-hunt offered an optimum use of time because of the certainty of sport, according to Taplin. When the carted stag was released, he would run straight across the country, inciting some fast, bold riding. Although Taplin could not deny fox-hunting's hold on the public imagination, and sang its praises as a patriotic discipline, as we have seen, he couldn't resist a jibe at those who could devote themselves so wholly to sport, without the pressure of business, that they could afford to spend a bad or blank day with hounds. All those hours on horseback in a wet and dirty country, and nothing done!

Alternatively, it remained possible to have a day out from London or from many provincial towns with a subscription pack of foxhounds. Subscription packs began to be formed in the 1760s, and even the masters of some private packs began to take a handful of subscriptions to support hounds, including, in 1761, Hugo Meynell of the Quorn.[10] These subscription hunts constituted a form of high-status leisure, dignifying provincial towns. In Peter Borsay's terms, they were merely one more instance of the worlds of 'town' and 'sport' 'becoming increasingly bound together in a marriage of mutual convenience', a trend that was firmly established by the end of the eighteenth century.[11]

In remoter districts, hunting continued unimproved. Nothing could have been further from the elegant ceremony and racing pace of Leicestershire than the informal hare- and fox-hunting praised by Cobbett in the 1820s. Thomas Rowlandson's *The Return* of 1788 captures something of this older atmosphere. The 'cheerful exhaustion of the hounds'[12] and horses implies a good run, the waving fox's brush a good day (except for the fox), the presence of women riders, at least one of whom seems to have been actually hunting since her horse looks suitably exercised,[13] a companionable sociability of both sexes in the field. *The Chase*, part of the same series, confirms women's participation, with one dashing female flying a gate (Fig. 8). This is not high heroic hunting or snobbish hunting,

Figure 8 Thomas Rowlandson (1756–1827), *The Chase* (*c.* 1788). By courtesy of the Witt Library, Courtauld Institute of Art, London.

but old-fashioned provincial hunting ironized as rustic, yet portrayed sympathetically as country contentment. In its bold-riding Yorkshire form provincial foxhunting impressed even Nimrod: 'One might also imagine there was something in the physical power of the climate that produced such a host of sportsmen – for every man there is one.'[14] It has been observed even of Surtees, bred in County Durham, who took 'the Yorkshireman' as his persona in the Jorrocks stories, that it was 'round the scratch packs' that his 'most cordial work was written'.[15]

The pleasures of this old-fashioned kind of hunting included a tradition of rural hospitality, and festive eating and drinking were part of the hunting calendar. In November 1802 Coleridge recorded in his notebook the following bill of fare at a hunt dinner near Penrith, which was hearty but not elegant, and contrasts significantly with the more meager walkers' cuisine:

A large round of Beef   Sirloin of Beef – a Ham 4 Geese – 4 Fowls – a Hare – 2 Giblet Pies, 1 Veal Pye – 12 Puddings – Vegetables of all sorts / 1 s,, 6 d a head, at an annual Hunt-Feat at Culthwaite, 7 miles from Penrith/close [crossed out] on the Thursday before Martinmas, 28 persons present – [?Ale/All] inc.[16]

Ale and giblet pies, one of Dorothy Wordsworth's specialities, did not feature in the diet of the sporting set at Melton Mowbray, where it was all champagne and claret, or 'pale brandy', and 'choice Hollands gin and soda "out of glasses the size of stable buckets" ', to wash down the occasional anchovy toast.[17]

Pleasures of the grass

By the 1830s, what the Quorn country of Leicestershire provided were unequaled opportunities for galloping over a sea of grass, with the fashionable bonus of Melton Mowbray as a social center. This formerly champion country was punctuated by four-foot, cut-and-laid, stiff thorn fences, sometimes topped with one rail, sometimes with two (the challenging double oxer), but always with a ditch on one side. These had to be taken at speed. James II as Duke of York was reputedly a bold rider across country, but he would have jumped fences off his horse's hocks, from a near standstill, and after craning his neck to see what lay on the other side, or risked castration on the high pommel of his saddle, since flat racing or hunting saddles did not appear until the early eighteenth century.[18] Well-bred horses were required that could maintain a near-racing pace for as many miles as a fast hunt, 'a quick thing', might last.[19] The famous Billesdon Coplow run of 1800 took place in the last year of Hugo Meynell's mastership, which had begun in 1753. After 28 miles, covered in 2 hours and 15 minutes over numerous obstacles, the fox beat both hounds and horses.[20]

Eighteenth-century hunting had been a much slower, more haphazard affair. Not only were hounds often taken out to put up whatever game might be available, rather than strictly 'entered' to fox, but hunting began early in the morning and could take all day. Making a hunt last rather than having a quick thing was considered showing sport in the eighteenth century. On 26 January 1738/39, the 'greatest Chase that ever was' in the records of the Charlton hunt in West Sussex, which was the Melton Mowbray of its day, lasted just over 10 hours, from a quarter to eight in the morning until ten minutes to six in the evening, and covered just over 24 miles.[21] This much slower pace, up and down steep hills and through bits of forest and heath, including such tricky-sounding places as 'Nightingale bottom' and 'my Lady Lewkner's buttocks', defeated all the field except the Master, the Duke of Richmond, a hunt servant, Billy Ives, and Brigadier Hawley, who weighed 17 stone. Flying leaps were sufficiently unknown for the Duke of Devonshire's leaping of a five-barred gate to be memorialized in verse in 1737, and still talked about in 1910:

> Northward, and riseing close above the Towne,
> another Mountain's Known, by Leving Downe;
> a Pirenean path, is Still there seen,

1 John Wootton, *Hare Hunting on Salisbury Plain* (c. 1700). Reproduced by kind permission of the Duke of Beaufort. Photograph: Photographic Survey, Courtauld Institute of Art, London.

2 George Morland, *Tavern Interior with a Sportsman Refreshing* (c. 1790). By courtesy of John Barrell. Photograph: Cambridge University Press.

3 James Seymour (1702–52), *A Coursing Party* (1738). By courtesy of the Richard Green Gallery, London.

4 William Henry Davis (c. 1786–1865), *Colonel Newport Charlett's favourite Greyhounds at Exercise* (1831). By courtesy of the Richard Green Gallery, London.

5 John Wootton, *Lady Henrietta Cavendish Holles, Countess of Oxford, hunting at Wimpole Park* (1716). By courtesy of private collection. Photograph: The Paul Mellon Centre for Studies in British Art, London.

6 John Ferneley (1782–1860), *Sir John Palmer on his favourite mare with his shepherd John Green and prize Leicester long wool sheep* (1823). By courtesy of Leicestershire Museums, Arts and Records Service.

7 Thomas Barker of Bath (1769–1847), *The Woodman and his Dog in a Storm* (c. 1790). By courtesy of (copyright) Tate, London 2001.

> where Devons Duke, full Speed, did drive his well
> bred Courser down, and flying, leap't five barrs;
> incredible the Acte! but still 'twas fact.[22]

Such feats of horsemanship would become commonplace in the modern hunting field of the nineteenth century.

The biggest difference between hunting in England in 1735 and 1830 might be characterized as the difference between riding in order to hunt, as Robert Bloomfield's stiff-limb'd foot follower no doubt longed to do, and hunting in order to ride. As Surtees's John Jorrocks, the Cockney grocer and provincial Master of Fox Hounds, observed, the fashionable hunting countries of the 1830s were full of would-be thrusters and 'Cut-'Em-Down Captains' perpetually home on leave.[23] 'Meltonians hunted to ride rather than rode to hunt.'[24] Lord Alvanley summed up the Meltonian spirit when he pronounced, 'What fun hunting would be, were it not for these damned hounds.'[25] A Master of the Pytchley, Sir Charles Knightley, lamented that hunting just wasn't what it used to be because of large numbers of thrusters:

> Formerly five or six men used to ride hard, and if they knew but little of hunting, they generally knew when hounds were on scent and when not. At present everybody rides hard, and out of three hundred, not three have the slightest notion whether they are on or off scent . . . If hounds were let alone and not ridden upon, they would rarely miss a day's sport.[26]

Modern fox-hunting has from its beginnings often been about something other than hunting. Roger Scruton has recently argued that hunting in Britain today is like nothing so much as British football, 'a sport of followers', though the '"team" you follow, and whose triumphs and disasters you share, is not a group of human beings but a pack of foxhounds'.[27] In the light of this history, in which showing sport in modern fox-hunting has come to mean, more than anything else, providing entertainment for the mounted field, replacing hunting a fox with following a 'drag' scent laid in advance seems a logical outcome, though it would be anathema to devotees of hunting.

Equestrian pleasures

Technological improvement in the shape of horses and hounds was necessary for this quickening of the pace. Horses were the first species to be highly selectively bred in England, with an infusion of Eastern blood imported through the Levant trade resulting in both in-and-in and cross-breeding. As Thirsk speculates, for want of proof, 'one can only hazard guesses when exactly the lessons learned from horse breeding influenced breeders of other livestock'.[28] There was ample choice among the many unimproved breeds of cattle, sheep and pigs, which, she concludes, 'satisfied for a long time'.[29] The increasing demand for meat in an expanding population which was also becoming more urban stimulated changes in sheep and cattle breeding, sacrificing quality for quantity of meat. Pigs, the mainstay of cottagers' and laborers' larders, were not considered commercial meat producers until the mid-nineteenth century.[30]

If market pressures finally transformed even pigs, gentleman's recreations – first on the champion plains, later on the new enclosed pastures – had a similarly stimulating effect on horse breeding. As Thomas Lister, Lord Ribblesdale, put it, 'An open galloping country makes a fast horse, and a fast horse makes a fast hound.'[31] He examined the evidence of sporting pictures to establish how as early as the second decade of the eighteenth century, imported Eastern blood had produced a fine fast stamp of English hunter:

> I have two or three pictures at Gisburne of members of my family on horseback, coursing and hunting, painted about 1720 or so. The horses are exceedingly well bred and full of character, but narrowish and not up to more than twelve stone. They *all* show much more Eastern blood in their heads – the bump on the forehead, the full eye, which we only find here and there in an individual now. They were, of course, much nearer the blends of Arab and Arabian blood to which we are indebted for everything we prize most in horseflesh. J. Ward's best horses are quite a different stamp. By that time we had somehow or other got more substance and size into our breed.[32]

Like other exotic products of the Eastern trade, Eastern equine blood could be naturalized on English soil.[33] '[C]areful attention to feeding on the new grasses' gave size and substance to the imported desert stock.[34] The Eastern blood horse, especially the pure-bred Arab, was the foundation of the modern thoroughbred, traceable to three foundation sires, the Byerley Turk, the Darley Arabian and the Godolphin Arabian. In addition to producing racing stock, the infusion of Eastern blood improved English horses for hunting and riding. Edward, Lord Harley, after 1724 the second Earl of Oxford, having been sent the famous Bloody Shouldered Arabian in 1719–20 by his uncle Nathaniel Harley, a merchant in Aleppo, 'was frequently prevailed upon by friends who begged for the privilege of using this stallion for breeding'.[35] John Wootton probably painted more portraits of this horse than of any other;[36] where a 'leap' or service by such a horse was not possible, his painted image might be acquired, for he represented an ideal type (Col. Pl. 4). After hunting on Arabian, thoroughbred or near-thoroughbred horses became commonplace, the sport was irrevocably altered. The naturalistic pleasures of riding out to put up whatever quarry showed itself took second place to achieving blinding speed.

Talking of hounds

Similar views were expressed about the English foxhound, who was produced by in-and-in-breeding to be fast and to hunt tirelessly.[37] Less is heard about the music of hounds, more about their speed and stamina. Somervile advocated that sportsmen 'A different hound, for every diff'rent chase, / Select with judgment' (*The Chace* 1: 225–6). And the pattern was very much a tough and speedy one, though hounds were meant to endear themselves to their masters and to please the eye as well as to have keen noses in order to 'own' a scent and hunt a line for miles across country. The ideal hound was no less than a work of art, an object worthy the connoisseurship of a gentleman possessed of a certain classical learning and a neoclassical taste:

> His glossy skin, or yellow-pied, or blue,
> In lights or shades by nature's pencil drawn,
> Reflects the various tints; his ears and legs,
> Fleckt here and there, in gay enamell'd pride
> Rival the speckled pard; his rush-grown tail
> O'er his broad back bends in ample arch;
> On shoulders clean, upright and firm he stands;
> His round cat foot, straight hams, and wide-spread thighs,
> And his low-dropping chest, confess his speed,
> His strength, his wind, or on the steepy hill,
> Or far-extended plain; in every part
> So well proportion'd, that the nicer skill
> Of Phidias himself can't blame thy choice.
>
> (1: 242–54)

Even while describing hound conformation in a manner that will be continuously quoted by subsequent writers, Somervile manages to insinuate the patriotic rightness of this particular stamp of hound. In describing the acceptable variety of colors of hounds' coats, Somervile echoed Pope while instituting a comparison between English hound and exotic foreign leopard: a moment of georgic inscription of English hunting in an imperial, global frame. Pope had written of the vegetable bounty, both wild and cultivated, in *Windsor-Forest* under Queen Anne's lax enforcement of forest law:

> Here blushing *Flora* paints th' enamel'd Ground,
> Here *Ceres'* Gifts in waving Prospect stand,
> And nodding tempt the joyful Reaper's Hand.
>
> (ll. 38–40)

Windsor's fields of flowers and hounds' coats are both enamelled ground – artistic but suggestive of martial maneuvers. Hounds' arching tails suggest plenitude as richly as waving plots of corn. John Chalker is right to claim for Somervile a high degree of 'literary self-consciousness', and to suggest that the tone of passages such as this one is 'likely to be over-complex rather than over-simple'.[38] We are following the line first run by Denham and Pope. Breeding an excellent stamp of hound, who might be considered as British as the bulldog, is tantamount to cultivating and defending the nation.[39]

Sartorial and linguistic pleasures

The pleasures of the chase as they developed between the 1730s and the 1830s were self-professedly imperial, patriotic, progressive and discriminating, like the peers and gentlemen who enjoyed them. They were increasingly technological pleasures, products of the agricultural revolution, but they were accompanied by social pleasures in a similar state of constant 'improvement'. We have already remarked the gradual evolution of a uniform dress for fox-hunters that was in effect, as Merrily Harpur notes, the fox's livery – red coat, white bib and black extremities. Masters and hunt servants wore hunting caps, and the field top hats, beginning early in the nineteenth century, though some hunts could be 'flung into confusion when the field wore caps and could not be distinguished from the hunt servants'.[40] Certainly by 1850, according to Raymond Carr, 'the tight-fitting scarlet coat' 'adorned with five brass buttons' was ' "correct" ' for men, though black coats were also worn by members of both sexes.[41] Red coats had been the coming thing for a long time – since before 1820, and even as early as the 1770s, if Stubbs's paintings are any indication. They were most definitely to be distinguished from the green coats of hare-hunters. By mid-century, '[p]articular care was lavished on boots and breeches', with Oxford undergraduates affecting to clean the mahogany or brown tops of their boots with 'champagne and apricot jam, or port and blackcurrant jelly'.[42] Hunting boot fetishism is epitomized by Colonel Hesmon in Siegfried Sassoon's *Memoirs of a Fox-Hunting Man* (1929), who every autumn visits, 'with the utmost solemnity, an illustrious bootmaker in Oxford Street', with the result that he has accumulated 27 pairs of 'chronological boots' with their 'mahogany, nut-brown, and salmon-coloured tops'.[43]

As with clothes, so also with language: modern fox-hunters were increasingly on the lookout for a solecism. Somervile's language in the passage about hounds is richly freighted with allusions and imperial georgic aspirations, but there are certain discrepancies between his diction and that of the writers who formalized the discourse of modern hunting in the 1820s and 1830s. In Apperley and Surtees, we will hear no more of hounds' 'tails', any more than we will hear of foxes' brushes being 'tipp'd with white', as Somervile has it (3: 58). Peter Beckford too, confident that 'perfect symmetry' was necessary for a hound, or he would 'neither run fast, nor bear much work', opined that a hound's 'tail' should be 'thick, and brushy; if he carries it well, so much the better'.[44]

One of the principal pleasures of nineteenth-century hunting would seem to have been the initiation into a coterie language that grew ever more specialized and refined. In Nimrod and Surtees and subsequent writers, hounds have 'sterns', not tails; foxes' brushes are always 'tagged', never tipped with white. Somervile wrote that a hound 'with deep-opening mouth' made 'the welkin tremble' (1: 324–5) and reports that 'The welkin rings, men, dogs, hills, rocks, and woods / In the full concert join' (2: 157–8). Tediously ever after, the 'welkin' will 'ring' with hounds' voices.

After 1820, the sporting code of gentlemanly conduct was no longer enough to guarantee inclusion. Despite the apparent openness of fox-hunting as a vermin chase, not a game chase, Cobbett's notion of hunting as democratizing was fast disappearing in a new exclusivity. One needed to dress and sound like the right sort of sportsman to be taken seriously as one. This is nowhere more clearly illustrated than in Apperley's novel of 1832, *The Life of a Sportsman*. The protagonist, young Francis Raby, yearns to be a sportsman in the modern fashion. Unlike Apperley himself, who was not born to money, Frank is fortunate in having means as well as birth, and a father who is a master of hounds. The language lessons begin as soon as sport is imminent:

> 'I hunt with you tomorrow, papa,' said Frank to his father, as soon as he had made his escape from what he called 'Egerton's botheration about Hannibal and Cicero.' 'I hope we shall find as good a buck hare as that which Mr. Gibbon's shepherd soho'd for us last time we met at the same place.' 'Frank,' said Mr. Raby, 'I must now be your tutor, and, in this instance, can do more for you than Mr. Egerton. You have made use of two terms not used in hare-hunting, and it becomes every person to adapt their language to their subject. A male hare, in hunting, is called a jack hare; and the word *tan-ta-ra*, not *soho*, denotes one espied in its form. The terms you have applied are peculiar to coursing.'[45]

Frank soon grows impatient with even the best harriers, so that reports of an eight-mile burst by his father's huntsman lead him to say dismissively, ' "[I]t was all very fine, but he had taken his leave of thistle-whipping." '[46] Mr Jorrocks is similarly superior regarding 'staggers' (stag-hunters) and 'muggers' (hare-hunters) in *Handley Cross*:

> Talk of stag-'untin! might as well 'unt a hass! – see a great lolloppin' beggar blobbin' about the market-gardens near London, with a pack of 'ounds at

its 'eels, and call that diwersion! . . . Puss-'untin' is werry well for cripples, and those that keep donkeys . . . I wouldn't be a master of muggers for no manner of money!'[47]

Jorrocks's bluster and Frank Raby's *hauteur* would prove decisive. Apperley's refinements of hare-hunting language, however, as voiced by Frank's father, did not catch on. C.E. Hare's *The Language of Sport* (1939), a compendium of earlier as well as modern usages, reports categorically, 'The male is called "the buck", and the female "the doe" ', and 'One of the commonest exhortations was the term *Soho*! used in hare hunting and coursing.'[48] We hear nothing of tan-ta-ras and jack hares.

That a specific discourse not only of hunting terms but of polite exchanges at the covert side had developed early in the nineteenth century is clear from Surtees's satire on its violation by metropolitan citizens. His Yorkshire persona, having a day with the Surrey foxhounds, rides round at the meet eavesdropping and being astonished at what he hears – and doesn't hear:

'What is that hound got by?' No. 'How is that horse bred?' No. 'What sport had you on Wednesday?' No. 'Is it a likely find to-day?' No, no no; it was not *where the hounds*, but what *the consols*, left off at; what the four per cents., and not the four horses, were *up to*; what the condition of the money, not the horse, market. 'Anything doing in Danish bonds, sir,' said one. 'You must do it by lease and release, and levy a fine,' replied another. 'Scott *v*. Brown, crim. con., to be heard on or before Wednesday next.'[49]

The true sportsman would be intent upon discovering the pedigrees of hounds and horses and determining what sort of hunting he could expect. This was acceptable conversation at a meet. Talking shop was beyond the pale. Once hounds 'moved off', however, the true sportsman wished to hear no chatter. 'Coffee-housing' was a crime, though the less than sporting continued to commit it. When hounds were 'at fault', after having 'found', and then lost the scent, Beckford wished not a word to be said. Yet this was precisely the moment when the less than sporting would fall to gossiping:

A propos, Sir, a politician will say, – What news from America? – *A propos*, – Do you think both the Admirals will be tried? – Or, *a propos*, – did you hear what has happened to my grand-mother? . . . Amongst the antients, it was reckoned *an ill omen* to speak in hunting – I wish it were thought so now. – *Hoc age*, should be one of the first maxims in hunting, as in life; and I can assure you, when I am in the field, I never wish to hear any other tongue than that of a hound.

(*Thoughts upon Hunting*, p. 152)

Hunting was meant to constitute an immersion in the world of hound and fox, to present an alternative to human society, not an extension of it. 'I know many gentlemen, who are excellent sportsmen,' Beckford remarked, 'yet I am sorry to say, the greater number of those who ride after hounds, are not . . . I consider many of them as gentlemen riding out, and I am never so well pleased, as when I see them ride home again' (pp. 181–2). Even in the later eighteenth century, there were those who hunted in order to ride, and thought they were sportsmen though they were ignorant of hunting procedures.

One of the clearest signs of modern hunting's status as an invented tradition in Hobsbawm's and Ranger's sense, as a practice with significantly modern origins, is this very obsession with implicit codes and definitive terminology, treated as sacred and unquestionable, as if they had been handed down from time immemorial. Yet within the canon of hunting writing recognized by nineteenth-century writers, within which Somervile's *The Chace* and Beckford's *Thoughts upon Hunting* were the most frequently cited works, there was no such fixed lexicon.

The pleasures of Somervile

In *The Seasons*, James Thomson gave a 'ludicrous account of foxhunting', putting the sporting ideals of earlier generations into doubt. The eighteenth-century hunting lobby responded to Thomson's challenge. Writing on behalf of the old sporting culture, in 1735 Somervile countered Thomson's urbane metropolitan views by representing hunting as an activity both requiring technical knowledge and possessed of moments of sublimity worthy of an imperial nation. Somervile replied most pointedly to Thomson by arguing that it was *natural* for humankind to hunt (it was vegetarianism which was unnatural):

> . . . for the green herb, alone,
> Unequal to sustain man's labouring race,
> Now every moving thing that lived on earth,
> Was granted him for food.
>
> (1: 64–7)

Somervile was successful in pleasing contemporary readers. Even Samuel Johnson admitted, 'To this poem praise cannot be totally denied.'[50] *The Chace* appeared in at least 13 editions during the eighteenth century, with Dublin, Glasgow and Birmingham as well as London imprints, and continued to be reissued in the nineteenth century.[51] The poem evinces how, when technical precepts might otherwise have seemed tedious to eighteenth-century readers, the georgic mixed style was designed to make them palatable – rather than unpalatable, as modern readers have sometimes protested. The poet presented himself as an elderly man, now retired to his elbow chair, but eager to impart his stock of hunting knowledge and enthusiasm for the chase. The 'preceptive' parts, he instructs us in the 'Preface', will be 'intermixed' with so many 'descriptions, and digressions, in the Georgick manner' that he hopes 'they will not be tedious'.[52] Somervile announced that he took comfort from being unlikely to 'trespass upon' the patience of his gentleman readers 'more than Markham, Blome, and the other prose writers upon this subject' (p. xviii). So we are invited to compare Somervile's advice with Gervase Markham's and Richard Blome's, those two standbys of a country gentleman's library. By invoking Xenophon, Pliny and Galen, and poetry by Oppian, Gratius, Nemesianus, Ovid and Virgil, Somervile strikes in the 'Preface' a gentlemanly, Oxonian note, entirely suitable for a former pupil of Winchester and fellow of New College. Virgil, Somervile observed, didn't say as much about hunting as we might have expected, giving us only ten verses on it in the third book of the *Georgics* and just touching upon it in the fourth and seventh books of the *Aeneid* (pp. xix–xx). But what all the ancients reveal is how different modern, improved, scientific hunting is from what hunting with hounds was in antiquity:

> It does not appear to me, that the ancients had any notion of pursuing wild beasts, by the scent only, with a regular and well-disciplined pack of hounds; and therefore they must have passed for poachers amongst our modern sportsmen. The muster-roll given us by Ovid, in his story of Actaeon, is of all sorts of dogs, and of all countries. (p. xx)

Hunting by scent, not sight, with a 'regular' (well-matched, specially bred) and 'well-disciplined pack of hounds': here is Somervile's brief for the improvements hunting has undergone in Britain. Even the ancients were poachers by contrast!

In 1735 both Thomson and Somervile were part of the self-styled Patriot Opposition, who looked to Frederick, Prince of Wales, for future leadership and were hostile to his father George II and Sir Robert Walpole, the chief minister. Somervile managed to construct his poem in such a way that it had both a political and a social agenda; the political agenda is the one commented on by modern critics, so we will concentrate on the social agenda here.[53]

Somervile simultaneously addressed the Prince, infusing the greenness of hunting into Patriot Opposition discourse, and he addressed those headstrong youths who were the scions of wealthy landed families and members of a former warrior class, for whom hunting was the 'Image of war, without its guilt' (1: 15). The appeal to Frederick was Whiggish and jingoistic, willing the prince to be the guardian of Britain's commerce and freedom of the seas, going to war if necessary to defend them (a criticism of Walpole's comparatively pacific policy towards Spanish shipping). Should some foreign invader threaten Britain's interests, Somervile assures the Prince that the hunting fraternity would support him:

> Thy hunter-train, in cheerful green array'd,
> A band undaunted, and innured to toils,
> Shall compass thee around, die at thy feet,
> Or hew thy passage through the embattled foe [.]
>
> (1: 25–8)

Fresh from the greenwood, sportsmen would rally round. Chalker remarks on the 'romantic mediaevalism' of this passage, and Christine Gerrard links the green-clad hunters with a masque of 1731 and a portrait Frederick commissioned of himself and companions in hunting costumes the following year.[54] The appeal to Frederick as a patron of the chase climaxes during the stag-hunt in Book 3, when Somervile dares to incorporate the kind of hyperbole that could so easily be read ironically or even taken as burlesque, another instance of the stylistically over-complex rather than over-simple:

> . . . But who is he,
> Fresh as a rose-bud newly blown, and fair
> As opening lilies, on whom every eye
> With joy and admiration dwells? See, see!
> He reins his docile barb with manly grace.
> Is it Adonis, for the chase array'd?
> Or Britain's second hope? . . .
>
> (3: 383–9)

This camp portrait of Frederick manages to make him an endearing rather than overbearing figure. Like so much else in the poem, it strives to excess for ambitiously complex effects, working political and social agendas concurrently. As Gerrard notes, Frederick's reception of *The Chace* must have been 'gracious enough' to inspire Somervile to dedicate his later *Field Sports* (1742) to the Prince.[55]

Apologizing neither for the social status attached to hunting nor its expense, Somervile compared hunting in Britain with the royal pastimes of rulers in India, Africa and Arabia, directing his poem to:

> Ye vigorous youths, by smiling fortune bless'd
> With large demesnes, hereditary wealth,
> Heap'd copious by your wise forefathers' care,

> Hear, and attend; while I the means reveal
> To enjoy those pleasures, for the weak too strong,
> Too costly for the poor: . . .
>
> (1: 103–8)

Somervile's benevolent sportsman is meant to be a man of taste and sensibility, who could appreciate verse as well as kennel management and hound work. Hunting offered an acceptable outlet for that instinct for blood among the warlike upper classes, guaranteeing good health and preventing melancholy and other diseases induced by the best of British weather:

> In vain malignant steams, and winter fogs,
> Load the dull air, and hover round our coasts;
> The huntsman, ever gay, robust, and bold,
> Defies the noxious vapour, and confides
> In this delightful exercise, to raise
> His drooping head, and cheer his heart with joy.
>
> (1: 97–102)

Thus Britons might have inherited a 'malignant' and 'noxious' climate, but it was one best dealt with by hunting, for which the 'highly favour'd isle' had been specially designed by heaven, since it produced both the fleetest horses and perfect hounds (1: 84–96). Such advice would be echoed nine years later by Armstrong in *The Art of Preserving Health*, though he would add walking to the exercise list.

Our hunting fathers

Somervile wrote to elevate existing hunting practice into an official gentlemanly discourse of hunting. He did so by following georgic precedent and projecting an ideal past as well as present for British hunting, claiming that Britain was not only supreme in modern hunting methods but had also been so in antiquity. According to Oppian, Britain was the source of the best hounds, with the best noses, 'this island having always been famous, as it is at this day, for the best breed of hounds, for persons the best skilled in the art of hunting, and for horses the most enduring to follow the chase' (p. xxi). Britain both then and now possessed the best in sporting talent and genetic inheritance. This comment will be slightly offset by the admission in Book 1, lines 74–83, that not until William the Conqueror did British hunting become properly formalized, with the use of the hunting horn and proper methods of breeding, feeding and disciplining hounds becoming established. It would seem that even then the French were better at organization, but that the native talents of Britons, whether human, canine or equine, were superior nevertheless.

The 'Preface' climaxes with a disingenuous query about why, since Britain has been famous for its hunting for so long, there has been no previous English poem devoted to it:

> It is, therefore, strange that none of our poets have yet thought it worth their while to treat of this subject; which is, without doubt, very noble in itself, and very well adapted to receive the most beautiful turns of poetry. Perhaps our poets have no great genius for hunting. Yet, I hope, my brethren of the couples, by encouraging this first, but imperfect essay, will shew the world they have at least some taste for poetry. (p. xxi)

Already by 1735, the mainstream of English literature had diverged from the tastes of the rural gentry. A dominantly urban form of literary culture had elbowed aside country house taste.

Somervile's query, however, works in at least two ways. His brethren of the chase have not been such keen followers of poetry either. They have oftener had in their hands 'couples', leather collars linked by chains for pairing hounds on the way to a meet, than books. He hopes to win them over to studious pursuits, particularly on blustery or stormy days, or days of hard frost, unsuitable for hunting, and so to alter the image of the hunting man as a 'bookless, sauntring youth, proud of the skut / That dignifies his cap, his flourish'd belt / And rusty couples gingling by his side' (1: 394–6). It is imperative for Somervile that hunting folk be unmistakably 'Well bred, polite' and a 'Credit' to their calling (1: 392–3). We are invited actively to scorn the bookless youth who lives to hunt: 'See! how mean, how low,' the poet intones; 'Be thou of other mould' (1: 393, 397). The anti-intellectualism of the modern English gentry would seem to possess a long pedigree. But then, such institutions of gentlemanly learning as Eton and the universities were also infected with sporting spirit, sometimes to the exclusion of any higher learning. Foxhunting society at Melton Mowbray has been described as 'a horse-centred continuation of the all-male social life of Library at Eton or rooms in Christ Church or Trinity'.[56]

The overarching structure of the poem, designed to instruct as well as please, neatly packages the primary excitements of hare-, fox- and staghunting presented in Books 2 and 3 between two books of technical detail and advice. The georgic nature of *The Chace* allows Somervile to recommend particular kennel designs and procedures for the keeping of hounds – 'O'er all let cleanliness preside; no scraps / Bestrew the pavement, and no half-pick'd bones, / To kindle fierce debate' (1: 49–51). Nearly 50 years later, Peter

Beckford would quote Somervile at length on hounds and kennel management, claiming that where he thought the same as the poet, his audience 'had better take it in his words, than mine' (p. 40). But there is little doubt that the poem's chief appeal, then and now, lies in descriptions of various forms of chase in Britain, modulated with views of fiercer beasts of chase abroad, suggesting an imperial outlook of global proportions.

The sadness of the creatures

There are, however, complications within this scheme of even-handedly narrating each of the different forms of chase. We might read these complications as suggestive of how hunting in Britain was undergoing changes during the mid-eighteenth century. Like Wilfrid Scawen Blunt's narrator of 1914, 'The Old Squire', the poet of *The Chace* prefers hare-hunting as the most traditional pastime of country gentlemen, the quintessence of British sport. Blunt's nostalgic poem memorializes how early rising, keeping to one's own fields and keeping faith with one's ancestors distinguished hare-hunting from other field sports:

> I like the hunting of the hare
>   Better than that of the fox;
> I like the joyous morning air,
>   And the crowing of the cocks.
> . . .
> I like the hunting of the hare;
>   New sports I hold in scorn.
> I like to be as my fathers were,
>   In the days ere I was born.
>
> (ll. 1–4, 65–8)[57]

Blunt might well be harking back to Somervile here, bucking a good 150 years of fashionable fox-hunting tradition. Even such a classic of fox-hunting literature as Surtees's *Handley Cross* begins with a chapter on 'The Olden Times', in which hare-hunting was the neighborhood's delight, before riding to hounds and the faster fox chase were established (pp. 1–5). The later novel *Hawbuck Grange* (1847), featuring the 'Goose and Dumpling' harriers, 'contains more hound-work than any other of Surtees' novels',[58] and in *Hillingdon Hall* (1845), Jorrocks and his huntsman James Pigg enjoy hare-hunting, for Surtees, 'with the appropriate sentiment of the cynic for what is genuine, respected the hare-hunter for his freedom from snobbery and his genius for good fellowship'.[59]

As we have seen, Somervile dismisses the ancient sport of coursing as base, lower-class and bloody – hardly a sport at all: 'nor the timorous hare / O'ermatch'd destroy, but leave that vile offence / To the mean, murd'rous, coursing crew, intent / On blood and spoil' (1: 226–9). Coursing just does not fit in with his brief for modern hunting as requiring a disciplined *pack* of hounds who hunt by scent, not sight. But having shown us the hare as a 'timorous' victim, Somervile finds himself in difficulties. It is as if he were beginning to have doubts about this sport, despite his own affection for it, and was quite fearful that his readers might be disdainful of it. By 1796, Aikin thought hare-hunting made for dull reading and was impressed by Somervile's ability to bring it off: 'It would be scarce possible even in prose to describe the hunting of the hare with more exactness than is here done; yet the language throughout is sufficiently elevated, and some of the passages are truly poetical.'[60]

The Chace provides evidence that the dominant taste may well have been on the turn as early as 1735. Pursuit of vermin – the fox – was gaining ground in the public imagination, both because of the faster pace set and the straighter line across country taken by the quarry, and because much less sentiment attached to the fox than to the hare. A four-legged poacher could be ruthlessly hunted when the harmless hare, so human in its shrieks, and so contested as a quarry in the light of the game laws, was beginning to trouble some consciences.

By Beckford's day, this shift had definitively happened. Already by 1781 Beckford could assume that his correspondent, and by extension his wider audience, would be primarily interested in fox-hunting: 'I Thought I had been writing all this time to a fox-hunter,' he comments in Letter 10, when his friend has asked him many questions about hare-hunting (p. 134). Never 'by inclination' a hare-hunter, and living in a country where the hare-hunting is bad, Beckford is respectfully dismissive of this sport, claiming that 'if I could have persuaded myself to ride on the turnpike road to the three-mile stone, and back again, I should have thought I had no need of a pack of harriers' (p. 134), and 'I always thought hare-hunting should be taken as a ride after breakfast, to get us an appetite to our dinner' (p. 140). He admits that in a good country, it provides a greater certainty of sport than any other chase, as well as the best opportunity for watching hounds work, but that is not enough to redeem it from a fox-hunter's point of view (p. 161). It is just too slow a sport for the 1780s, and besides, the hare, unlike the fox, is an object of compassion or sympathy:

> I HOPE, you agree with me, that it is a fault in a pack of harriers to go too fast; for a hare is a little timorous animal, that we cannot help feeling some compassion for, at the very time when we are pursuing her destruction: we

should give scope to all her little tricks, nor kill her foully, and overmatched. (p. 139)

Beckford does not dwell on the timorousness of the hare and how it inspires human compassion, but this sentiment informs his conviction that the hare should be given a fair chance to display her wiles and not simply be outpaced.

Like Gray's narrator in the 'Elegy', claiming absolute possession of the rural scene when the gathering twilight leaves 'the world to darkness and to me' (l. 4), Somervile has a moment of possessive sympathy with the hare. The effect of Gray's line is to insist upon *my* scene, my village churchyard, my previously unhonored dead, my poem. Somervile likewise inserts himself into his own narrative as if he could not help doing so, and at the very moment at which the hare has hidden herself from view – from all eyes, that is, except the poet's:

> Ah, there she lies! how close! she pants, she doubts
> If now she lives; she trembles as she sits,
> With horror seized! The wither'd grass, that clings
> Around her head, of the same russet hue,
> Almost deceived my sight, had not her eyes,
> With life full beaming, her vain wiles betray'd.
> At distance draw thy pack, let all be hush'd,
> No clamour loud, no frantick joy be heard,
> Lest the wild hound run gadding o'er the plain,
> Untractable, nor hear thy chiding voice.
> Now gently put her off; see how direct
> To her known meuse she flies! Here, huntsman, bring,
> But without hurry, all thy jolly hounds,
> And calmly lay them on.
>
> (2: 137–50)

'Almost deceived my sight,' he writes: his intrusion of himself into the action is also an act of possession. The poet and the hare have a special relationship. Her eyes, so full of life, stare into his. Her gaze betrays her, in spite of her deceptive powers; her wiles are all in vain. The hare is deserving of sympathy, especially when she is reeling with tiredness and black with sweat, over 100 lines later: '[W]ith infant screams / She yields her breath, and there reluctant dies!' (2: 271–2). But she is still an agent, as well as a sentimental subject; her tricky, devious course has somehow not only made the chase possible but legitimated it.

The good old hare, like other good old causes, was fading from fashionability. This bold puss is somehow the poet's own special object, though he seems unconvinced his audience will be satisfied with it, or her:

> Thus the poor hare,
> A puny, dastard animal! but versed
> In subtle wiles, diverts the youthful train.
> But if thy proud aspiring soul disdains
> So mean a prey, delighted with the pomp,
> Magnificence, and grandeur of the chase,
> Hear what the Muse from faithful records sings.
>
> (2: 295–301)

The hare is a questionable quarry because she is incapable of fighting back to defend herself, and cowardly in that she always runs away. That she is a game animal, protected by the game laws from pursuit by the unqualified sportsman, is not mentioned. She is now classifed as 'So mean a prey' – a poor, puny, dastard animal rather than a valuable prize for the table. British readers, the poet seems to fear, may find this abject object not to their taste. Yet Somervile devoted 286 exciting lines to the pursuit of the hare; fox- and stag-hunting will receive 191 lines and 264 lines, respectively, and otter-hunting a mere 124.

It may be that the temper of the times, imperial aspirations and mercantile ambitions, or the pomp and circumstance of hunting abroad had influenced Britons to prefer some more imperial kind of chase, some nobler prey. From the 'Argument' we know that the Indian hunting scene that follows the hare hunt in Book 2 is drawn from 'Monsieur Bernier, and the History of Gengis Cawn the Great'. The moral of this history is that when the women of Aurengzebe's harem ask him to spare the lives of the wild beasts who survive the first fierce pitched battle, he obeys them and the animals flee to the hills to be hunted again another day. Somervile then extracts a further moral precept from this imperial Indian example:

> Ye proud oppressors, whose vain hearts exult
> In wantoness of power, 'gainst the brute race,
> Fierce robbers, like yourselves, a guiltless war
> Wage uncontroll'd; here quench your thirst of blood:
> But learn, from Aurengzebe, to spare mankind.
>
> (2: 519–23)

Here Somervile sounds more like Thomson and other poets of the Patriot Opposition. As a hymn to Whig compassion and love of freedom, *The Chace* subtly criticizes the machinery of what John Brewer has identified as 'the fiscal-military state'.[61] The chase is once again a proving ground for the difference between good and bad government. And that difference is a matter of sparing humans and pursuing animals instead, since this is a comparably guiltless activity. Once again, the poem insists, hunting is the 'Image of war, without its guilt'.

Not surprisingly, Book 3 features the fox chase, that pursuit of a 'subtle, pilfering foe' and 'felon vile,' a 'conscious villain' (3: 24, 37, 55). The poet juxtaposes fox-hunting with the trapping of lions and elephants in Africa, and the chasing of wild boar in Arabia, before giving a lengthier view of royal stag-hunting at Windsor. Colonial scenes are inserted within the British domestic frame, an emblem of imperial imaginings characteristic of the eighteenth-century georgic. Reading Book 3 alongside Thomson's 'Autumn', a pattern of implicit dialogue with Thomson reveals itself. Thomson, we recall, had satirically depicted an English fox chase before going on to deplore the possibility of British women taking to the hunting field. Somervile's rendition of fox-hunting echoes Thomson's but contradicts its tone, giving fox-hunters the benefit of the doubt about the heroism and justness of their cause. And his version of stag-hunting actively invites and celebrates the participation of women. Turning the tables on Thomson, Somervile hopes to reclaim hunting for a properly heroic, rather than mock-heroic, inscription within the tradition of British georgic verse.

An uncanny similarity to the line taken by Thomson's fox governs Somervile's as well:

> Far o'er the rocky hills we range,
> And dangerous our course; but, in the brave,
> True courage never fails: in vain the stream
> In foaming eddies whirls; in vain the ditch,
> Wide-gaping, threatens death: the craggy steep,
> Where the poor dizzy shepherd crawls with care,
> And clings to every twig, gives us no pain;
> But down we sweep, as stoops the falcon bold
> To pounce his prey: then up the opponent hill,
> By the swift motion slung, we mount aloft.
> So ships, in winter seas, now sliding, sink
> Adown the steepy wave, then, toss'd on high,
> Ride on the billows, and defy the storm.
>
> (3: 87–99)

Here are the same steep and rocky hills, stream in raging torrent and lethal-looking ditch to be jumped, as in Thomson. But there is none of Thomson's mock-heroic posturing and bragging. Every aspiration of the eighteenth-century georgic is present – at least in so far as Somervile wishes to put hunting alongside agriculture as evidence of British progress and scientific improvements. Whipping up bravery in his readers, Somervile leaves the poor shepherd of pastoral and agricultural georgic behind, explicitly displacing him from the scene. The mounted field goes thundering past where shepherds, dizzy with the height, have feared to tread. The riders ride the country just as British ships, those great symbols of a mercantile empire, ride the waves – dangerously, but defiantly. Aikin believed that Somervile had 'laboured' the fox chase 'more *con amore* than any other', and justifiably so, since it was 'the capital scene of action to the English sportsman'.[62]

Somervile via Beckford

This description of a good galloping run was approvingly quoted by Beckford. And since Surtees's Mr Jorrocks cribbed unabashedly from Beckford for his 'Sporting Lectors', it was also quoted in *Handley Cross*. And since Surtees has been read by everybody interested in hunting literature, the afterlife of Somervile's description has been a long one. This passage, like Beckford's Letter 13 in which it appears, has been internalized and refracted in subsequent hunting writing right into the twentieth century, whether in first-person memoir (Willoughby de Broke), fictionalized memoir (Ralph Greaves, Siegfried Sassoon) or fiction (Rudyard Kipling):[63]

> How musical their tongues! – Now as they get nearer to him, how the chorus fills! Hark! he is found. – Now, where are all your sorrows, and your cares, ye gloomy souls! Or where your pains, and aches, ye complaining ones! one halloo has dispelled them all. – What a crash they make! and echo seemingly takes pleasure to repeat the sound. The astonished traveller forsakes his road, lured by its melody; the listening ploughman now stops his plough; and every distant shepherd neglects his flock, and runs to see him break. What joy! what eagerness in every face! (p. 167)

This is the music of hounds transposed from georgic poetry to elegant epistolary prose: the poetry of hunting journalism. Once again the countryside has become a social space in which a community is constituted through hunting. The generic shift itself is democratizing, since Beckford's audience was a wider, more socially varied one than Somervile's.

Curiously, although he quoted so much about fox-hunting from *The Chace*, Beckford overlooked adjacent passages in which, as the horses begin to tire, cruelty breaks out. The fox chase in Somervile is both smelly and gory: the fox tries to cover his tracks by running through 'every jakes' (3: 156), and the fox-hunters do not spare their horses:

> . . . see yon poor jade;
> In vain the impatient rider frets and swears,
> With galling spurs harrows his mangled sides;
> He can no more; his stiff unpliant limbs,
> Rooted in earth, unmoved and fix'd he stands,
> For every cruel curse returns a groan,
> And sobs, and faints, and dies! who, without grief,
> Can view that pamper'd steed, his master's joy,
> His minion, and his daily care, well clothed,
> Well fed with every nicer cate; no cost,
> No labour, spared; who, when the flying chase
> Broke from the copse, without a rival led
> The numerous train; now, a sad spectacle
> Of pride brought low, and humbled insolence,
> Drove like a pannier'd ass, and scourged along!
> While these, with loosen'd reins and dangling heels,
> Hang on their reeling palfreys, that scarce bear
> Their weights; another, in the treacherous bog,
> Lies floundering, half ingulf'd.
>
> (3: 117–35)

Because these lines bespeak plenty of action in a fox chase, we might have expected Beckford to quote them, but he does not. Why not? Beckford wished above all to promote fox-hunting as a properly gentlemanly sport, and so they do not suit. The temper of the times by the last two decades of the eighteenth century was much less stiff-upperlippish about cruelty to horses than it had been earlier in the century, as witnessed by numerous publications on the care of horses and the prevention of cruelty to animals generally.[64] Horses' flanks running with gore and horses dropping dead on the hunting field were not acceptable sights. If the fox chase was all about defying danger from the rider's point of view, by Beckford's day fox-hunting had to be shown to be safely negotiable by equine participants.

Beckford was keen to give his contemporaries an aura of politeness and their sport an air of dignity which Somervile's graphic descriptions would disturb, for, Beckford insists:

> [T]he profession of fox-hunting is much altered since the time of Sir John Vanbrugh; and the intemperance, clownishness, and ignorance of the old fox-hunter, is quite worn out . . . – Fox-hunting is now become the amusement of gentlemen; nor need any gentleman be ashamed of it. (pp. 176–7)

Beckford was hoping to banish forever the ghosts of characters like Vanbrugh's Sir Francis Headpiece, who by the age of 42 had drunk 'two and thirty ton of ale', and spent the rest of his time

> in persecuting all the poor four-legged creatures round, that wou'd but run away fast enough from him, to give him the high-mettled pleasure, of running after them. In this noble employ, he has broke his right arm, his left leg, and both his collar bones – Once he broke his neck, but that did him no harm; a nimble hedge-leaper, a brother of the stirrup that was by, whipt off his horse and mended it.[65]

With only comic chaps like these to emulate, the hare-hunter Wetenhall Wilkes's dislike of fox-hunting in 1748 was understandable: 'Fox-hunting is too violent and dangerous a Pursuit for my Delight; being cautious, neither to venture my own Neck, nor to break my Horse's Heart.'[66] In 1781, Beckford had some residual resistance to fox chasing to overcome, indicated by even his friend's adherence to the traditional quarry.

The complexity of Somervile's achievement

Somervile's hunting georgic thus performs some complex cultural work in the fox-hunting passage. Somervile both counters Thomson's satirical mockery of fox-hunting – by representing it as honorable and brave – and perhaps also gets his revenge regarding the rising popularity of fox-hunting by portraying its bloodier, dirtier side. Exciting it may be, and legalistically justified by the fox's characterization as a poaching felon, but, the poet implies, only in this chase do horses suffer such torments. Once again Somervile's text seems ahead of its time. Apperley will have Mr Somerby, a keen fox-hunter from Leicestershire, discourse on the problem of horses that were not clean-bred in *Life of a Sportsman*. They could not stand the pace set by the new Meynellian hounds on the line of a fox, especially after a summer spent turned out at grass, with no concentrated food and no exercise. Both purer blood and year-round conditioning would be necessary if horses were not to drop dead in the field:

'– Mr. Meynell, and some other masters of foxhounds, have brought them to the very highest pitch of perfection of which their nature I believe is capable, both as to high breeding and condition; whilst the state of the horses that follow them is left very nearly where it was.' (p. 58)

Thus Nimrod sought to answer the charge of cruelty in 1832.

Somervile's transition to stag-hunting offers a great modulation in tone as well as content. Royal panegyric and praise of Britain's fairer sex join in an ecstatic evocation of hunting at Windsor. Aikin observed that this scene was 'made to partake of the polish and splendour of a court', though it was 'vastly inferior in magnificence to that of the Indian hunting before described'.[67] Here all is beauty and excitement, the only pain that of erotic desire:

> How melts my beating heart! as I behold
> Each lovely nymph, our island's boast and pride,
> Push on the generous steed, that strokes along
> O'er rough, o'er smooth, nor heeds the steepy hill,
> Nor falters in the extended vale below;
> Their garments loosely waving in the wind,
> And all the flush of beauty in their cheeks:
> While at their sides their pensive lovers wait,
> Direct their dubious course; now chill'd with fear,
> Solicitous, and now with love inflam'd.
>
> (3: 443–52)

So much for Thomson's sermon on the unseemliness of women hunting. It would seem to be the most attractive thing a woman could do. Between the advantages of being dashingly mounted on horseback, and the aerobic exercise that puts color into their cheeks, the women who follow hounds in the royal train could hardly advertise their charms more successfully. Accompanied by 'solicitous' lovers, who are both fearful for their safety and inflamed by the sight of them galloping fearlessly, pushing on their horses, these women are for Somervile the very 'boast and pride' of Britain. Aikin opined that 'much as we must admire the graceful form of the huntress, the *pensive* lover at her side makes rather an insipid figure'.[68] Somervile had turned the tables on Addison's and Thomson's criticisms of Amazons by feminizing their lovers. Aikin also regretted that stag-hunting was less egalitarian than other chases, so that 'the ardour and animation congenial to the Chace when partaken of by equals, is somewhat kept down', though royalty had its privileges: '[A] kind of awe and respect for the exalted personages' had to substitute here for 'the sportsman's rapture' (p. 18).

The only pain in this socially elevated form of chase was to be experienced by the stag, but even that could be alleviated by royal decree, according to Somervile. Like Aurengzebe in Book 2, George II spares the life of the hunted animal, exercising the royal prerogative of mercy:

> The tears run trickling down his hairy cheeks;
> He weeps, nor weeps in vain. The king beholds
> His wretched plight, and tenderness innate
> Moves his great soul. Soon, at his high command,
> Rebuked, the disappointed, hungry pack
> Retire, submiss, and grumbling quit their prey.
>   Great Prince! from thee, what may thy subjects hope;
> So kind, and so beneficent to brutes?
>
> (3: 594–601)

By the end of the day, it seems, both stag and hounds will have suffered, but only mildly. Hounds may grumble, but they submit, and the stag lives to be hunted again another day. A compromise has been reached between rewarding the pack with blood, always a graphic exhibition but occasionally necessary to keep hounds happy, and emphasizing that hunting need not always end in a kill. A certain enlightened sensibility can be discerned in Somervile's conception of the chase here. Being 'beneficent to brutes' – in this case the stag, not the hounds – is a properly royal attitude; the King's human subjects may expect similar benevolence. Praise of the King enables a further ennobling of the stag chase, as well as an admission of women into it, in direct refutation of Thomson. Aikin, however, found sparing the stag's life 'too trivial' an occasion to 'justify the pomp of the sentiment' (p. 19). And what of those grumbling hounds? We may not be able to avoid the suspicion that Somervile prefers hare-hunting to all other forms of chase, and to the pageantry of stag-hunting in particular, because it is most likely to end fairly for the hounds, in a kill or not depending on the sagacity of their noses.

Otter-hunting is saved for Book 4, as if to leaven briefly this otherwise heavy book largely concerned with hound diseases, especially the dreaded 'madness' – rabies. Otter-hunting is a summer amusement and thus a diversion from the hot weather complaints most likely to affect hounds. It also rounds off the hunting calendar, which begins in autumn with hare-hunting, moves on to winter fox-hunting and finishes up in summer with royal stag-hunting. Following the spirit if not the letter of his own advice to 'A different hound for every chase / Select with judgment' (1: 225–6), Somervile himself kept two packs of hounds: about twelve couple of foxhounds, 'rather rough and wire-haired', with which he hunted the fox and an occasional buck, and some of which he used for hunting the otter in summer; and about twelve couple of harriers, a cross between the old Southern hound (large and slow) and the old Cotswold hound (small, almost a beagle), which he kept solely for hare-hunting.[69]

As if this sport needed some particular justification, Somervile stresses the private property question. Otters deserve hunting because of the number of fish they eat; like foxes, they are 'felons', poachers (4: 386, 463), and so meet with an especially grisly fate. In this respect, the poet is in agreement with the anonymous author of a pamphlet from 1770, who argued that the otter was a more destructive predator than the fox: 'The quantity of Fish, that are sometimes destroyed in a pond or river, in one night, by a single Otter is scarcely to be believed.'[70] Yet, as something of a minority sport, practiced when no other form of hunting may be available, otter-hunting receives the most limited treatment in, even as it concludes, *The Chace*. In his survey of rural sports in 1801, the Rev. Daniel will have even less to say of otter-hunting, which 'was formerly considered as excellent sport,' only mentioning it on account of the otter's 'being so inveterate a foe to the fisherman's amusement; for the Otter is as destructive in a pond, as a Polecat in a hen-house'.[71] The young Frank Raby in *Life of a Sportsman* will sing Somervile's praises even while questioning whether a pedestrian activity like otter-hunting suits his sporting tastes.[72]

When in 1781 Beckford came to write his hunting letters, he could name only Somervile as a worthy predecessor. Wishing that Somervile, as an 'elegant poet', had answered all his friend's questions, Beckford opined modestly that he would have deferred entirely to Somervile's authority (and, implicitly, to the higher genre of poetry): '[Y]ou then would have received but one letter from me – to refer you to him' (p. 327). There was a distinct lack of hunting knowledge among men of letters, Beckford regretted, because they now spent their time 'chiefly in town'; the only knowledgeable parties were either servants who could not write or country gentlemen who would not 'give themselves the trouble' (pp. 327–8). But this would soon cease to be the case, as, following Beckford's sterling example, hunting discourse accrued capital. Those who might make a profession of writing about it began to do so, and certain country gentlemen gave themselves the trouble of writing about it too.

8
The Pleasures of Surtees

> On they go – now trotting gently over the flints – now softly ambling along the grassy ridge of some stupendous hill – now quietly following each other in long-drawn files, like geese, through some close and deep ravine, or interminable wood, which re-echoes to their never-ceasing hollas – every man shouting in proportion to the amount of his subscription, until day is made horrible with their yelling. There is no pushing, jostling, rushing, cramming, or riding over one another; no jealousy, discord, or daring; no ridiculous foolhardy feats; but each man cranes and rides, and rides and cranes, in a style that would gladden the eye of a director of an Insurance office.
>
> Robert Smith Surtees, 'The Swell and the Surrey',
> *Jorrocks' Jaunts and Jollities* (1838)

Surtees' novels, and the Jorrocks books in particular, have long been regarded as the essence of fox-hunting literature, but in their day they were not lacking in critical edge. Virginia Blain has remarked how ironic it is that it should have been Surtees, the 'demythologizer', who transmitted the myth of fox-hunting as a ' "glorious ideal" ' to later generations.[1] But how ironic is it, in fact? Norman Gash describes Surtees as an example of '"the large ironic English mind" on which the fashionable moral and intellectual forces of the age were seeking to impose themselves'.[2]

When he initially appeared, Jorrocks was not a success. First published serially in the *New Sporting Magazine*, *Handley Cross* was far from popular when it appeared in 1843, and it was not until *Mr Sponge's Sporting Tour* appeared ten years later, in 1853, that Surtees had a publishing success.[3] As one critic has remarked, Surtees's vulgar Coram Street grocer, who announced that ' "all time is lost wot is not spent in 'unting – it is like the hair we breathe – if we have it not we die – it's the sport of kings, the image of war without its guilt, and only five-and-twenty per cent. of its danger" ',[4] was designed to alienate

196 *The Invention of the Countryside*

the MFH who fancied himself the most important representative of the landed interest, and such social pretensions were firmly attached to hunting by 1843. 'Had Surtees ridiculed Jorrocks all might have been well,' Frederick Watson argues, 'because it was obvious that the dignity of the office of Master would have been preserved. But when he laughed not *at* him, but only *with* him the sense of social indiscretion' was impossible for the snobbish to ignore.[5] Yet subsequent generations would embrace this irreverent view of hunting's pretensions, and treat Surtees's send-ups as amusing but reliable conduct manuals for behavior in the field.

Of many possible examples of this reverent consumption, two stand out, one fictional, one historical. In Kipling's short story ' "My Son's Wife" ' (1913), Frankwell Midmore, an intellectual and member of the London Immoderate Left, inherits a small country property from a maternal aunt, is converted to riding, hunting, shooting and country life – but mainly hunting – and falls in love with a booted and spurred attorney's daughter, learning to recognize as he progresses the specimens of rurality he meets from the 'natural-history books by Mr. Surtees':[6]

> It was a foul world into which he peeped for the first time – a heavy-eating, hard-drinking hell of horse-copers, swindlers, matchmaking mothers, economically dependent virgins selling themselves blushingly for cash and lands: Jews, tradesmen, and an ill-considered spawn of Dickens-and-horse-dung characters (I give Midmore's own criticism), but he read on, fascinated, and behold, from the pages leaped, as it were, the brother to the red-eyed man of the brook, bellowing at a landlord (here Midmore realised that *he* was that very animal) for new barns; and another man who, like himself again, objected to hoofmarks on gravel. Outrageous as thought and conception were, the stuff seemed to have the rudiments of observation. (p. 113)

Soon Midmore is hunting three, sometimes four, days a week, owns three hunters and wonders 'quite sincerely why the bubbling ditches and sucking pastures held him from day to day' (p. 121), until the great flood brings about an *éclaircissement* with Connie Sperrit, the attorney's daughter, who has been hunting since she was seven.

The romance of fox-hunting for Siegfried Sassoon in the thinly disguised autobiography of the George Sherston memoirs is homosocial rather than heterosexual but it is a romance, nevertheless. Brought up on Surtees by his aunt in Kent, and taught to ride by her sporting groom-gardener Dixon, Sassoon's persona is keen but anxious on his first day out with Lord Dumborough's hounds (in life, Lord Abergavenny's Eridge pack):[7]

My first reaction to the 'field' was one of mute astonishment. I had taken it for granted that there would be people 'in pink,' but these enormous confident strangers overwhelmed my mind with the visible authenticity of their brick-red coats. It all felt quite different to reading Surtees by the schoolroom fire.[8]

At twelve, Sherston is eager to bring credit to Dixon, his servant, but also his only live model of a hunting man, and his schoolmaster in all things equestrian. So it is with anxious eyes that he compares himself unfavorably with a well-dressed and utterly self-possessed boy of his own age at the meet, Denis Milden (modeled on Norman Loder).[9] Sassoon's understated homoeroticism makes a triangular relation between Sherston, Dixon and Milden inevitable. 'I felt what a poor figure I must be cutting in Dixon's eyes while he compared me with that other boy,' comments Sherston (p. 33). And then after arriving home and reporting on Milden to his aunt, he adds, 'It was the first time that I experienced a feeling of wistfulness for someone I wanted to be with' (p. 39). Nervous in the face of Milden's self-assurance, Sherston drops his 'clumsy unpresentable old hunting-crop', and has to dismount to retrieve it, scrambling back onto his Welsh cob, Sheila, with difficulty (p. 33). But when hounds are running, he inexplicably follows the famously hard-riding Mr. Macdoggart over the biggest place in a hedge, tumbling off but arousing the interest of the Master, Lord Dumborough, who sees ' "quite a young thruster" ' in the making (pp. 35–7). This sporting male society excites Sherston greatly, but he is never entirely at home in it. The first time Sherston views a fox out hunting, he embarrasses himself in front of Milden again by speaking from a protective and entirely unsporting impulse:

> It was the first time I had ever seen a fox, though I have seen a great many since – both alive and dead. By the time he had slipped out of sight again I had just begun to realize what it was that had looked at me with such human alertness. Why I should have behaved as I did I will not attempt to explain, but when Denis stood up in his stirrups and emitted a shrill 'Huick-holler,' I felt spontaneously alarmed for the future of the fox.
> 'Don't do that; they'll catch him!' I exclaimed.
> The words were no sooner out of my mouth than I knew I had made another fool of myself. Denis gave me one blank look and galloped off to meet the huntsman, who could already be heard horn-blowing in our direction in a maximum outburst of energy. (p. 49)

As Jane Ridley has observed, Sassoon's book is one of the few hunting memoirs enjoyed by non-hunting as well as hunting people.[10] Coupled with its skilled representation of the mixed feelings of panic and euphoria induced by hunting, and its painterly descriptions of weather, effects of light and landscape, the book's characterization of Sherston as a fallible, foolish, confused young man endears itself to many readers.

Years later, after Cambridge, young Sherston will again meet Milden, who has become Master of the Packlestone country (a thinly disguised version of the nearly-shire pack, the Atherstone), and 'Surteesian' observations will once again serve him well:

> Foxes were plentiful, except in parts of the Friday country; but there was no shortage anywhere as regards rich-flavoured Surteesian figures. Coming, as I did, from afar, and knowing nothing of their antecedents and more intimate prospects, I observed the Packlestone people with peculiar vividness. I saw them as a little outdoor world of country characters and I took them all for granted on their face value. How privileged and unperturbed they appeared – those dwellers in a sporting Elysium! (p. 269)

Now an initiate, Sherston looks to 'cram' his horse at as many fences as possible (p. 255).

What distinguishes the Surteesian social landscape and infects so many people with a wish to become part of it, if they can? The generic transposition from Somervile's complex georgic verse, with its incorporation of burlesque elements into an heroic idiom, to the comic novel of Surtees, was literally performed, as we have seen, via Beckford's *Thoughts*. In *Handley Cross*, much of Beckford, and thus much of Somervile, is directly lifted and recycled in the new MFH's 'Sporting Lectors'. Surtees's hunting fields are both generically and historically layered, retailing elements of an heroic sporting past for present consumption.

Surtees's hunting fields are also socially complex. Figures from the shires do crop up. The 'Swell', 'like a monkey on a giraffe – striding away in the true Leicestershire style', is a novel amusement in the Surrey field.[11] Although varied in terms of their rurality or proximity to a town, Surtees's sporting scenes remain resolutely provincial. Fox chases are interrupted by hares; packs of harriers live cheek-by-jowl with packs of foxhounds. The fox-hunting Jorrocks has the odd day with the Surrey staghounds, interrupted by the carted stag leaping into a 'return post-chaise from the Bell, at Seven Oaks', from which his head sticks out of the side window 'with all the dignity of a Lord Mayor'.[12]

'Smoke-dried cits'

Metropolitan hunting is the preserve of 'smoke-dried cits',[13] who discuss business between energetic bursts of sport and have adopted the cautious style of older generations in the field. The 'Swell' may look a fool, but anti-Meltonian irony is complicated by anti-metropolitan irony, as evidenced by this chapter's epigraph. The Surrey foxhounds' country, unlike Leicestershire with its sea of grass, is a highly picturesque country, replete with flints to lame a horse, 'interminable' woods and 'close and deep' ravines, though so close to London, and rendered 'horrible' – residually sublime, but also just plain awful – with ever-prudent business and tradespeople's anxious hollers. Money talks, literally. Each man contributes to the chorus in proportion to his subscription. The 'holloa' is meant to be reserved for actually viewing a fox – impossible when following one behind the other in a close and deep ravine. But the Surrey field are nothing if not a communicative field, full of fellow-feeling – a mutual reassurance society. Surtees spared no one who was not properly sporting: neither Leicestershire snob, gutless citizen, nor fancier of picturesque scenery. He even appropriated the language of the picturesque in order to satirize certain sorts of citified behavior in the field.

Hunting and the spa town

Handley Cross emblematizes provincial hunting remote from London but urbanized in precisely Peter Borsay's sense. The spa town must have its pack of hounds, and Jorrocks is imported, along with his fine teas, to be the Master of it. When the best of provincial hunting is laid on for the Nimrod figure in Surtees's fiction, Pomponius Ego, consisting of a jolly unsporting drag scent of 'aniseed and red-herring over some of the best of their country', with a bagged fox 'turned down' (set loose) 'at the far end, in some convenient unsuspicious-looking place' (p. 466), all goes much better than expected:

> Nothing can be finer than the line! Large grazing grounds, some forty, none less than twenty acres, are sped over, and twice Dribbleford Brook comes in the way for those whose ambition is waterproof. What a scene! – what blobbings in and scramblings out! what leavings of hind legs and divings for whips, sticks, and cigar cases! (p. 482)

The name Dribbleford suggests the puniest of brooks, but even so, the Handley Cross field are unaccustomed to exerting themselves to jump any sort of water. Far from demonstrating battlefield nerves on their willing chargers, these followers 'blob' awkwardly in and scramble awkwardly out, stopping to retrieve their precious personal effects.

In spite of the inglorious spectacle thus presented, Jorrocks is thrilled with the pace and the show of fine turf and water. He fancies that Leicestershire itself could offer nothing finer. ' "If this don't 'stonish old Hego, there arn't no halligators!" ' thinks Jorrocks to himself. ' "Come hup, you hugly beast," he adds to his horse, again spurring and kicking him into a canter' (p. 482). Mr Bagman performs as wished:

> A stranger in the land, the fox goes stoutly down wind, with the hounds too near to give him much chance for his life. As if anxious for the promotion of the sport, he makes for the vale, and the pack come swinging down the hill in the view of the field planted below. Fresh ardour is caught at the sight! Those who ridiculed the cast are now loudest in its praise. They reach the bottom, and fox and hounds are in the same field. Now they view him! How they strain! It's a beautiful sight. Old Priestess is tailed off, and Rummager falls into the rear. Ah age! age! Now Vanquisher turns him, and races with Dexterous for the seize! Who-hoop! Fox and hounds roll over together! (p. 486)

Despite the pitiful plight of the bagged fox, most unsporting of gimmicks, echoes of Somervile can plainly be heard. The pleasures of the chase – viewing hound work, straining sinews of horse and rider, and the hound-pleasing closure of a kill – are all present in this passage. But the governing irony is that poor old Jorrocks and his huntsman, James Pigg, demonstrate all too clearly that the name of the game is showing sport and providing the entertainment of an exciting run for the mounted field, rather than actually hunting a fox with hounds.

For the Handley Cross field, the pace has been 'truly awful' (p. 481), with James Pigg, taciturn and iron-nerved Newcastleman, flying stiff fences that stop Pomponius, who is always hoping for a gap (pp. 480–1). Here Meltonian pomposity about bold jumping at racing pace is comically undermined. Pomponius, doubtful of scent and sport on such a frosty day, and 'inwardly rejoicing at the thoughts of a check' (p. 481) – funking it, in other words – manages to miss most of the excitement of the run made possible by the secret drag and reports in print only that 'After a good deal of cold and slow hunting, we at last worked up to our fox, and Mr. Jorrocks most politely presented me with the brush' (p. 491). Thus will the provinces, by definition, always fail to excite the true Meltonian, who believes the only proper sport is to be had in the shires.

Other days will not be so exciting in the Handley Cross country: Surtees always implies the possibility that far more people hunt because they think they ought to than because they like it. Jorrocks himself will often funk a run as badly as Pomponius, and he will never be a lightweight jockey or one who makes light of leaping fences. But although he may not be keen on thrusting riding, Jorrocks is genuinely keen on hunting. He has taken to heart such old-fashioned Beckfordian advice as 'NEVER take out your hounds on a very windy, or bad day',[14] and 'the best way to follow hounds across a country, is to keep on the line of them, and to dismount at once, when you come to a leap, which you do not choose to take.'[15] Hoping for a quiet 'bye day', Jorrocks is more than ever disinclined to exert himself in galloping and jumping, and he harbors a secret desire that the fox whom hounds have unaccountably found may have gone to ground. But when a fox shows himself, Jorrocks simply cannot contain his ecstasy, all thoughts of going home early forgotten:

> 'Tally-ho!' now screamed Jorrocks, as a magnificent fellow in a spotless suit of ruddy fur crossed the ride before him at a quiet, stealing, listening sort of pace, and gave a whisk of his well-tagged brush, on entering the copsewood across. 'Hoop! hoop! hoop! hoop!' roared Mr. Jorrocks, putting his finger in his ear, and halloaing as loud as ever he could shout . . . (p. 237)

Fox-hunting has its rituals, and its cult-like procedures. It also has aroused some peculiarly conflicting emotions regarding the fox, from Sassoon's immediate sympathy to Jorrocks's bloodlust on behalf of his hounds.

For fox-hunting is in some sense a fox cult, leading to fox *kitsch* and fox sentiment, but above all to preoccupation if not obsession with foxes (Col. Pl. 5). Jacques-Laurent Agasse's *Sleeping Fox* of 1794 suggests that 'admiration, almost affection' for foxes David Macdonald has found typical of fox-hunters and other countrymen of the old school. The painting exhibits natural-historical detail but also elicits affection, and a desire to protect the fox, who appears so peaceful and so vulnerable.[16] We are back in the world of hunting as paradox. As Frederick Cotton wrote of the fox in the popular nineteenth-century poem, 'The Meynell Hunt', 'Though we all want to kill him we love him'.[17] This refrain has sometimes been refashioned – in the New forest, for example – as 'We don't want to kill him, we love him!'[18]

Whether metropolitan would-be thrusters, in whom he inspired hope of future prowess or opportunity, or faithful followers of less fashionable provincial packs, Surtees's readers were likely to feel they needed initiation manuals of some sort into such arcane rituals, and the more amusing they were, the better. Surtees preserves, however ironically, and often by means of noting their absence, some trace of the traditional pleasures of hunting. He has kept alive for posterity the essence of Somervile's *Chace* and Beckford's *Thoughts*, which, so long as hunting exists as a discourse as well as a sport, may continue to find readers. Somervile's verse may have been surgically reshaped as well as framed by prose narrative, but literate followers of hounds have internalized it, nevertheless, as their music, their poetry, their art, their pleasure.

Part III
Walking in the Countryside

9
The Pleasures of Perambulation

> I took a walk to day to botanize . . . the green is covered with daiseys & the little Celadine the hedge bottoms are crowded with the green leaves of the arum were the boy is peeping for pootys with eager anticipations & delight –
>
> John Clare, Natural History Letter IX (25 March 1825)

During the later eighteenth century, walking ceased to be merely walking and became self-conscious pedestrianism. This tradition is perpetuated today by the Ramblers' Association and allied organizations. What were, and what continue to be, the pleasures of walking?

Gear

If hunting had a coterie language as well as, increasingly by the 1820s, color-coded uniforms, walking had its own equipment and apparel. Plumptre's Pedestrians dressed as sailors, Hucks and Coleridge wore trousers and carried knapsacks. The Rev. Richard Warner and his companion on a first walk in Wales had specially customized coats to carry their 'single change of raiment, and some other little articles for the comfort of the person', as well as maps, compass and drinking-horns.[1] Warner's companion, 'C –', had side-pockets added to his coat, and Warner himself had 'a neglected *Spencer*, which, though somewhat threadbare and rusty, may still make a respectable figure in North-Wales', fitted up by a tailor 'with a sportsman's pocket, that sweeps from one side to the other, and allows room sufficient for all the articles necessary to be carried' (pp. 3–4). These bulky garments, however ingenious, were put to shame by gear they observed carried by three gentlemen pedestrians in Cardiganshire:

> a handsome leathern bag, covered with neat net-work, which, being suspended from the right shoulder by a strap, hung under the left arm, in

the manner of a shooting-bag. This was occasionally shifted from one shoulder to the other, and at the same time that it proved a most convenient conveyance for linen, &c. was no inelegant addition to the person; at least, it gave the wearer much less the appearance of a pedlar than attached to us, from the enormous side-pockets of my companion, and my own swoln Spencer. (p. 4, n.)

The fetishism of gear for walking comfortably, and even with a degree of elegance, while carrying necessities, began early in the history of pedestrian tourism.

A few months after his ascent of Sca Fell in 1802, on a six-week journey in Wales with Tom Wedgwood, Coleridge lovingly entered in his notebook a description of ideal fell-boots and procedures for caring for them:

N.B. Have two Lasts made exactly the shape of my natural foot – the Boots to have a sole less on the hollow of the foot – Mutton suet 1. Hog's Lard 2. Venice Turpentine 1/2 – all mixed & melted – always put on warm, Shoe or boot being held to the fire, while it is being rubbed in – The middle sole of the Boot covered with Cobbler's wax – or still better, steeped thoroughly in the above Composition/the Leather of the Boot should be stout Horse leather – if none to be had, Cow-leather/a piece of oil Silk 6 inches above the Heel, 2 inches wide with a back strap to the Boots.[2]

The back strap would aid in pulling them on, the oil-silk cuff six inches above the ankle would be elegant and waterproof. Richard Holmes has described Coleridge's enthusiasm here as that of the 'true fell-addict'.[3]

But in these early days records of gear fetishism were comparatively rare. Other pleasures took precedence. Unlike fox-hunters, walkers were likely to take their pleasures in comparative solitude – alone, or with one or two companions. Unlike hunting, pedestrianism was not so much about immersion in the animal world, supplemented by hearty good fellowship in the field or around the dinner table. It was about a different kind of absorption in nature.

Botanizing, etc.

When Clare wrote, 'I took a walk today to botanize', he expressed the same purposeful analytical and classificatory relationship to nature that White's writing exemplified.[4] Clare made 'an exact catalogue of the locations of 16 species of orchid he knew from Helpston', and based on the accuracy of his descriptions in verse and the 'conviction' his poetry carries of being 'based on personal experience', he 'has been granted 65 first county records for birds and more than 40 for plants'.[5] Such amateur naturalism was neither an exclusively male nor middle-class pastime in the eighteenth and early nineteenth centuries. Before the professionalizing of the sciences in the nineteenth century, there was room for working-class men as well as women in botany.[6] Systematic botanical fieldwork began with a livery company, the Society of Apothecaries, which established the Chelsea Physic Garden in 1673. From 1620 to 1834, 'herbarizings' took place about once a month between April and September, for the instruction of apprentices. In July the 'General Herbarizing' would be celebrated with a haunch of venison. 'These excursions, without any doubt, were the major seminal influence in the establishment of the great field tradition that forms the core of modern natural history.'[7]

Herbalist traditions of the seventeenth century, and the flower and vegetable shows that 'became part of the recreational repertoire of working men' in the eighteenth century, laid a foundation for working-class interest in botanical science, as Anne Secord has shown in her work on the artisan botanists of Lancashire.[8] The popularity of the Linnaean system of classification enabled certain exchanges of information and specimens between artisan botanists and gentlemen. Although women had also been acknowledged herbalists for centuries, the evidence of working-class women's participation is scantier than the evidence for men, signifying how powerful gender distinctions could be in determining who could be publicly acknowledged to possess technical skill.[9]

In *Mary Barton* (1848), Elizabeth Gaskell observed of the handloom weavers of Oldham that among them were botanists 'equally familiar with the Linnaean or the Natural system, who know the name and habitat of every plant within a day's walk from their dwellings', and who, 'tying up their simple food in their pocket-handkerchiefs, set off with a single purpose', as well as entomologists who set off with nets and dredges for raking 'green and slimy pools'; both groups pored over every new specimen 'with real scientific delight'.[10] In his botanical guide to the environs of Manchester (1849), Richard Buxton reported that he preferred 'carefully to observe living plants on the spots where they grew, and not to make great collections of dried specimens', which looked to him like 'pallid corpses'. His was a conservationist mentality: 'Besides a dislike to dead plants, I did not like to take away and destroy living things, which might be enjoyed by others as well as myself.'[11] Buxton invited his fellow workmen 'to go into the green fields and fresh air of the country; and, whenever they can, to take their wives and children with them'.[12] He also advocated the keeping open of footpaths by landowners.

In this scientific and hunting-gathering mode, the new pedestrianism contributed to grounding ecological thinking and practice. Long-distance rambling, along with hunting and other field sports, helped form the basis for naturalistic fieldwork and transformed pedestrian labor into scientific pleasure.

The Rev. William Bingley, dedicated botanist, sometimes climbed to perilous heights in search of rare plants. His ascent of Clogwyn Du'r Arddu, in the company of a local clergyman, was a case in point:

> For a short time we got on without much difficulty, but we were soon obliged to have recourse to our hands and knees, and clamber thus from one crag to another. Every step now required the greatest care, for even the mere laying hold of a loose stone might have proved fatal. I had once taken hold of a piece of the rock, and was about to trust my whole weight upon it, when it loosened from it's bed, and I should have been sent headlong to the bottom had I not instinctively snatched hold of a tuft of grass, which grew close by it, and was so firm as to save me. When we had ascended a little more than half-way, I was much afraid we should have been doomed to return, on account of the masses of rock over which we had to climb, beginning to increase in size; we knew, however, that a descent would have been attended with infinite danger, and being urged on partly by eagerness in our pursuit, but more from a desire to be at the top, we determined to brave every difficulty. This we did, for in about an hour and a quarter from the time of our beginning the ascent, we found ourselves on the top of this dreadful precipice, and in possession of some very uncommon plants which we had picked up during our walk.[13]

What is most remarkable about this passage is Bingley's recalling of strenuous climbing in such dramatic and psychologized detail – the treacherous loose rock, his instinctive reach for a tuft of grass that will save him, the triumph of getting it right. Bingley here has all the makings of one of John Buchan's athletic outdoor heroes. Robin Jarvis finds an 'element of surprise, even of bathos' in Bingley's mention of the plants at the end of this exciting escape from 'infinite danger'; according to Jarvis, the botanical eye 'belongs properly to a different discursive tradition' from the discourse of recreational tourism, 'with the emphasis on strenuous physical activity', which Bingley's text evokes.[14] This separation of kinds of writing, however, was evidently not one which had come fully into being when Bingley wrote about his 1798 tour, which explicitly combined the interests of a naturalist with a delight in pedestrianizing and advertising walking's pleasures.[15] Between the 1770s and the 1830s, walking often became what hunting and shooting had been, an absorption in the natural world with scientific requirements and outcomes, accompanied by a descriptive discourse that sought to advertise and justify the pleasures that would be found therein. Given the history of the game laws, walking took on some of the cultural meanings of hunting and shooting's shadow practices, trespassing and poaching.[16]

As Joseph Hucks effused on his walking tour of Wales with Coleridge:

[E]very object in nature is interesting, and wherever nature is, I felt similar sensations; mountains and valleys, rivers and rivulets, nay the smallest plants that are trodden under our feet, unseen or unregarded, are inexhaustible sources, to a contemplative mind, of gratification and delight.[17]

One member of his party was a keen botanist, alerting Hucks to what he might otherwise have trodden underfoot without any sense of gratification whatsoever.[18] Undergraduate walkers like Hucks were likely to be diverted from gathering natural historical knowledge by more aesthetic concerns. The science of the greenwood in its traditional form, as practiced by foresters, gamekeepers and poachers, and epitomized by Thomas Barker of Bath's painting, *The Woodman and his Dog in a Storm* (*c*. 1790) (Plate 7) was easily lost sight of by educated pedestrians.

'Hunting the borough'

Pedestrian travel encourages looking at what can be seen from a ground-level point of view, at least until the attaining of the summit from which a prospect might unfold. Ramblers recapture something of the old perspective common to villagers such as Clare, who grew up in the late eighteenth and early nineteenth centuries in the open-field parish of Helpston in the Soke of Peterborough in what was then Northamptonshire. A landscape divided between limestone heath and level fen, Helpston was organized, as open-field parishes were, in a circular fashion, inward-looking rather than open to the outside world before undergoing Parliamentary enclosure between 1809 and 1820.[19] Clare saw as one who regularly walked the fields and woods around Helpston without having internalized the panoramic sense of landscape-in-general attending the prospect views of Denham and Pope and Thomson and the paintings of Claude, and without being versed in the horseback traveler's commanding eye for a landscape that seemed to come naturally to William Marshall or Arthur Young, agriculturalists who made a profession of improvement. Particularly when he first began writing poetry, Clare had to confront the fact that, apart from what he gained by reading, he possessed only the knowledge specific to his parish. It was only so long as Clare was in Helpston that his knowledge 'was valid, was knowledge', as John Barrell puts it; 'the names he knew for the flowers were the right names as long as the flowers were in Helpston'.[20] Clare gloried in his knowledge, recreating himself on his walks, and comparing what he saw with what he had read. He enacted his own private beating of the bounds, undertook a solitary perambulation.

Beating the bounds, which was known as 'hunting the borough',[21] was a form of hunting a country without hunting it.

Walkers, like hunters, experience the displacement of the ordinary and habitual. Walking in Wales, Joseph Hucks observed: 'I cannot, when I am upon the summit of a mountain, with a beautiful and fertile country widely extending upon the sight, think of any thing but the prospect before me.'[22] A concentrated focus upon the now, the immediate present, unbinds the subject from everyday concerns. A more literal kind of displacement also occurs, in that the walker's immersion in the walked landscape causes a kind of forgetting of worldly knowledge, of other landscapes, other places. William Hazlitt briefly mentioned this Clare-like immediacy of perception in his essay on the pleasures of walking:[23]

> In travelling through a wild barren country, I can form no idea of a woody and cultivated one. It appears to me that all the world must be barren, like what I see of it . . . All that part of the map that we do not see before us is a blank.[24]

He could say this, he could enter into a Clare-like sense of locality, but it was for him, as he observed of foreign travel, 'too remote' an experience from his 'habitual associations' to be his dominant experience of the world, and 'like a dream or another state of existence', this perspective 'does not piece into' his 'daily modes of life' (p. 189). Indeed, he admitted that a cartographic sense of prospect views giving way imaginatively to whole counties, which in turn led to a view of the entire kingdom of Great Britain, was only momentarily suppressed by the pedestrian perspective:

> All that part of the map that we do not see before us is a blank . . . It is not one prospect expanded into another, county joined to county, kingdom to kingdom, lands to seas, making an image voluminous and vast; – the mind can form no larger idea of space than the eye can take in at a single glance. (p. 187)

Thus a Clare-like, hedge-bound vantage point was imaginatively possible, but Hazlitt also always knew better. He had the map in his head, though he could forget it while walking. The contraction of view to one of absolute locality, in which Clare *knew* he knew the names of the wildflowers in his own parish, but not necessarily the names of those outside it – without reference to some other authority, that is – was, for Hazlitt, 'an animated but a momentary hallucination' (p. 189).

This contraction of view to one of absolute, if temporary, locality, to no prospect more extensive than the very spot in which one finds oneself at the moment, was, for Hazlitt, intensely liberating. Is this not reminiscent of

Mr Jorrocks's ecstasy at finding a fox and pursuing that fox's line across country? 'Nothing can be finer than the line!' The line taken by any given fox at any given time will serve this purpose. Hazlitt, however, who walks to go a journey rather than to climb stupendous heights, has abandoned 'the image of war without its guilt, and only five-and-twenty per cent. of its danger'. Walkers' risk factor: nil.

Walkers' palimpsests

Gilpin advertised landscape tourism precisely as a new form of hunting, as we have seen. And when readers accustomed to a taste for landscape went walking abroad, judging other people's properties, they felt entitled to 'a kind of Property' in everything they saw, as Addison had put it in 1712. The gathering of landscape images, stored for future contemplation and use, constituted a kind of hunting and gathering that left hardly any trace behind to mark the surface of the land being thus consumed. As Anne Wallace says of what she terms peripatetic poetry, 'It is a miracle of loaves and fishes, a bottomless pitcher, continually renewing the resources it consumes.'[25]

In 'An Invitation to Selborne' (composed before 1793; published in 1813), Gilbert White alluded to his pedestrian habits in a way that could serve as a partial paradigm for future ramblers' pleasure:

>   SEE Selborne spreads her boldest beauties round,
> The varied valley, and the mountain ground,
> Wildly majestic! . . .
>   . . .
>   Arise, my stranger, to these wild scenes haste;
>   . . .
>   Oft on some evening, sunny, soft, and still,
> The muse shall lead thee to the beach-grown hill,
>   . . .
> Romantic spot! from whence in prospect lies
> Whate'er of landscape charms our feasting eyes;
> The pointed spire, the hall, the pasture-plain,
> The russet fallow, or the golden grain,
> The breezy lake that sheds a gleaming light,
> Till all the fading picture fail the sight.
>   . . .
>   Now climb the steep, drop now your eye below,
> Where round the blooming village orchards grow;
> There, like a picture lies my lowly seat,
> A rural, shelter'd, unobserved retreat.
>
>   (ll. 1–3, 7, 19–20, 27–32, 73–6)[26]

John Barrell rightly objects that it is not clear where White is standing in these lines, and in this respect the poem does not conform to the pictorial conventions of eighteenth-century topographical verse derived from landscape painting. The structure that served Claude and Thomson has become 'meaningless' for White: '[A]fter a leisurely and over-observant progress through the intervening ground,' Barrell complains, we have to wait an inordinately long time to reach the gleaming lake, and even then White 'is not clear whether he is over *here*, seeing the lake gleam in the distance, or over *there*, feeling the breeze blow, or watching it make ripples on the lake's surface'.[27]

I would describe White's departure from a stable vantage point as an innovation rather than an error in judgment. White seems to be everywhere, all over the landscape surrounding Selborne, because, at one time or another, he has walked in all these places. He takes his time getting us to the lake through 'a leisurely and over-observant progress through the intervening ground' because he is representing the experience of walking that ground, and not simply the experience of viewing it from a fixed vantage point. He is also inviting readers to 'feast' upon both the prospect and the many walks offered by the prospect. Simply to address himself to the environs of the parish of Selborne from a single spot on the highest attainable ground is too meager a feast, too reductive a form of landscape consumption. The different views and beauty spots appear in the poem as a kind of palimpsest, a layering or overlaying of walks and views, all constituting the experience of feasting upon 'Selborne', which is distinguished, we should note, by its old-fashioned architecture as well as its unspoiled natural and agricultural beauty. The medieval-sounding 'hall' and the church's 'pointed spire' signal community, just as the 'beach-grown hill', mixture of fallow and grain-growing fields, and 'blooming village orchards' suggest an anciently enclosed and unmodernized landscape.

For walking writers the palimpsest of tracks trod and views absorbed encapsulated the hoarding of visual experiences they eagerly sought, and sought to disseminate to their readers. In seventeenth- and eighteenth-century usage, a palimpsest referred to paper, parchment or other writing material specially prepared for writing on and then erasing for re-use (*OED*). By the nineteenth century, this definition had become largely obsolete, as writing paper became cheaper and much more common. In nineteenth-century usage, the palimpsest came to designate a layered repository, a key to the past, since new chemical discoveries had enabled nineteenth-century archeologists to recover the layered past from pieces of vellum by reconstituting successive generations of lost inscriptions.[28] English writers were not slow to appropriate the palimpsest's metaphorical possibilities. As Josephine McDonagh has noticed, the palimpsest 'provided a model for subjectivity, history, and epistemology' for writers such as Coleridge, Thomas Carlyle, Elizabeth Barrett Browning, and

G.H. Lewes, as well as Thomas De Quincey, the subject of her study. For De Quincey, according to McDonagh, 'the model served to represent the construction of human consciousness through the indelible traces of the past'.[29] In *Suspiria de Profundis*, De Quincey described the process of recollecting or reconstructing the past through a palimpsest as peculiarly violent:

> the footsteps of the game pursued, wolf or stag, in each several chase, have been unlinked, and hunted back through all their doubles [.][30]

'Figured as the pursuer of game, the palimpsest editor becomes a hunter and the original text an animal, which, when discovered, reconstituted and read, will be slaughtered,' as McDonagh puts it.[31] The coincidence of language between De Quincey's description of the palimpsest editor and Gilpin's description of the devotee of picturesque theory – '[S]hall we suppose it a greater pleasure to the sportsman to pursue a trivial animal, than it is to the man of taste to pursue the beauties of nature?' – is no coincidence. Both convey the devouring, consumptive sense of their respective pursuits – of the past, of picturesque views – by comparing them with hunting.

Walking as trespassing, but also writing: Wordsworth

If a footpath can be proved to have been used 'as of right' for 20 years or more, it shall be considered 'public'.[32] Wordsworth was nothing if not dedicated to the right to roam, for he 'championed public footpaths with a spirit that should warm the hearts of modern walkers and conservationists'.[33] Wordsworth has been called the National Trust's 'Patron Saint',[34] and he has been credited with inspiring the British National Park Movement.[35] Accompanied by John Taylor Coleridge on a walking visit to Lowther in 1836, Wordsworth, when the path through the Glenridding Walks ended abruptly at a wall:

> attacked the wall as if it were a living enemy, crying out, 'This is the way, an ancient right of way too', and passed on. That evening after the ladies had left the room, Mr. Justice Coleridge said to Sir John Wallace who was a near resident: 'Sir John, I fear we committed trespass today; we came over a broken-down wall on your Estate'. Sir John seemed irate and said that if he could have caught the man who broke it down, he would have horsewhipped him. The grave old bard at the end of the table heard the words, the fire flashed into his face and rising to his feet he answered: 'I broke your wall down, Sir John, it was obstructing an ancient right of way, and I will do it again. I am a Tory, but scratch me on the back deep enough and you will find the Whig in me yet.'[36]

By defending ancient rights of way in this manner, Wordsworth enacted privately the collective beating of bounds of the parish or country he inhabited. When the keeper-opener of footpaths is also a writer, who in writing about his walking keeps footpaths open in imagination as well as reality, 'even private lands along his path are metaphorically unenclosed', as Anne Wallace puts it.[37] '[W]e become tenants in common, enriched by the poetic fruits the walker gathers, enjoying individual access to a mutual estate' (p. 117). Thus the private ramble can have public consequences. In Wallace's view, walking was for Wordsworth serious business, replacing agricultural cultivation as the basis for poetry in the georgic mode (pp. 67–118). The new georgics of walking, rather than of farming, established what Wallace terms a 'peripatetic' mode or aesthetic (pp. 8–9), especially in Wordsworth's 'excursive walking' poems, poems about journeys on foot (p. 119).

Wordsworth's pedestrian verse, including 'Resolution and Independence' and the Snowdon ascent in *The Prelude*, has inspired generations of ramblers. What is perhaps less well known is the way in which the rhythm of walking in a sheltered and confined space, such as a garden or small grove of trees, was intrinsic to Wordsworth's method of composition. He seems to have walked his meter, or associated pedestrian movement with metrical generation.[38] This form of walking – pedestrianism without travel, or walking as repetitive movement without going a journey, to use Hazlitt's phrase – distinguishes Wordsworth from the rambling tradition. This form of walking is akin to modern walking for exercise, where the walks are short ones, if not mere supplements to car journeys undertaken to arrive at somewhere pleasant to walk in.

That Wordsworth's walking could be subordinated to his poetic composition could not have been more clearly stated than in 'When first I journeyed hither' (composed 1800–4). In Wordsworth's day the confinement of modern urbanized life was unthinkable, except perhaps by analogy with going to sea. In this poem, Wordsworth bonds with his brother John (1772–1805), home from his naval adventures, through their mutual liking for walks in a particular fir grove. The Wordsworthian narrator relates that he had always found the grove too closely planted to be entirely satisfactory to walk in because the trees:

> Had by the planter been so crouded each
> Upon on the other, and withal had thriven
> In such perplexed array that I in vain
> Between their stems endeavoured to find out
> A length of open space where I might walk
> Backwards and forwards long as I had liking
> In easy and mechanic thoughtlessness.

> And, for this cause, I loved the shady grove
> Less than I wished to love a place so sweet.
>
> (ll. 32–40)

Wordsworth wished to walk without thinking about walking, in 'easy and mechanic thoughtlessness', concentrating on metrical composition while his body performed the rude mechanics of walking, and this appeared impossible to effect in the closely planted grove.

The sailor brother, however, accustomed to the confinement of his ship, possessed a better eye for a confined space. John treads a pleasing line through the trees, responding to their natural order, and leaving behind him:

> A hoary pathway traced around the trees
> And winding on with such an easy line
> Along a natural opening that I stood
> Much wondering at my own simplicity
> That I myself had ever failed in search
> Of what was now so obvious. With a sense
> Of lively joy did I behold this path
> Beneath the fir-trees, for at once I knew
> That by my Brother's steps it had been traced.
> My thoughts were pleased within me to perceive
> That hither he had brought a finer eye,
> A heart more wakeful: that more loth to part
> From place so lovely he had worn the track,
> One of his own deep paths! by pacing here
> With that habitual restlessness of foot
> Wherewith the Sailor measures o'er and o'er
> His short domain upon the Vessel's deck
> While she is travelling through the dreary seas.
>
> (ll. 57–74)

Recognizing his brother's footwork, the poet writes this poem to memorialize the walk as a bond between them. Only now can the poet love the grove as it has always deserved to be loved, and only now can he appreciate that his Brother is a 'silent Poet!' (l. 88). For the Wordsworthian narrator, the ultimate bonding experience lies in the mutual metrical walking society founded by them after the Brother has returned to sea:

> ... when Thou,
> Muttering the verses which I muttered first

> Among the mountains, through the midnight watch
> Art pacing to and fro' the Vessel's deck
> In some far region, here, while o'er my head
> At every impulse of the moving breeze
> The fir-grove murmurs with a sea-like sound,
> Alone I tread this path, for aught I know
> Timing my steps to thine, and with a store
> Of indistinguishable sympathies
> Mingling most earnest wishes for the day
> When We, and others whom we love shall meet
> A second time in Grasmere's happy Vale.
>
> (ll. 105–17)

One brother mutters the other's verses, one brother treads the other's path. The imagined simultaneity forges an equation between the walking and the verse-making or reciting. The bond between the brothers, their 'indistinguishable sympathies', is paced out by both as the shared palimpsest of their much-loved fir grove returns to the mind of each. Or so it appeared to the poet, if not the sailor brother. The projected reunion of the brothers was not to be. John Wordsworth, who had joined in 1789 the East India Company's ship, the *Earl of Abergavenny*, drowned when this ship went down on 5–6 February 1805.

This poem represents a particularly intense instance of the imagining of walking as mutual enjoyment, or rather, enmeshment. Such bonding through walking, through a shared palimpsest of walks, sights, sounds and smells, is an important feature of all walking writing. The greatest walking poems and prose descriptions imply or construct just such an intimate, mimetic relation between the walker-writer and the audience. But however much the activity of walking might have meant to Wordsworth as a political gesture or a source of pleasure or social bonding, the walking was always subordinated to the writing. Composing verse was Wordsworth's obsession. Walking was ancillary to literary production.

Walking as hunting: Clare

If Wordsworth's paradigm was forward-looking, influential for future generations, Clare's was more deeply interfused with seventeenth- and eighteenth-century influences, especially that of Thomson, of whom, as a poet of nature, Clare wrote, 'and my favourite Thomson shall not yield to any one, either ancient or modern, in my opinion – only mine perhaps. See how he paints the white hyacinth . . .'[39] Clare did not chop himself free from eighteenth-century taproots. His was a retrospective paradigm, but one in touch with E.P. Thompson's 'rebellious traditional culture'. As an agricultural and gardening laborer, Clare sought recreation on Sundays by walking in the fields and wastes, composing verses and making notes on the flora and fauna he observed. Even at work, he would sometimes be overcome by the urge to note something down and would 'drop down behind a hedge bush or dyke and write down my things upon the crown of my hat'.[40] More consistently and conscientiously than Wordsworth's, Clare's walking constituted a perambulation or beating of the bounds in protest against improvement and enclosure.[41] Clare's pedestrianism was solitary, but he walked and wrote on behalf of his fellow parishioners and even fellow creatures whose liberties were being curtailed by agrarian improvement. Clare's walking was crucial for his writing, but it was also a form of imaginative political and ecological practice.[42]

In Clare's aesthetics of perambulation, the walking writer performs a solitary beating of the bounds, reclaiming territory from agrarian improvement and enclosure, often hunting out botanical specimens and natural historical observations. His miming of hunting while walking is a kind of reinscription of fox-hunting – a rewriting that is also a reinventing. Just as Surtees appropriated the discourse of the picturesque in the service of hunting writing, Clare appropriated the discourse of hunting for walking. He thus transformed modern mounted fox-hunting back into the pedestrian activity it had once been. Although no sportsman, Clare was attracted by the spectacle of the hunting field. He couldn't help thinking of fox-hunters in their red coats as he went for an autumnal walk in 1841. More surprisingly, he actually imitated the prose of fox-hunting reports in the sporting press while describing walking:

> – here is a drove leads us on its level sward right into the flaggy fens shaded on each side with whitehorn hedges covered with awes of different shades of red some may be almost called red-black others brick red & others nearly scarlet like the coats of the fox hunters – now we have a flaggy ditch to stride which is almost too wide for a stride to get over – a run & jump just lands on the other side & now a fine level bank smooth as a bowling green curves & serpentines by a fine river whose /wood of/ osiers & reeds make a pleasant rustling sound though the wind scarcely moves a single branch – how beautifull the bank curves on like an ornament in a lawn by a piece of water the map of ploughed field & grass ground in small alotments on the left hand with an odd white cottage peeping some where between the thorn hedges in the very perfection of quiet retirement & comfort & on the right hand the clear river with its copices of reeds & oziers & willow thickets . . .[43]

The use of scarlet coats to describe the red of some whitethorn hawes introduces the hunting theme. Readers are invited to follow the line in the best sporting tradition. At our peril we go at a ditch which is wider than it at first seems, and we have to take a running jump at it. We cross the country as if hunting it, negotiating obstacles by leaping them, but attending too to the naturalistic details and picturesque beauties of the scene. A bank covered in fine grass, 'smooth as a bowling green', makes a pleasing surface for walking or running on, a miniature Leicestershire sea of turf that gets the blood up for galloping on – and on. Meanwhile, the bank itself offers an aesthetically satisfying serpentine line, enhancing the fine prospect of the river. This is a landscape of social goods as well as natural beauties – a cared for and worked landscape, but not a harshly modernized and improved one. No ruined cottages mark the passing of a former commoners' community; rather, the white cottage peeping between thorn hedges embodies the essence of rural peace, a settled continuity – the desired object of so many country walks. The cottage emblematizes 'the very perfection of quiet retirement & comfort'. As we continue to view the country at hunting pace, we see 'on the right hand' the rich wildlife habitat of the riverbank reflected in the river's purity, a residue of wild wetland undisturbed by improvement.

Clare thus naturalized fox-hunting and fox-hunters, making them part of his picturesque and botanically specific landscape. Their colors contributed to his palette, and their sport enlivened his imaginings while walking. The sanguinary language of hunting he contained within the hunters' wearing of the fox's livery, those entirely naturalized red coats. Fox-hunting *kitsch* of the curio and table-mat variety was already on the horizon by 1841.

At the end of the eighteenth century, walking, whether more influenced by picturesque theory or the impulse to botanize, was grounded in a displacement of hunting. When modern fox-hunting as it had come into being by the 1830s, with its cut and thrust semiotic of competitive riding, coupled with a frequent not-knowing if hounds were in fact hunting or not, became the dominant form of hunting, and shooting became all about big bags and battues or game drives, field sports in their earlier incarnation of 'Countrey Contentments', replete with naturalistic fieldwork, were already dying out. At this historical juncture long-distance walking in pursuit of picturesque views, climbing in search of sublime experiences, and scouring the countryside for botanical specimens, became alternative forms of hunting. Once hunting had been displaced by these alternatives, it could be dispensed with, and forgotten. The alternative game bag was a rich aesthetic repository of overlaid images absorbed at ground level.

When written up, perambulations might well inspire imitation in the form of actual walks taken and other walking writing. Whole travelers' guidebooks might be written in response, sometimes in verse. And the written perambulations themselves often represented the experience of walking and 'feasting' as a richly layered palimpsest of images, views and momentary hallucinations. The risk of 'infinite danger' which Bingley incurred in order to botanize, and which modern fox-hunting epitomized, would be less and less often on the menu.

10
'This Lime-Tree Bower' as Walking Poem

> Coleridge has told me that he himself liked to compose in walking over uneven ground, or breaking through the straggling branches of a copsewood; whereas Wordsworth always wrote (if he could) walking up and down a straight gravel-walk, or in some spot where the continuity of his verse met with no collateral interruption.[1]

As the anecdote from Hazlitt indicates, Coleridge constantly compared himself with Wordsworth, usually anxiously and emulatively. It was, after all, *Coleridge's* description of his own and Wordsworth's walking habits that Hazlitt reported. Because Coleridge suffered such great anxiety regarding verse composition in particular and public reception in general, his notebooks are a particularly rich source of comparatively uncensored thoughts. In a notebook entry of 24 October 1803, describing a walk through Borrodale into Watendlath with Robert Southey and Hazlitt, Coleridge seems to have entered into a strange fantasy of out-Wordsworthing Wordsworth by writing a poem according to the number of strides necessary to walk from one place to another. The parodic quality of the entry illustrates brilliantly the difference between Wordsworth and Coleridge as walking poets. Coleridge began the entry by sounding more like Wordsworth than himself: 'Of course it was to me a mere walk; for I must be alone, if either my Imagination or Heart are to be excited or enriched.'[2] Compare this disclaimer with what Hazlitt reported of his own experience of Coleridge's gift of invention while walking long distances:

> ... I accompanied him six miles on the road. It was a fine morning in the middle of winter, and he talked all the way ... In digressing, in dilating, in passing from subject to subject, he appeared to me to float in air, to slide on ice.[3]

In October 1803 Coleridge seems to have been preoccupied by the need to write a reply to Wordsworth's *The Prelude*: 'O surely I might make a noble Poem of all my Youth nay *of all my Life.*'[4] Coleridge's plan for doing so in the notebook, never fulfilled so far as we know, was to build a cairn, marking for future travelers where they should leave a well-trodden track in order to receive the best view of the Lake of Keswick. Coleridge would then write a poem by pacing the distances within this scene. The competition with Wordsworth is explicit. If the tone of the entry were slightly different, we might suspect that Coleridge was parodying Wordsworth in order to mock him:

> This is /I have no hesitation in saying it –/ the best, every way the best & most impressive View in all the Lake Country – why not in all the Island? – . . . How could Wordsworth think otherwise? – Go & build up a pile of three, by that Coppice – measure the Strides from the Bridge where the water rushes down a rock in no mean cataract if the Rains should have swoln the River – . . . from this Bridge measure the Strides to the Place, build the Stone heap, & write a Poem, thus beginning – From the Bridge &c repeat such a Song, of Milton, or Homer – so many Lines I [will] must find out, may be distinctly recited during a moderate healthy man's walk from the Bridge thither – or better perhaps from the other Bridge – so to this Heap of Stones – there turn in – & then describe the Scene.[5]

Raimonda Modiano notes that this notebook entry indicates Coleridge's poetic rivalry with Wordsworth in all things to do with representing nature.[6] It also indicates his rivalry with Wordsworth in turning footfalls into metric feet: a rivalry of perambulation in shared landscapes.

As evidenced in Hazlitt's anecdote, Coleridge longed to thrust himself into physically demanding circumstances. From tackling uneven ground and breaking through straggling copse-woods would inspiration come, not from re-treading well-worn paths. David Craig has described Wordsworth as the first person to fuse the very different approaches to hill-walking and climbing assumed by local Cumbrian shepherds and by tourists, while Coleridge was almost entirely a tourist, pursuing 'intensity through risk' and always on the lookout for extreme experiences.[7] Coleridge's ascent and, more hair-raisingly, his descent of Sca Fell in August 1802 make thrilling reading, whether in his notebook entries or Sara Hutchinson's transcriptions of his letters. 'When I find it convenient to descend from a mountain,' wrote Coleridge to Sara, his 'Asra', 'I am too confident & too indolent to look round about & wind about 'till I find a track or other symptom of safety; but I wander on, & where it is first *possible* to descend, there I go.'[8] David Craig has described this letter as the first piece of climbing prose in English, recording the ascent and descent of Sca Fell for the first time. Coleridge found himself descending Broad Stand, a series of 10- and 15-foot steps and smooth rectangular corners which rise up from the col between Sca Fell and Scafell Pikes. Craig observes of Coleridge's feat, 'It is now graded Moderate and I've seen notices in Lake District shops warning walkers not to climb up it without a rope.'[9] Coleridge reported that his 'violent exertions' on this occasion produced both a moment of sublimity and a heat rash:

> ... the stretching of the muscle[s] of my hands & arms, & the jolt of the Fall on my Feet, put my whole Limbs in a *Tremble*, and I paused, & looking down, saw that I had little else to encounter but a succession of these little Precipices – . . . I lay upon my Back to rest myself, & was beginning according to my Custom to laugh at myself for a Madman, when the sight of the Crags above me on each side, & the impetuous Clouds just over them, posting so luridly & so rapidly northward, overawed me / I lay in a state of almost prophetic Trance & Delight – . . . so I began to descend / when I felt an odd sensation across my whole Breast – not pain nor itching – & putting my hand on it I found it all bumpy – and on looking saw the whole of my Breast from my Neck [to my Navel] – & exactly all that my Kamell-hair Breast-shield covers, filled with great red heat-bumps, so thick that no hair could lie between them . . . It was . . . a startling proof to me of the violent exertions which I had made.[10]

Coleridge would have made a thruster in the hunting field – had he been able to ride without being perpetually thrown and getting boils on his backside, as happened to him in the army;[11] and had he been able to approve of a ritual chase that must sometimes end in a kill.

Walkers for pantisocracy

Coleridge's 'This Lime-Tree Bower My Prison' (composed 1797, first published 1800, in the second volume of Southey's *Annual Anthology*)[12] epitomizes the walking poem as both palimpsest and perambulation. The poem was produced from layered memories of previous walks superimposed upon a walk taken by the poets' friends in an enactment of community. This particular walk, on the 'unenclosed Quantocks hilltops on which one may roam unrestricted by fences', as Tim Fulford notes, produces an experience of 'land shareable in common' because it *is* common land.[13] The poem makes evident the extent to which the consumption of landscape was a form of consumption – a displacement of hunting by consuming bloodlessly and nonviolently – and an appropriation of nature that hardly left a trace, except in writing. Some erosion on hill tracks, a few stones dislodged from a scree, an impression of a human form in the flattened heather: these are likely to be all the traces left behind by Romantic walkery – except for the written record of the walk. This procedure is what Coleridge captured so quintessentially in 'This Lime-Tree Bower'. The poem also evidences something of that split we have observed coming into being between walking for gathering scientific knowledge of individual plants and animals, and walking as primarily aesthetic experience (picturesque landscape consumption, metrical composition, momentary hallucination).

Having mapped out an itinerary for a walk with his visiting friends, Charles Lamb and the Wordsworths, based upon his own extensive knowledge of the district of Nether Stowey, Holford and Alfoxden, Coleridge was prevented from walking with them by an accident – the unpleasantness at the cottage, by now infamous, of having had boiling milk spilled on his foot by Sara Fricker Coleridge, his wife. The poem, addressed to 'Charles Lamb, of the India House, London', is framed by the following anecdote:

> In the June of 1797 some long-expected friends paid a visit to the author's cottage; and on the morning of their arrival, he met with an accident, which disabled him from walking during the whole time of their stay. One evening, when they had left him for a few hours, he composed the following lines in the garden-bower.[14]

This is a poem about a walk not undertaken in the flesh, but imaginatively, from memory and projection. Coleridge found, to use Addison's terms, 'a secret Refreshment in a Description' of a prospect he had often possessed visually in person.[15] As Richard Holmes puts it, '[The poem] vividly evokes the bare "springy" heath of the upper hills, the damp "ferny" dark of Holford Combe and its waterfall, the "deep radiance" of the evening sunlight on Poole's garden of limes, walnut trees and ancient ivy.'[16] By the poem's end, Coleridge had composed himself from feeling self-pity into feeling 'a kind of Property in every thing' he saw.[17] Imagining a community of fellow walkers and readers into being thus, Coleridge granted them a certain moral, social and political authority arising from 'the common sharing, rather than exclusive possession' of the landscape, in Fulford's terms.[18] This was a great, though fragile, achievement in a single poem, and significantly it was Coleridge and not Wordsworth who produced it.

> WELL, they are gone, and here must I remain,
> This lime-tree bower my prison! I have lost
> Beauties and feelings, such as would have been
> Most sweet to my remembrance even when age
> Had dimm'd mine eyes to blindness! They, meanwhile,
> Friends, whom I never more may meet again,

> On springy heath, along the hill-top edge,
> Wander in gladness, and wind down, perchance,
> To that still roaring dell, of which I told;
> The roaring dell, o'erwooded, narrow, deep,
> And only speckled by the mid-day sun;
> Where its slim trunk the ash from rock to rock
> Flings arching like a bridge; – that branchless ash,
> Unsunned and damp, whose few poor yellow leaves
> Ne'er tremble in the gale, yet tremble still,
> Fann'd by the waterfall! and there my friends
> Behold the dark green file of long, lank weeds,
> That all at once (a most fantastic sight!)
> Still nod and drip beneath the dripping edge
> Of the blue clay-stone.
>
> (ll. 1–20)

The speculation 'and wind down, perchance, / To that still roaring dell, of which I told' is disingenuous precisely because Coleridge sounds so confident the walkers will have followed his instructions to the letter so as not to miss anything, especially not a parenthetically protected view – '(a most fantastic sight!)' – so exclusive that only their friend Coleridge could have found it and recognized it for what it was.

A rare vision of animate creation, of vitality in natural objects, the dell offers a glimpse of the one life flowing through all things, a naturalist's supranatural find.[19] The ash's yellow leaves, the dripping and nodding 'dark green file of long, lank weeds', inspirited by the spray from the waterfall, appear as animate beings. These particular long, lank 'weeds', which still grow plentifully in Holford Combe near Alfoxden, fascinated Coleridge. He inserted a geological reference to secure their specificity: it is blue clay-stone country, that holds the moisture upon which these lank but moving plants thrive. Coleridge was as ecstatic at their finding as any naturalist coming upon a rare species or any poacher upon game lying up, ready for bolting, but it was because of what the plants signified, not because of what they were. Coleridge wasn't sure what they were, or what to call them. Neither he nor the Wordsworths would acquire their copies of William Withering's *An Arrangement of British Plants* (1796) until sometime between the autumn of 1800 and the early spring of 1801.[20] And when the Withering volumes did arrive, Coleridge would find himself in difficulties.[21]

This particular long, lank, but graceful plant had figured in an earlier poem by Coleridge, 'Melancholy: A Fragment' (first published in *The Morning Post*, 12 December 1797, but written several years earlier).[22] In spite of an absence of moving air, *Asplenium scolopendrium*, or Hart's Tongue, had fluttered against

Melancholy's cheek as she lay dreaming on a mouldered abbey's wall, barely touched by 'the flagging sea-gale weak' (l. 7): 'The long lank leaf bow'd fluttering o'er her cheek' (l. 8). But Coleridge had made a botanical error in 'Melancholy' and named the plant 'Adder's Tongue' (*Ophioglossum vulgatum*). The true Adder's Tongue, an atypical fern that is stiffer and more upright than the Hart's Tongue fern, resembling a lily or other member of the arum family, would not have comported itself in this way.[23] But the name Adder's Tongue struck Coleridge as more poisonously melancholy, and thus more 'poetical', than the trace of courtly blood sports signified by the Hart's Tongue. Although in 1802 he defended 'a force & distinctness of Image' in these lines, 'ill as they are written',[24] the poem was spoiled for Coleridge by this misnaming. When 'Melancholy' appeared in *Sibylline Leaves* (1817), Coleridge added a note: 'A botanical mistake. The plant I meant is called the Hart's Tongue, but this would unluckily spoil the poetical effect. *Cedat ergo Botanice.*'[25] Reluctantly, Coleridge retired from the field of botanical identification. He was more cautious in 'This Lime-Tree Bower', calling the plants simply 'weeds', but he could not resist appending an equivocating note, drawing attention to his own confusion and his continuing desire to depart from botanical nomenclature: 'The *Asplenium Scolopendrium*, called in some countries the Adder's Tongue, in others the Hart's Tongue, but Withering gives the Adder's Tongue as the trivial name of the *Ophioglossum* only.'[26]

Not for Coleridge, then, any acquisition of 'a poetics of the botanically exact', in which Charlotte Smith, Dorothy Wordsworth or Clare might take pride.[27] Dorothy Wordsworth demonstrates the kind of systematic study necessary for such exactness. At Grasmere on 28 May 1802 Dorothy noted in her journal:

> There is yet one primrose in the orchard. The stitchwort is fading. The wild columbines are coming into beauty. The vetches are in abundance, Blossoming and seeding. That pretty little waxy-looking Dial-like yellow flower, the speedwell, and some others whose names I do not yet know.

As Dorothy's editor Mary Moorman comments, '[S]he identified it two days later', writing on the first fly-leaf of the Wordsworths' copy of Withering, volume 1: 'Lysimachia Nemorum, Yellow Pimpernell of the Woods Pimpernel Loosestrife May 30th 1802.'[28] In defending Coleridge's commitment to accuracy in his recording of natural phenomena, including the notes concerning the Hart's Tongues he wished were Adder's Tongues, Kelvin Everest remarks, 'The unease that Coleridge felt over his inability to fit his "poetical effect" ("adder" helped to intensify the atmosphere of Gothic menace in *Melancholy*) into the external framework of botany, is most suggestive.'[29] According to Everest, 'Coleridge was genuinely anxious to get the detail right' in both poems.³⁰ But Coleridge, however anxious for accuracy, would remain stubbornly resistant to learning botanical classification. He wished for a command of the scientific language of the natural world, but could never internalize it. He would write in October 1803 of the walk through Borrodale into Watendlath with Southey and Hazlitt:

> – What was the name of that most vivid of all vivid green mosses by the side of the falling water, as we clomb down into Watendlath! – that red moss, too and that blood-red Fungus? – . . . – O surely I might make a noble Poem of all my Youth nay of *all my Life* – One section on plants & flowers, my passion for them, always deadened by their learned names.³¹

This entry illustrates the difficulty of writing about natural phenomena without an exact language in which to do so. Coleridge's 'passion' for plants is 'deadened' by their 'learned names', but sometimes, as in the case of the mosses and fungus above, there are no other names.

In 'This Lime-Tree Bower', perhaps his most successful poem at recording natural phenomena, Coleridge thus displays both his strengths and weaknesses. As a botanist he may be lacking, but as a walking poet he is sublime. Holmes notices how profoundly the topography of the upland Quantocks affected Coleridge's imagination during his time at Nether Stowey.³² In this poem particularly, Coleridge conducts his friends in his mind's eye on a perambulation of the bounds of his new-found country, managing to reassure himself that they are wandering happily in his foot steps, especially Charles Lamb, a fugitive from London:

> Now, my friends emerge
> Beneath the wide wide Heaven – and view again
> The many-steepled tract magnificent
> Of hilly fields and meadows, and the sea,
> . . . Yes! they wander on
> In gladness all; but thou, methinks, most glad
> My gentle-hearted Charles! . . .
>
> (ll. 20–8)

In his sense of Charles as having 'hunger'd after Nature', 'in the great City pent' (ll. 29–30), Coleridge was self-deceived.³³ But even within the poem there is anxiety that the landscape might fail to inspire Charles as it should. In a supreme effort to convey his sense of the rich pleasures of the Quantocks as a palimpsest of views, Coleridge asks the landscape to perform for his friends. As we shall see in relation to early Dartmoor walking in the final chapter, conceiving of weather and the creation of atmosphere as a theatrical performance necessary to the experience of place is typical of the aesthetic of perambulation. In hoping that Charles particularly will be gladdened out of his urban misery, Coleridge loses confidence in the landscape's ability to work its magic unassisted. He conjures a command performance:

> . . . Ah! slowly sink
> Behind the western ridge, thou glorious sun!
> Shine in the slant beams of the sinking orb,
> Ye purple heath-flowers! richlier burn ye clouds!
> Live in the yellow light, ye distant groves!
> And kindle, thou blue ocean! So my Friend
> Struck with deep joy may stand, as I have stood,
> Silent with swimming sense; yea, gazing round
> On the wide landscape, gaze till all doth seem
> Less gross than bodily; and of such hues
> As veil the Almighty Spirit, when yet he makes
> Spirits perceive his presence.
>
> (ll. 32–43)

Remembered topography is lovingly described, but by means of imperatives. We might call this a form of interactive picturesque. A moment of divine revelation is briefly hallucinated. And the writer's own happiness is restored: 'A delight / Comes sudden on my heart, and I am glad / As I myself were there!' (ll. 43–5). Sometimes a vicarious walk, the literary composition or, by extension, consumption of a walk, can have the same effect as a walk itself. Unbinding the subject to a state 'Less gross than bodily', and storing up images of beauty for future use, the aesthetic of perambulation conflates the practices of pedestrian travel and print culture in a self-ratifying circuit of pleasure as consumption. But it is a form of consumption less gross than bodily, a bloodless hunting and tracking and gathering of virtual specimens for subsequent analysis. As if to nail down this implicit departure from earlier modes of human bonding through rural rituals, Coleridge replaces both the georgic lens focused on agriculture and the sights of the gun emblematic of field sports with his own perambulatory emphases:

> . . . Henceforth I shall know
> That Nature ne'er deserts the wise and pure;
> No plot so narrow, be but Nature there,
> No waste so vacant, but may well employ
> Each faculty of sense, and keep the heart
> Awake to Love and Beauty! and sometimes
> 'Tis well to be bereft of promised good,

> That we may lift the Soul, and contemplate
> With lively joy the joys we cannot share.
> My gentle-hearted Charles! when the last rook
> Beat its straight path along the dusky air
> Homewards, I blest it! deeming, its black wing
> (Now a dim speck, now vanishing in light)
> Had cross'd the mighty orb's dilated glory,
> While thou stood'st gazing; or when all was still,
> Flew creeking o'er thy head, and had a charm
> For thee, my gentle-hearted Charles, to whom
> No sound is dissonant which tells of Life.
>
> (ll. 59–76)

Coleridge takes a bead on the bird, but the gun is missing. Agricultural revolution – 'no waste so vacant' but it might employ men to cultivate it – is actively displaced by the cultivation of the senses. There is a strange tension in these lines between an unnatural, but seemingly inevitable solitude – regret at missing those 'joys we cannot share' – and a yearning for close companionship, for bonding to be enacted through pedestrianism – 'My gentle-hearted Charles!' As with William and Dorothy at Hart-Leap Well, or William and John in the fir-grove, the solitary status of the sensitive observer of nature is temporarily overcome by a shared experience of a profound Nature. One lesson of the lime-tree confinement is that temporary deprivations can be illuminating and uplifting. ' 'Tis well to be bereft of promised good' if the result is increased empathy – with friends and fellow creatures, extending even to less-loved birds such as rooks.

Rooks' 'creeking' or raucous cawing voices signified backwoods rurality from Restoration drama onwards. In Sir George Etherege's play *The Man of Mode* (1676), when Harriet contemplates a return to the country from London, she laments, 'Methinks I hear the hateful noise of rooks already – kaw, kaw, kaw!'[34] Rooks were not beautiful songsters nor valued for their plumage nor considered particularly good to eat. They were neither ornamental pastoral furniture nor game. Scientifically, they were recognized to be the farmer's friend in the eighteenth century, but popular prejudice was against them. As Thomas Pennant remarked in *The British Zoology* of 1766, because the rook thrusts its bill 'into the ground in search of the *erucae* of the Dor-beetle', the rook 'should be treated as the farmer's friend; as it clears his grounds from caterpillars, that do incredible damage by eating the roots of the corn. Rooks are sociable birds; living in vast flocks; crows only go in pairs.'[35] In the 1790s Gilpin and Richard Payne Knight appropriated the rook for picturesque aesthetics. Gilpin opined that forest scenes were better adapted to the rook than to any other bird, and that 'Among all the sounds of animal nature, few are more pleasing than the cawing of rooks.'[36] Knight echoed both Pennant and Gilpin, characterizing the rook as agriculturally useful as well as politically instructive in *The Landscape*. Rooks, 'little politicians', demonstrated 'The ills, that from a social compact flow' (1: 349–50), but also asserted the rule of law, high among the oak trees that symbolized British liberty on gentlemanly estates (356, 338). Knight appended a naturalistic note to these verses, complaining once again of farmers' prejudice:

> The farmers, when they see the rooks feeding on the fields that are newly sown, are apt to imagine that they are eating the seed-corn, and thence endeavour to destroy them; whereas they are in reality digging up the worms and slugs, and by that means doing the most essential service. The large white grub with a brown head, which, after lying three years in the ground, becomes the common brown beetle or caterpillar, and which is so destructive to the roots of grass and corn, while in this embryo state, is a favourite food with them; – whence those insects seldom appear near to rookeries.[37]

Here we have another instance of the tenets of picturesque theory aiding and abetting ecological knowledge by contradicting common agricultural practice. The rook population was regulated by annual shooting in the nineteenth century. This spring event was particularly looked forward to by laborers, prevented from shooting other birds at other times; sometimes young birds were poached from nests in anticipation.[38]

Coleridge thus followed a highly significant rook across the sky in his mind's eye: this was an unnaturally solitary, would-be sociable, too often unacknowledged to be ecologically beneficial, bird. The poet's eye follows the rook as a man taking aim with a gun – or a crossbow – would. But instead of bonding through the ejaculation of gunpowder and falling birds, the poet joins with his friend by keeping a bead on the rook's singular flight against the sun, creaking across the eye of eternity.[39] They are united not in manly rituals of bloodshed but in appreciating the dissonance of the rook's call. Walkers for Pantisocracy.

This displacement of the gentlemanly sporting ideal through the rejection of shooting, like the displacements of hunting we have observed in other walking poems, now reads as merely conventional. But in its day, it was a richly overdetermined moment marking a seismic shift in English culture.

11
Dartmoor Visible

> Devonshire is certainly the worst hunting country I ever was in; yet, strange to say, there are more hounds kept in it than in any other three counties in England. Independent of the established packs of stag and fox-hounds (of which there are one of the former, and four of the latter), nearly half of the resident gentlemen, and the greater part of the yeomanry, keep what they call '*a cry* of dogs;' and a friend of mine, who resides among them, told me he had hunted with seventy-two packs!
>
> Charles James Apperley, *Nimrod's Hunting Tours* (1835)

The county of Devon in southwestern England is the home of provincial hunting *par excellence*. Still largely agricultural, Devon lies between two coasts and two great moors, the National Parks of Exmoor and Dartmoor. Since 1995, the National Parks of Britain have been attempting 'conservation and enhancement' of cultural as well as natural 'heritage' within their boundaries.[1] Equivocating about what may or may not constitute 'a genuine Dartmoor tradition', the Dartmoor National Park Authority have recently hazarded the example of 'common land management with its ancient origins, pony drifts, hunting and field sports'.[2]

This chapter endeavors to account for Dartmoor's emergence as a tourist destination in the late eighteenth and early nineteenth centuries because that phenomenon exemplifies the arguments of this book. As a royal forest, Dartmoor had been a hunting country for centuries, and its wildness continued to attract sportsmen and deter other visitors. This hunting history inflected the writing of early pedestrians, who reproduced, however unconsciously, something of a hunting frame of mind. The chapter is also a personal reflection, because for the past ten years, I have been living for several months of the year in a Dartmoor village, and thus the moor has provided a lens through which to view both the present and the past. (This lens is nothing like a Claude Glass, I might add – rather more like the focusing space provided by a cap's brim, when eyes are screwed up against the weather.)

Walking culture put Dartmoor on the tourist map. It was as if travelers could only properly appreciate moorland scenery or even, in a sense, bring themselves to *look* at it, to see it as a sight worth seeing, once they had become deliberate pedestrians. The link between walking and liberal freedom observed by Langan apparently found its most satisfactory expression beside the granite tors and peat bogs of moorland wastes such as Dartmoor. Modern conservationists who are dedicated ramblers speak of finding freedom in wild and solitary places. Conservationists' preference for moorland over the ancient enclosed English landscape, remarked upon by Marion Shoard, may well have something to do with moorland barrenness; ecologically speaking, a moor is a form of 'biological desert', as well as an escape route from other people.[3] British moorland is not wilderness in the sense of primaeval forest as North American wilderness is. Most British moorlands, Shoard observes,

> are relatively recent landscapes, created at most 4000 years ago through the destruction of forest to provide wood, charcoal or sheep runs. What is more, most moors rely on Man's activities – burning and the grazing of his animals – for their continued existence: left to itself, heather moorland reverts to scrub or woodland within about 60 years. (p. 58)

Only the wilderness of moorland, it would seem, can satisfy many conservationists' psychological requirements for a truly '*therapeutic*' experience of landscape, as Kate Ashbrook of the Ramblers' Association describes her attitude to moorland.[4] 'Wildness, openness, asymmetry, homogeneity, height, freedom for the rambler to wander at will, and the absence of what is obviously human handiwork' are the characteristics of moorland that satisfy many conservationists.[5]

The moor looks natural rather than manmade, though this is an illusion. Its openness suggests a realm of freedom. And its weather and topography are both physically challenging and forbidding – potentially sublime. Moorland devotees thus described sound remarkably like urban picturesque tourists of the late eighteenth and early nineteenth centuries. Dartmoor fulfills the criteria recommended by picturesque theorists. Under the rubric of preserving 'a "wilderness" experience which is unique in southern Britain',[6] Dartmoor conservationists have been urging in recent years that isolated moorland cottages be left to decay into ruins rather than be made habitable 'because of the effects on the surrounding landscape of ancillary facilities such as drives, gardens, power lines etc'.[7] The aim of the policy appears to be increasing the 'wilderness' effect by achieving a landscape with a look of human depopulation, a landscape returned to nature from human cultivation.

Moorland conservationists, according to Shoard, stress the personal pleasures of walking as therapeutic solace, not naturalistic field study. They may pay no more attention to individual plants and animals than do less knowledgeable walkers:

> Spectacular birds like ravens, merlins, hen harriers, even peregrine falcons could flit past unnoticed or at least undistinguished. Nor do they bend down and admire, close up, a sprig of heather or gorse, lichens, mosses, fungi or a blade of grass . . . The curlew and the buzzard, through their weird cries, are perceived as background noise, a feature of the moors rather than of the birds themselves. Flowers are perceived *en masse* through their smell and through their colour.[8]

The conservationist Malcolm MacEwen takes pleasure in this 'background noise' of birds in remote places as a kind of *'silence with sounds'*.[9] In this respect the moorland advocates have different priorities from the preservers of ancient woodlands and 'veteran' trees.[10] Scientific knowledge or even interest in individual plants and animals, their presence or absence, what they might indicate about the health or fragility of the local ecosystem, have been supplanted by aesthetic pleasure and what we might call the unbinding of the subject in an experience of liberal freedom. Thus the Dartmoor National Park *Plan* speaks of striving to accommodate heavy tourist use of the moor at certain designated sites, while leaving 'room for solitude and personal discovery elsewhere'.[11]

It is surely significant that Dartmoor failed to appear in travelers' itineraries until late in the eighteenth century. Early notices of Devonshire often ignored Dartmoor entirely. Drenched by heavy rain, Celia Fiennes rode the 20 miles from Launceston in Cornwall to Okehampton in Devon, where she 'met with a very good Inn and accomodation, very good chamber and bed', and proceeded the next day through Crockernwell to Exeter.[12] In sight of the looming mass of Dartmoor for much of the way, she never once mentions seeing it. Defoe similarly traveled from Exeter to Plymouth, round the southern end of the moor, without once alluding to it.[13] He did cite William Camden's *Britannia* when referring to Exmoor – 'Cambden calls it a filthy, barren, ground, and, indeed, so it is' – but quickly turned his gaze to the lower course of the river Exe and 'Devonshire in its other countenance, viz. cultivated, populous, and fruitful'.[14] Thomas Fuller said much the same thing in 1662, contrasting South Hams in South Devon as 'so fruitful, it needs no art' with the barrenness of 'Dart-more', which 'will hardly be bettered by art'.[15]

Even William Marshall, whose profession it was to ride over and survey a country with an eye toward agricultural improvement, and who claimed in

1796 to have 'had opportunities of seeing almost every square mile' of the surface of 'Dartmore, and its Uncultivated Environs', and of 'observing its natural characters, in different and distant parts', did so at a *distance*, from several different vantage points on high ground.[16] Not surprisingly, he grew particularly excited when, deep in the moor, he spied 'HERBAGE! – greensward!' even on 'the highest bleakest hills; frequently intermixed, however, with HEATH'.[17] The sight of good pasturage was the most reassuring view he could find amongst the heather and granite outcrops. In 1795 J. Laskey had gone in search of Cranmere Pool, but failed to find it. He went with an eye toward possible enclosure and cultivation, but was not impressed by what he found, remarking that the

> rage for improvement of poor lands seems of late to have been carried too far, and instances are not wanting of furze-brakes in particular being cleared and grubbed up, and, in the course of a few seasons, suffered to return to their pristine form (furze-brakes) again.[18]

By 1799 it had become possible to contemplate walking on the moor as part of a recreational itinerary, but the idea still stimulated anxiety about the wildness of the place. The Rev. Richard Warner dithered in September of that year about whether or not to take his landlord's advice in Lydford and go across Dartmoor to Two Bridges: 'The idea of travelling twelve miles over a desolate moor, wild as the African Syrtes, without a single human inhabitant or regular track, has something in it very deterring, I must confess.'[19] On the 14th he met 'a brother pedestrian' on the road and was persuaded to go with him across 'the wilds of Dartmoor' between Okehampton and Chagford, by way of Gidleigh Park and Kestor Rock (though he refers to the latter as 'the *Spinster* or old-maid rock'):[20]

> This expressive name, which happily conveys the ideas of desolation and barrenness, solitude and individuality, will suggest to you in a moment the appearance and situation of the object in question. It is a vast natural acervation of granite rock, naked and rugged, forming the apex of one of the hills of Dartmoor, which rises in solitary sadness on the edge of the waste ... the many rock-basons on the top, and circular arrangements of small stones on the sides, may be considered as vestiges of the holy mummery which was heretofore practised on this consecrated hill. The prospect from it is immense. (p. 172)

For Warner the only thing that made 'this wide extent of naked barrenness' palatable was a horseman's imaginings of its sporting possibilities (p. 169).[21] Contemplating stag-hunting over such 'a vast tract of unobstructed country',

Warner opined that 'Perhaps, indeed, no English hunting can be more animating than that of Dartmoor', since such a country gives 'a variety and beauty to the chace, that no inclosed country can possibly afford' (p. 170). And this from a man who earlier in his tour deplored the sounds of echoing guns near Glastonbury, concurring with Thomson that field sports were a 'falsely cheerful barbarous game of death' (p. 7).

The folklore of Dartmoor contains many hunting stories. Thomas Fuller reported that a gentleman called Child from Plimstock, 'hunting in Dartmore, lost both his company and way in a bitter snow. Having killed his horse, he crept into his hot bowels for warmth; and wrote this with his blood;

> "He that findes and brings me to my Tombe,
> The Land of Plimstock shall be his doom."[22]

Frozen to death nevertheless, Child was found by the monks of Tavistock, who promptly claimed possession of the manor of Plimstock.

In fact, though red deer were 'anciently common' on Dartmoor, as the Rev. Sabine ('Onward Christian Soldiers') Baring-Gould remarked, by the mid-nineteenth century, they had disappeared, apart from occasional incomers from Exmoor.[23] The Forest of Dartmoor had been a mainstay of the king's larder in medieval times, and commoners had to build cornditches – deep ditches with the dug earth thrown and mounded behind stone fascias – to discourage the deer from entering farm enclosures.[24] Because the deer could not see over these walls, they were deterred from leaping over them, but if they did, getting back was easier still, and 'that suited the King very well'.[25] As forerunners of the ha-ha, Dartmoor's cornditches evidence once again the connection between the medieval deer park and the eighteenth-century landscaped park. But because the deer were 'mischievous to the crops of the farmer, and to the young plantations', according to Baring-Gould, the 'farmers, yeomen, and squires combined to get rid of them from Dartmoor'.[26] Unlike the farmers of Exmoor, for whom 'the spell of the hunt', in Ted Hughes's phrase, operated to stay their prejudices against the agricultural damage caused by deer, the commoners of Dartmoor refused this compromise, with the result that the deer were exterminated. Thus it was 'for fox, hare, and otter hunting' that the sportsman went to Dartmoor during the nineteenth century, and not for red deer.[27]

Between the 10th and the 16th of September 1799, while Warner wandered around Gidleigh and Chagford, thinking about stag-hunting, Coleridge and Southey were also on a walking tour in Devon. From Exeter they went to Paignton, Dartmouth, Totnes and across part of Dartmoor back to Exeter.[28] From Southey's lodgings in Exeter, Coleridge wrote to Tom Poole, supplier of the cottage and lime-tree bower at Nether Stowey:

Here I am, just returned from a little Tour of five days – having seen rocks; and waterfalls; & a pretty River or two; some wide Landscapes; & a multitude of Ash-tree Dells; & the blue waters of the [']ROARING Sea!' as little Hartley [Coleridge's son] says . . . — The Views of Totness & Dartmouth are among the most impressive Things, I have ever seen / but in general, what of Devonshire I have lately traversed is tame to Quantock, Porlock, Culbone, & Linton. — So much for the Country – now as to the inhabitants thereof, they are Bigots; unalphabeted in the first Feelings of Liberality . . .[29]

These particular inhabitants, who supported Church and King and the slave trade, were largely Coleridge's own extended family (he had been born in Ottery St. Mary), with whom he claimed to have 'neither Tastes or Feelings in common'.[30] He had become acclimatized to Somerset and North Devon, with their more free-thinking, dissenting populations. As in seventeenth-century Wiltshire – the Royalist chalk downlanders so different from the Parliamentarian and litigious folk of the wooded cheese country, as John Aubrey had opined – so also in eighteenth-century Devon:[31] a conservative south and a more intellectual and radically influenced north coexisted within a single county.

Like Plumptre's Pedestrians, Coleridge and Southey were interested in the manners and ideas of fellow human beings, but it was rocks, waterfalls, rivers, wide landscapes and secret places, such as ash tree dells, that stimulated them to make their tour. Dartmoor became visible for the same reasons that walking became a recreational activity. Both were consequences of 'improvement' which kept a certain fear of the wild at bay: moorland could now be aestheticized because of its very difference from intensive agricultural cultivation, and even in remote areas, anxieties regarding highwaymen and robbers dispersed with better regulation of roads and transport. Famous for 'their enthusiastic adherence to the old festive culture, and for drinking and general debauchery', Dartmoor's tin miners had long been a law unto themselves, since they were 'exempt from national and county taxes, had the right to be mustered in their own militia units, and could not be tried by the county courts for any crimes except wounding and murder'.[32] All other offenses were dealt with by the tinners' own local Stannary court, and the Great Court or Tinners' Parliament legislated upon all matters concerning the tinners. This 'self-governing state within the state', notable for the toughness of the men involved, seems to have deterred travelers from casually venturing into its moorland recesses.[33] However, by the early nineteenth century, although many men were once again employed in mining – tin, but also copper, arsenic, lead, silver and zinc – prompted by continuing war with France and population increases, the tinners were only a remnant of their former numbers.[34] The Napoleonic campaign also brought about the first recorded use of Dartmoor for military exercises, a tradition which still continues in the National Park, and the prison at Princetown was built to house French prisoners of war.[35] For so long a wasteland, Dartmoor 'could now, it seemed, be penetrated, tamed and exploited'.[36] The prospect of a wilderness or upland heath was newly pleasing precisely because of its unimproved, unenclosed and uncultivated qualities, especially if they could be contemplated free from fear.[37]

In the nineteenth century, Dartmoor poems as well as guidebooks began to be written. Despite existing verse tributes, an assumption persisted that the moor remained undiscovered country. William Browne of Tavistock had begun his *Britannia's Pastorals* of 1613–16 with lines identifying him as a Dartmoor man:

> I that whileare neere *Tavies* straggling spring,
> Unto my seely Sheepe did use to sing,
> And plaid to please myselfe, on rusticke Reede,
> Nor fought for *Baye*, (the learned Shepheards meede,)
> But as a Swaine unkent fed on the plaines,
> And made the *Eccho* umpire of my straines:
> Am drawne by time, (although the weak'st of many)
> To sing those layes as yet unsung of any.
>
> (ll. 1–8)[38]

Because this part of Devon must have seemed sufficiently remote to readers to warrant explication, these lines were accompanied by an explanatory note:

> *Tavie* is a River, having his head in *Dertmore* in *Devon*, some few miles from *Marie-Tavy*, and falls South-ward into *Tamar*: out of the same Moore riseth, running north-ward, another, called *Taw* . . .

Michael Drayton began *Poly-Olbion* (1622) with Cornwall and Devon, naming the '*Dert*' and the '*Ting*' among other rivers.[39]

In 1795 Ann Batten Cristall included a poem, 'Written in Devonshire, Near the Dart', in her *Poetical Sketches*.[40] Cristall's single volume has been praised by Jerome McGann for its 'commitment to (not a mere belief in) expenditure and ecstasy'.[41] McGann suggests that we learn from Cristall's obscurity, her single volume of work, and the single surviving copy known to us, that 'For her poetry should not be defined by measures of fame or endurance. A thing of beauty is not a joy for ever, it is a joy for now'.[42] Cristall strives for fleeting ecstasy by writing, conscious of her fragile, mortal state. Not so much attuned to a 'silence with sounds' as listening for individual birds' songs, she makes nature notes with feeling: 'Hark! where the shrill-ton'd thrush, / Sweet whistling, carols the wild harmony!' (ll. 51–2). The poet would have her own verse do the same. Thrilling to this 'lov'd society', Cristall anticipated many later writers, who have wondered, as their minds filled with sensations stimulated by birds, animals and vegetation, 'Ah! where again' would they 'such pleasures find?' (ll. 60–2). In the Dart valley, where 'wandering waters join with rapid force', the poet was struck by the 'varying prospects' of 'Rough-hanging woods' rising behind 'cultur'd hills' (ll. 16–19). Dartmoor was a presence, striking in its promise of wild energy, but only just visible, looming in the background.

Ignoring the moor's poetical past, the Royal Society of Literature advertised in 1820 a contest for the best poem on Dartmoor and, out of a field of entrants that included Joseph Cottle, awarded the prize of 50 guineas to Felicia Hemans the following June.[43] When publishing his *Dartmoor: A Descriptive Poem* in 1826, N.T. Carrington was quick to point out that his poem 'was not one of those which were rejected on that occasion', but had been written at the suggestion of W. Burt, Esq., Secretary of the Plymouth Chamber of Commerce, who had contributed the preface and the notes.[44]

Among guidebooks, Samuel Rowe's *A Perambulation of the Antient and Royal Forest of Dartmoor* of 1848 initiated a trend.[45] Baring-Gould, given Rowe's book as a birthday present when he was a boy, said of it:

> It arrested my attention, engaged my imagination, and was to me almost as a Bible. When I obtained a holiday from my books, I mounted my pony and made for the moor.[46]

An entire moor-exploring industry sprang up that remains vital to the district's economy to this day.

Of all the early writing about Dartmoor, none was more plainly an eye-witness account of the strenuous pleasures of moorland walking than that composed by Sophie Dixon. A resident of Princetown, in the vicinity of Dartmoor Prison, Sophie Dixon published a collection of poems in 1829, identifying herself as Dartmoor's truest native poet:[47]

> Ye mighty Tors, – ye regions of the cloud!
>   Dartmoor! in chant of Bard too long unsung!
> So lofty, so remote, so passing proud,
>   Ye tower on high, the scenes of song among.
>
> None e'er hath looked upon you, as fit theme
>   For verse – nor e'er in kindred feeling warm,
> Deemed there might mantle one inspiring beam
>   O'er giant halls of solitude and storm.
>
>   ('Stanzas. – Written on Dartmoor', ll. 1–8)

So far as Dixon was concerned, to give voice to Dartmoor, in its 'solitude and storm', was the essence of poetry. This particular landscape encoded the turbulent, striving human psyche but also surpassed it in grandeur. In 'Sonnet', she wrote:

> THERE is a glory on the dark rough hill,
>   When the low sun his setting radiance throws;
> . . .
> Oh Earth! Air! Ocean! wherefore should we seek
>   Language, save yours? – The Eternal's glorious fane,
> Where oracles of Heaven around us speak!
>
> (ll. 1–2, 12–14)

The very topography of the moor, dark and rough, and as vast as the ocean, lends the poet a language as enduring and beyond human control as the geological forces that formed that topography. A glimpse of glory, as fleeting as the radiance of a setting sun, is all that can be hoped for. In 'On Longaford Tor – one of the hills of Dartmoor', Dixon waxed both more topographically specific, by naming a particular tor, and more intimate in her relation to the moor than she had yet done. Dartmoor was a 'Desert', unvisited by most people, but all the more dear to her for that:

> What dreams are ours, thus pondering mid
>   The Desert all around us spread!
> Half seen in light, in shade half hid,
>   Dusk vales below, rocks overhead;
>   And where the cataract flashing dread
> Boils up in its tremendous glee, --
>   By the blithe crowd unvisited,
>   -- Yet sought and loved by me.
>
> (ll. 25–32)

Dixon had internalized the codes of the picturesque and the sublime along with the forms of contemporary verse; she quoted Beattie and Byron and echoed Coleridge and Wordsworth. The 'cataract flashing dread' is like nothing so much as Coleridge's Exmoor-inspired 'deep romantic chasm' (l. 12) complete with insurgent river in 'Kubla Khan' (1816), combined with his description of the poet who has acquired the forbidden knowledge necessary to represent such a landscape:

> And all should cry, Beware! Beware!
> His flashing eyes, his floating hair!

> Weave a circle round him thrice,
> And close your eyes with holy dread,
> For he on honey-dew hath fed,
> And drunk the milk of Paradise.
>
> (ll. 49–54)⁴⁸

Dixon is more modest than Coleridge in her claims for poetry. Like Cristall, she seems to have written to expend feeling as much as to capture or contain it. In the desert, she does not long for fame, but only to recompense the moor for the pleasure it has given her.

In Dartmoor, Dixon had found not only her native, but her ideal, subject matter. The apparently unchanging face of the moor, too rugged and remote ever to become a modern cultivated landscape, gave her a sense of a massive permanence beyond herself, or beyond any human capacity. The moor was a repository of vast energies and forces as well as a realm of freedom for wandering humans. Wild nature was now not to be feared so much as its loss in overcultivation. What would in earlier times have been a language of menace had become the language of desire.⁴⁹ The following lines from 'On Longaford Tor' could serve to gloss the appeal of Dartmoor for many people since Dixon, including many conservationists:

> Man may encroach – but never plough
>   Shall e'er thy craggy summit pass –
> His roofs may grow around – but thou
>   Untouched shalt lift thy mountain-mass;
>   And when all he hath wrought, must class
> With things gone by, – thy rugged brown
>   And its thin wreath of desert grass,
> Remain as they are now.
>
> (ll. 81–8)

Imagining the new roofs of an increasing population surrounding the moor, but the high moor itself remaining unchanged, a monument to geology rather than agriculture, Dixon recorded a sensation felt by many people since. The obscurity of Dixon herself, the few copies of her works that were printed, and the very few now in existence, all testify to the comparative insignificance of any single human endeavor on the scale of geological time. The slightness of Dixon's *oeuvre* is no measure of the significance of her achievement.

In the summer of 1830, Dixon made two journeys on foot from Princetown and published *A Journal of Ten Days Excursion on the Western And Northern Borders of Dartmoor* and *A Journal of Eighteen Days Excursion on the Eastern And Southern Borders of Dartmoor, and on the Western Vicinity of Exmoor; including*

Ilfracombe, Lynton, &c.[50] The Bodleian Library copy of these two journals contains the following manuscript note:

> The two Journals are seldom found together, and in this state are exceedingly rare as they emanated from different presses, & were not printed for sale.
>
> ---
>
> This diary, or memorandum book, or what you please to call it, is not to be despised for its untutored simplicity. On the contrary, I think that the writer has very fairly confined her pen to the work of recording what she actually saw, and that she really saw a good deal more than most people who are less observant.[51]

Dixon sought above all the experience of wild nature which would induce a feeling of staggering sublimity. To achieve this end, she was willing to endure incredible discomfort and fatigue. Dixon appears to have made nothing of a walk of 28 or 30 miles, 'the largest portion of which was over hills, where the wind blew with such violence that we had extreme difficulty to proceed at all,' reporting only being 'sufficiently tired' afterwards.[52] She might almost have been enjoying a few good days' hunting, given her dedication to moor-walking in the worst sort of weather, which Dartmoor was capable of providing even in May and July. Dixon concluded her first tour-journal thus, sounding pleased at:

> having obtained from the various objects of our tour a very considerable degree of satisfaction, and treasured up many recollections of beauty and grandeur, such as can only be derived from the observance of nature in her most sequestered recesses; – those haunts where the forms of art have not yet been intruded, and where man comes only as a visitor, to wonder, to admire, and, if his own perversity withstands not, to be instructed and rendered happier. (p. 32)

There is the language of a full bag of recollections here. Dixon's nature, like Gilpin's picturesque beauty, is a goddess worthy of venery, of eroticized pursuit culminating in penetration of her most secret parts. Consuming landscape has become therapeutic in precisely the ways hunting was once considered to be. The object of such a search is to be made happier, having hunted down 'beauty and grandeur' and gathered them for future use.

Attaining a view of nature's sequestered recesses in 1830 involved Dixon in being 'blown upon and rained upon and splashed from head to heels with mud', exactly the effects of fox-hunting as Virginia Woolf described them in 1926. According to Woolf, we recall, these consequences of rough weather and strenuous physical activity in crossing a country at speed had worked themselves 'into the very texture of English prose and given it that leap and dash, that stripping of images from flying hedge and tossing tree which distinguish it not indeed above the French but so emphatically from it'. Dixon captured these effects for pedestrian touring. On Thursday, 27 May, ascending 'Eastor or Yestor', Dixon wrote:

> While nearing the peak of the hill, the clouds suddenly collapsed, and resting on the Tor in a dense mass, the shower began to roll upon us with great impetuosity. During the first violence of the rain we found an imperfect shelter beneath the vast pile of rocks which crown the summit. The mist or cloud that enveloped us preventing any prospect of the surrounding country, which is in clear weather very extensive from this Tor.
>
> Deeming it hazardous to proceed further upon the hills, in a state of atmosphere so uncertain, we abandoned our intention of visiting great Kneeset, about four miles from hence to the south-eastward, and confined our excursion to Blackator, and the valley below it which forms the channel of the West Ockment. In descending the river side of this range of tors, we beheld a scene of the most magnificent character, when at the clearing of the showers, the clouds had collected on the ridge of the opposite steeps, rolling in gigantic columns of vapour, that seemed to station themselves successively on the very brow of the rugged hills over which they hung. A gleam of sunshine breaking through these vast mist-wreaths, they were suddenly clothed with a splendour of hue, and a diversity of light and shade, that rendered the whole prospect one of indescribable beauty and grandeur, the spirits of the storms unveiled their mighty forms, and arrested the eye with an impression so powerful and poetical, that no words can do it sufficient justice. We thought the encountering even of a tempest to behold this scene, would have cheaply bought the pleasure of that sublime recollection which it left behind. (p. 29)

We might have expected that phrase about abandoning the intention of visiting Great Kneeset in such a storm to have been followed by the admission that they returned to Okehampton, where they were staying. Not a bit of it! Dixon and her companions simply explored Blackator and then sought slightly lower ground in the West Okement valley; they did not go home. Their perseverance paid off. Like Mr Jorrocks, they found their fox in the form of a scene of such sublimity for future use in recollection that Dixon regarded it as cheap at the price of a long day's walking in soaked skirts. Of hunting itself, Dixon has nothing much to say, beyond praising a tapestry at Cotehele of a huntsman, a greyhound, and 'all the different varieties of sporting dogs; the bulldog, mastiff, bloodhound, wolfdog, pointer, &c., all of a particularly fine and natural character, and done in colours, which seem to retain all their first freshness and beauty' (p. 12).

Dixon's second excursion concluded with these lines of verse, which sum up the pleasures of perambulation as they had been appropriated from hunting in former times:

> Yet ends not here the pleasure we have won
> In wandering and beholding – nor the knowledge
> That from the beautiful and enduring
> Comes on our spirits – as the morning sunrays
> Brighten the dusk cloud with their early flame.
> Yes! ours hath been delight which shall not perish
> Though exiled from its source – but still renew
> In memory's meditation, what we felt;
> When by the rough rocks, and eternal billows,
> And solemn woods, and cataracts of foam
> Treading like native dwellers of the wild.
> We have heard other waves, and other rivers
> Than unto us belong – and heard the winds
> Play in the clefts that seam the ocean crag:
> Or where like hoary tresses, curl the breakers
> Around some rock that lifts its ancient head
> Amid the dread immensity of waters.
> We have beheld the waving forests lean
> Upon the rugged hills; and the bare cliffs
> Jaggy and stern, protrude against the sky.
> We have communed with these in their wild places
> And desolate abodes, and unto them
> Frequent in thought returning, will again
> At least in thought be present, and recall
> The beauty and the glory left behind.[53]

So long as memory serves, the intrepid venturer can return to these scenes of energy and power in imagination. The Miltonic syntax of 'Frequent in thought returning', with its Latin and Greek precedents, gives a sense of history to this movement, but also suggests its fragility in the scheme of time. The poet 'will again / At least in thought be present' at these scenes, but there is inevitably a sense of loss of what has been left behind.

Dixon had left behind more than she knew. 'Treading like native dwellers of the wild' had been for centuries the ambition of huntsmen and poachers, sportswomen and men, naturalists who prided themselves on their woodcraft.

Now, 'communing with these in their wild places' meant communing not with wild creatures, but with the spirits of seas, rivers, rocks, forests, hills and cliffs. If hunting animals exceeds simply communing with them, the traditions of woodcraft and knowledge of local ecology required to hunt nevertheless represent a loss. Neither seeking out beasts of chase nor studying closely the habits and habitats of any animals or plants figures on Dixon's agenda. The memory, not of a good day's hunting, but of walks walked and views viewed, would be sufficient recompense for what had been endured during such an immersion in an inanimate wild nature.

One question which modern conservationists must answer is which period of a particular landscape's history they wish to emphasize. How strict a definition of 'indigenous' species will be applied? Is the aim to restore the landscape to a prehistoric purity, or is it to acknowledge the historical layering brought about by centuries of importation and naturalization of foreign species? How old must a tree be to be called 'veteran' and thus worthy of preservation? How old is an ancient woodland?, and so on. The Woodland Trust and English Nature offer the definition of pre-1600 for trees and woodlands in England and Wales, and pre-1750 for those in Scotland.

The present preservation policies for Dartmoor have been inspired by the same picturesque aesthetic that animated the earliest writing about the moor by pedestrian tourists (1799–1830). The Dartmoor of these comparatively modern walkers is the wilderness that the conservationists seek to restore. Once again we have the invention of tradition. Dating not from prehistoric or medieval times, or even from 1600, the woodland preservationists' threshold of the ancient, but from the early nineteenth century, this Dartmoor is a modern, comparatively recent phenomenon. The landscape of Dartmoor, although it appears to be a wilderness, has derived from nothing so much as centuries of grazing, dwelling, felling, hewing, mining, military training and hunting. The conservationists' preference for a picturesquely inspired walker's Dartmoor over a sportsman's or woman's, a poacher's, a miner's or a farmer's Dartmoor can be explained, as we have seen, by the history of landscape aesthetics and the associated movement from hunting to walking. But this idea of the moor has material consequences for its future. Is this comparatively recent perception of moorland as an aesthetic object and a space of 'solitude and personal discovery' all that should be conserved?

If rural communities are to be preserved as well as ancient monuments, then a georgic rather than a pastoral sense of what it means to inhabit that modern invention, the countryside, will remain necessary. The notion of stewardship 'need not be an entirely anachronistic ideal', despite the fact that in the future 'Only in terms of land use will rural England be agricultural England', as Howard Newby has observed.[54] 'In all other senses – economically, occupationally, socially, culturally – rural England will be (and in many cases already has been) comprehensively "urbanized".'⁵⁵ And if preservation of the landscape in its historical layerings, from pre-history to the present, is the object of conservation today, then the pre-history of modern scientific ecology in hunting and poaching should not be forgotten or ignored. Among Dartmoor hill farmers there is a conviction that a ban on fox-hunting would almost certainly bring about the disappearance of the fox from many 'in country' areas surrounding the high moor, and a drastic reduction in the number of foxes living on the moor itself.[56] Certainly the hare and the deer have vanished from Dartmoor, not as animals hunted to death by hounds, but as casualties of agricultural improvement.[57] Needless to say, none of these animals figures anywhere in the official conservationists' policies and plans for Dartmoor.[58]

Modern hunting will have to change, of course, before it is worthy of not being banned. Let hunting be returned to its greener, naturalistic former self. Let it also be rethought in terms of ecology and non-instrumental relations with the natural world. Let walkers know themselves to be treading in the foot-steps of hunters and poachers as well as natural historians.

Notes

Preface

1. The Hunting Bill listed three options: an outright ban enforced by fines up to £5,000 (advocated by Deadline 2000); the Middle Way approach involving licensing of hunting and coursing; and self-regulation under the auspices of the Independent Supervisory Authority for Hunting (supported by the Countryside Alliance). On 17 January 2001, the Commons voted 387 to 174 for a ban, a majority of 213; licensing was rejected by 382 to 182, a majority of 200; and self-regulation fell by 399 to 155, a majority of 244; George Jones and Benedict Brogan, 'MPs Vote for Total Ban on Hunting', *Daily Telegraph*, Thursday, 18 January 2001. '[T]he Hunting Bill is certain to be blocked by the Lords and will not become law before the general election, expected in May'; Andrew Grice, Political Editor, *Independent*, Friday, 19 January 2001. See also Michael White, Political Editor, 'Tory Peers Accuse Blair of "Cynical Ploy" on Hunting Ban', *Guardian*, Friday, 19 January 2001. On 29 March, the House of Lords rejected the entire Bill. The outbreak of foot and mouth disease meant that the issue was politically too sensitive to return it to the Commons, and with a general election declared, the Bill collapsed.
2. Sue Everett, 'Conservation News: Right to Roam', *British Wildlife: The Magazine for the Modern Naturalist* 10: 4 (April 1999), p. 290. The Countryside and Rights of Way Act 2000 came into effect on 30 January 2001, but much of the acreage has yet to be re-mapped showing strictly defined 'right to roam' areas; Peter Hetherington, 'Landowners Still Have Time on their Side', *Guardian*, Saturday, 20 January 2001. The foot and mouth outbreak has closed off much of the countryside to the public as this book goes to press. On 18 September 2000, the Rural History Centre at the University of Reading held a conference on 'The Contested Countryside', organized by Jeremy Burchardt. For recent views both left and right, see Ingrid Pollard, 'Another View', *Feminist Review* 45 (Autumn 1993): 46–50; Paul Cloke and Jo Little, eds, *Contested Countryside Cultures: Otherness, Marginalisation and Rurality* (London and New York: Routledge, 1997); Anthony Barnett and Roger Scruton, eds, *Town and Country* (London: Jonathan Cape, 1998); and Michael Mosbacher and Digby Anderson, eds, *Another Country* (London: Social Affairs Unit, 1999).
3. *The Sporting Magazine; or Monthly Calendar of the Transactions of The Turf, The Chace, And every other Diversion Interesting to The Man of Pleasure and Enterprize* (London: Printed for the Proprietors, and Sold by J. Wheble) first appeared in 1792, and 'held its own until 1870, when it ceased; having twenty-two years previously absorbed the *Sporting Review* which at different dates had absorbed the *Sportsman* and *New Sporting Magazine*'; Sir Walter Gilbey, *Animal Painters of England from the year 1650*, 2 vols (London: Vinton and Co., 1899), 1: xii.
4. Raymond Carr, *English Fox Hunting: A History* (1976; London: Weidenfeld and Nicolson, 1986), p. 26.
5. Marion Shoard's *The Theft of the Countryside* (London: Temple Smith, 1980), 'an impassioned attack upon farmers for bringing about the environmental ruination of the countryside', as Howard Newby notes, had its weaknesses as a 'balanced

246 *The Invention of the Countryside*

analysis', but 'as a polemic it was an outstanding success', becoming an instant bestseller and shifting the terrain of the public debate from 'conservation through voluntary agreement to conservation via legislative control'; 'Epilogue', *Green and Pleasant Land? Social Change in Rural England* (1979; London: Wildwood House, 1985), pp. 283–305; this passage p. 296. Newby compares Shoard's book with Rachel Carson's *Silent Spring* (1963) in that it 'placed the farming lobby markedly on the defensive – itself a considerable achievement', p. 296. In this vein see, most recently, Graham Harvey, *The Killing of the Countryside* (London: Jonathan Cape, 1997).
6. Christopher Morris, ed., *The Illustrated Journeys of Celia Fiennes c. 1682–c. 1712* (London and Sydney: MacDonald, 1984), p. 36.
7. Sir John Denham, *Coopers Hill*, Draft I (1641), in Brendan O Hehir, *Expans'd Hieroglyphicks: A Critical Edition of Sir John Denham's Coopers Hill* (Berkeley and Los Angeles: University of California Press, 1969), lines 243–5.
8. Marchamont Nedham, *Mercurius Pragmaticus* (London, 1649); quoted in Nigel Smith, *Literature and Revolution in England, 1640–1660* (New Haven and London: Yale University Press, 1994), p. 68.
9. David W. Macdonald and Paul J. Johnson, 'The Impact of Sport Hunting: A Case Study', in V. Taylor and N. Dunstone, eds, *The Exploitation of Mammal Populations* (London: Chapman and Hall, 1996), pp. 160–207; see pp. 185, 194, 199.
10. Roger Scruton's *On Hunting* (London: Yellow Jersey Press, 1998) is a defense.
11. Charles Clover, Environment Editor, 'Hunting is Our Art, Our Life. We'll Fight: Labour Peer Warns Blair of Tragedy in the Country', *Daily Telegraph*, Friday, 11 July 1997, p. 4.
12. Patrick Wright, *On Living in an Old Country: The National Past in Contemporary Britain* (London: Verso, 1985), pp. 81–7.
13. Ibid., p. 87.
14. Nick Atkinson, *Dartmoor National Park Plan*, Second review (Bovey Tracey: Dartmoor National Park Authority, 1991), p. 36, and Dartmoor National Park Authority, *Dartmoor National Park Management Plan: Consultation Draft* (Bovey Tracey: Dartmoor National Park Authority, February, 2000), p. 12.
15. Anthea Hall, 'Lady who Refuses to be Outfoxed', *Sunday Telegraph*, 'Review', 22 December 1991, p. viii. All quotations from this page.
16. Macdonald and Johnson, 'Farmers and the Custody of the Countryside: Trends in Loss and Conservation of Non-Productive Habitats 1981–1998', *Biological Conservation* 94 (2000): 221–34; this passage p. 222.
17. Ibid., pp. 222, 232.
18. Ibid., pp. 233, 230, 233.
19. Macdonald and Johnson, 'The Impact of Sport Hunting', in Taylor and Dunstone, eds, *Exploitation of Mammal Populations*, p. 203. Arguing that the fox-hunting debate is complex and requires an 'interdisciplinary, non-partisan analysis', and steering a course between 'pathocentric' and 'ecocentric extremes' – minimizing foxes' suffering while accounting for the wider ecological picture – they recommend banning the use of terriers to dig out foxes that have gone to ground, making earthstopping illegal, curtailing the season before vixens have become heavily pregnant, and, ultimately, developing ways of making drag-hunting appealing, as, given the urban majority's finding of hunting animals for recreation 'morally unacceptable', 'This would seem the only course that is likely to preserve, and indeed potentially to enhance greatly, the traditions, skills, social infrastructure and employment associated with foxhunting', pp. 199–205.

Introduction

1. See Barrell, *The Idea of Landscape and the Sense of Place 1730–1840: An Approach to the Poetry of John Clare* (Cambridge: Cambridge University Press, 1972), pp. 1–3, and Janowitz, *England's Ruins: Poetic Purpose and the National Landscape* (Oxford: Basil Blackwell, 1990), pp. 3–4. Jeremy Burchardt makes a similar argument to mine, that 'the concept of "the countryside" came into being' when rural landscape began to be experienced as 'an object of consumption, rather than as a means of production', in *Urban Perceptions and Rural Realities: Attitudes to the Countryside in England 1800–2000* (forthcoming 2001). My thanks to him for letting me read the book in manuscript.
2. On the rivalry with France as a cultural solidifier of the English, Welsh, Scots and, most problematically, the Irish, see Linda Colley, *Britons: Forging the Nation 1701–1837* (New Haven and London: Yale University Press, 1992), and Gerald Newman, *The Rise of English Nationalism: A Cultural History, 1740–1830* (New York: St. Martin's Press, 1987).
3. As Raymond Williams observed, locating this stable rural society, this proverbial Golden Age of country life, has proved illusory, as we find ourselves on a kind of historical escalator – and yet, 'Old England, settlement, the rural virtues – all these, in fact, mean different things at different times, and quite different values are being brought to question'; *The Country and the City* (New York: Oxford University Press, 1973), p. 12.
4. G[ervase] M[arkham], *Countrey Contentments, In Two Books: The first, containing the whole art of riding great Horses in very short time, with the breeding, breaking, dyeting and ordring of them, and of running, hunting and ambling Horses, with the manner how to use them in their travell. Likewise in two newe Treatises the arts of hunting, hawking, coursing of Grey-hounds with the lawes of the leash, Shooting, Bowling, Tennis, Baloone &c. The Second intituled, The English Huswife* ... (London: Printed by J.B. for R. Jackson, 1615).
5. Paul Langford, *A Polite and Commercial People: England, 1727–1783* (Oxford and New York: Oxford University Press, 1992), p. 4.
6. Blackstone, *Commentaries on the Laws of England* (1765–69), 3: 326; quoted in Langford, *Polite and Commercial People*, p. 1. See also Robert Brenner, *Merchants and Revolution: Commercial Change, Political Conflict, and London's Overseas Traders, 1550–1653* (Princeton: Princeton University Press, 1993); and Peter Earle, *The Making of the English Middle Class: Business, Society and Family Life in London, 1660–1730* (London: Methuen, 1989).
7. Langford, *Polite and Commercial People*, p. 3.
8. Ibid., p. 5.
9. Paul Langford, *Public Life and the Propertied Englishman, 1689–1798* (Oxford: Clarendon, 1991), p. 9.
10. Ibid., p. 368.
11. For a summary of economic and social change during this period, see Gerald MacLean, Donna Landry, and Joseph P. Ward, 'Introduction: The Country and the City Revisited, c. 1550–1850', in MacLean, Landry, and Ward, eds, *The Country and the City Revisited: England and the Politics of Culture, 1550–1850* (Cambridge: Cambridge University Press, 1999), pp. 1–23. See also Peter Borsay, *The English Urban Renaissance: Culture and Society in The Provincial Town 1660–1770* (Oxford: Clarendon, 1989).

12. Howard Newby, *Country Life: A Social History of Rural England* (London: Weidenfeld and Nicolson, 1987), p. 34.
13. *The Farmer's Boy; A Rural Poem, In Four Books*, by Robert Bloomfield (London: Printed by T. Bensley for Vernor and Hood, et al., 1800), in Jonathan N. Lawson, ed., *Collected Poems (1800–1822) by Robert Bloomfield* (Gainesville, FL: Scholars' Facsimiles & Reprints, 1971).
14. Roger B. Manning, *Hunters and Poachers: A Cultural and Social History of Unlawful Hunting in England, 1485–1640* (Oxford: Clarendon, 1993), p. 17, and J.M. Neeson, *Commoners: Common Right, Enclosure and Social Change in England, 1700–1820* (Cambridge: Cambridge University Press, 1993), pp. 40, 170.
15. Manning, *Hunters and Poachers*, p. 61.
16. *Ibid.*, p. 61.
17. 'Petunt a rege ut omnes warennae tam in aquis quam in parco et boscis communes fierent omnibus, ita ut libere posset, tam pauper quam dives, ubicunque in regno in aquis et stagnis piscariis et boscis et forestis feras capere, in campis lepores fugare, et sic haec et hujusmodi alia multa sine contradictione exercere'; *Chronicon Henrici Knighton, vel Cnitthon, Monachi Leycestrensis*, ed. Joseph Rawson Lumby, 2 vols (London: Eyre and Spottiswoode, 1889–95), 2: 137; translated in Manning, *Hunters and Poachers*, p. 17.
18. David Underdown, *Revel, Riot, and Rebellion: Popular Politics and Culture in England 1603–1660* (Oxford: Clarendon, 1985), p. 160.
19. *Ibid.*, p. 160.
20. Paine, *Rights of Man*, Part II (1792), ed. Eric Foner (New York and London: Penguin, 1985), p. 232, n.
21. Pope, *An Essay on Man* (1733–34), Maynard Mack, ed., vol. 3: 1 (1950), *The Twickenham Edition of the Poems of Alexander Pope*, gen. ed. John Butt, 11 vols (London: Methuen; New Haven: Yale University Press, 1939–69). All quotations from Pope's poetry are from the Twickenham text. For convenience see John Butt, ed., *The Poems of Alexander Pope: A One-Volume Edition of the Twickenham Text With Selected Annotations* (London: Methuen, 1963).
22. Barrell, 'The Public Prospect and the Private View: The Politics of Taste in Eighteenth-Century Britain', in Simon Pugh, ed., *Reading Landscape: Country-City-Capital* (Manchester and New York: Manchester University Press, 1990), pp. 19–40; this passage p. 29.
23. Bob Bushaway, *By Rite: Custom, Ceremony and Community in England 1700–1880* (London: Junction Books, 1982), p. 82.
24. Thompson, *Customs in Common* (New York: The New Press, 1993), p. 98, n. 3.
25. Underdown, *Revel, Riot, Rebellion*, pp. 80–1.
26. Bushaway, *By Rite*, p. 25.
27. Thirsk, ed., *The Agrarian History of England and Wales, Volume V.ii., 1640–1750: Agrarian Change* (Cambridge: Cambridge University Press, 1985), p. 371.
28. *Ibid.*, p. 370.
29. Munsche, *Gentlemen and Poachers: The English Game Laws 1671–1831* (Cambridge: Cambridge University Press, 1981), p. 32.
30. On Whig versus Tory landscape aesthetics, see Nigel Everett, *The Tory View of Landscape* (New Haven and London: Yale University Press, 1994). On nationally connected county gentry versus locally minded parish gentry, see Lawrence Stone and Jeanne C. Fawtier Stone, *An Open Elite? England 1540–1880* (Oxford: Clarendon, 1984). On Gerrard Winstanley and the Diggers, see David Loewenstein, 'Digger Writing and Rural Dissent in the English Revolution: Representing England as a

Common Treasury', in MacLean, Landry and Ward, eds, *Country and City Revisited*, pp. 74–88.
31. On Walpole's use of the game laws for prosecuting political dissidents, see E.P. Thompson, *Whigs and Hunters: The Origin of the Black Act* (New York: Pantheon, 1975).
32. Stephen Deuchar, *Sporting Art in Eighteenth-Century England: A Social and Political History* (New Haven and London: Yale University Press, 1988), p. 156. On the poaching wars and man traps, see Munsche, *Gentlemen and Poachers*, pp. 62–75.
33. Nigel Duckers and Huw Davies, *A Place in the Country: Social Change in Rural England* (London: Michael Joseph, 1990), p. 11.
34. Rev. H.A. Macpherson, Hon. Gerald Lascelles, Charles Richardson, J.S. Gibbons and G.H. Longman, Col. Kenney Herbert, *The Hare* (London, New York, and Bombay: Longmans, Green, 1896), pp. 61, 85–6.
35. Fielding, *The History of the Adventures of Joseph Andrews, And of his Friend Mr Abraham Adams*, 2 vols (London: A. Millar, 1742; rpt Harmondsworth: Penguin, 1986), p. 219.
36. On pre-eighteenth-century agrarian complaint, see James G. Turner, *The Politics of Landscape: Rural Scenery and Society in English Poetry, 1630–1660* (Oxford: Basil Blackwell, 1978), and Andrew McRae, *God Speed the Plough: The Representation of Agrarian England, 1500–1660* (Cambridge: Cambridge University Press, 1996). On pre-eighteenth-century sympathy for hunted animals, see Keith Thomas, *Man and the Natural World: A History of the Modern Sensibility* (New York: Pantheon, 1983).
37. Thomas, *Natural World*, p. 181.
38. Ritvo, *The Animal Estate: The English and Other Creatures in the Victorian Age* (London and New York: Penguin, 1990), p. 1.
39. *Ibid.*, p. 2.
40. See Langford, *Propertied Englishman*, pp. 1–70, and Louis Althusser, 'Ideology and Ideological State Apparatuses (Notes towards an Investigation),' in *Lenin and Philosophy and Other Essays*, trans. Ben Brewster (New York and London: Monthly Review Press, 1971), pp. 127–86.
41. Joyce E. Salisbury, *The Beast Within: Animals in the Middle Ages* (New York and London: Routledge, 1994), p. 134.
42. *Ibid.*, p. 128.
43. Niccolò Machiavelli, *The Prince* (1532), trans. Angelo M. Codevilla (New Haven and London: Yale University Press, 1997), p. 65.
44. Patterson, *Fables of Power: Aesopian Writing and Political History* (Durham, NC and London: Duke University Press, 1991), pp. 15–16.
45. R. O[verton], *Mans Mortallitie* (Amsterdam, 1644); Erica Fudge, *Perceiving Animals: Humans and Beasts in Early Modern English Culture* (Houndmills and London: Macmillan and New York: St. Martin's, 2000), pp. 153, 156.
46. On the doctrine of similitudes, which preceded Enlightenment taxonomies of species and so on, see Michel Foucault, *The Order of Things: An Archaeology of the Human Sciences*, Eng. trans. of *Les Mots et les choses* (1966) (New York: Vintage, 1973), pp. 17–45.
47. Fudge, *Perceiving Animals*, pp. 124–5.
48. For example, Alexander Pope argued 'Against Barbarity to Animals', in *The Guardian*, No. 61, Thursday, 21 May 1713; Norman Ault, ed., *The Prose Works of Alexander Pope: Vol. I. The Earlier Works, 1711–1720* (1936; rpt New

York: Barnes & Noble, 1968), pp. 107–14. The clergyman James Granger offered remarkably similar arguments in *An Apology for the Brute Creation, or Abuse of Animals censured; In a Sermon on Proverbs xii. 10. Preached in the Parish Church of Shiplake, in Oxfordshire, October 18, 1772* (London: Printed for T. Davies and Sold by W. Goldsmith, 1772). Moira Ferguson has worked most extensively on connections between anti-slavery and anti-cruelty movements; see 'Breaking in Englishness: *Black Beauty* and the Politics of Gender, Race and Class', *Women: A Cultural Review* 5: 1 (1994): 34–52, and *Animal Advocacy and Englishwomen, 1780–1900: Patriots, Nation, and Empire* (Ann Arbor: University of Michigan Press, 1998).

49. See John Lawrence, 'On the Rights of Beasts', *A Philosophical And Practical Treatise On Horses and on the Moral Duties Of Man towards The Brute Creation* (London: T. Longman, 1796), pp. 117–63. Animal rights are satirized in Thomas Taylor's parodic *A Vindication of the Rights of Brutes* (1792); see Timothy Morton, *Shelley and the Revolution in Taste: The Body and the Natural World* (Cambridge: Cambridge University Press, 1994), pp. 30–3, and 13–56 for the wider context.

50. Coleridge, 'To a Young Ass, Its Mother Being Tethered Near It', in Ernest Hartley Coleridge, ed., *Coleridge: Poetical Works* (London: Oxford University Press, 1969), ll. 1, 26–8, and n., p. 75. David Perkins contextualizes the poem in 'Compassion for Animals and Radical Politics: Coleridge's "To a Young Ass" ', *English Literary History* 65: 4 (Winter 1998): 929–44. I disagree with his conclusion that Coleridge was only a half-hearted democrat, and that the 'poem laughs apologetically at its own silliness', p. 940. Coleridge anxiously feared reprisals against his Pantisocratic views, but that did not mean those views lacked conviction.

51. On the class-related aspects of anti-cruelty, see Robert W. Malcolmson, *Popular Recreations in English Society 1700–1850* (Cambridge: Cambridge University Press, 1973), pp. 152–7 and 172–3; and Ritvo, *Animal Estate*, pp. 130–53.

52. Granger, *Apology for the Brute Creation*, p. 12.

53. [Kendall], *The Canary Bird: A Moral Fiction, Interspersed with Poetry*, By the author of *The Sparrow, Keeper's Travels, The Crested Wren*, &c. (London: Printed for E. Newbery by J. Cundee, 1799), pp. 71–2.

54. Carr, *English Fox Hunting*, pp. 195–214, 246–56; and David C. Itzkowitz, *Peculiar Privilege: A Social History of English Foxhunting 1753–1885* (Hassocks, Sussex: Harvester Press, 1977), pp. 135–50.

55. Allen, *Enclosure and the Yeoman* (Oxford: Clarendon, 1992), p. 21.

56. Colin D.B. Ellis, *Leicestershire and the Quorn Hunt* (Leicester: Edgar Backus, 1951), pp. 14–15; Eric Kerridge, *The Agricultural Revolution* (London: George Allen & Unwin, 1967), pp. 311–25; and Thirsk, ed., *Agrarian History V.ii.*, pp. 577–8.

57. Elspeth Moncrieff with Stephen and Iona Joseph, *Farm Animal Portraits* (Woodbridge, Suffolk: Antique Collectors' Club, 1996), pp. 231, 234–5.

58. *Ibid.*, p. 234.

59. Thirsk, ed., *Agrarian History V.ii.*, pp. 577–8. On the early importation of 'arabs, turks, and barbs', see Peter Edwards, *The Horse Trade of Tudor and Stuart England* (Cambridge: Cambridge University Press, 1988), pp. 40–1.

60. See, out of many possible sources, the chapters on thoroughbred lineages derived from the three foundation sires, the Darley Arabian, the Godolphin Arabian and the Byerley Turk, in *Silk and Scarlet*, by The Druid [H. H. Dixon] (London: Vinton, 1859; new edn 1895), pp. 156–242. The best scholarly study of horses in East–West relations is the chapter, 'Managing the Infidel: Equestrian Art on its Mettle', in Lisa Jardine and Jerry Brotton, *Global Interests: Renaissance Art between East and West* (London: Reaktion and Ithaca, NY: Cornell University Press, 2000), pp. 132–85. For

'Orientalism', see Edward W. Said, *Orientalism: Western Conceptions of the Orient* (1978; London and New York: Penguin, 1995), who does not discuss horse-breeding.
61. Allen, *Enclosure and the Yeoman*, p. 21.
62. See, for example, David Levine, ed., *Proletarianization and Family History* (London and Orlando: Academic Press, 1984), and, for a useful critique, Rab Houston and K.D.M. Snell, 'Proto-Industrialization? Cottage Industry, Social Change, and Industrial Revolution', *The Historical Journal* 27: 2 (1984): 473–92.
63. Wordie, 'Introduction', in C.W. Chalklin and J.R. Wordie, eds, *Town and Countryside: The English Landowner in the National Economy, 1660–1860* (London: Unwin Hyman, 1989), pp. 1–25; this passage p. 19.
64. Repton, *Fragments on The Theory And Practice of Landscape Gardening* (London: Printed by T. Bensley and Son for J. Taylor, 1816), p. 94.
65. Bunce, *The Countryside Ideal: Anglo-American Images of Landscape* (London and New York: Routledge, 1994), p. 34.
66. *Ibid.*, p. 34.
67. *Ibid.*, p. 34.
68. *Ibid.*, p. 8.
69. Clare, [The Fens], in Eric Robinson and David Powell, eds, *Oxford Authors: John Clare* (Oxford and New York: Oxford University Press, 1984), ll.77–84. Quotations from Clare's verse will be from this edition unless otherwise noted.
70. Newby, *Green and Pleasant Land?*, pp. 17–18.
71. Eric Robinson, ed., *John Clare's Autobiographical Writings* (Oxford: Clarendon, 1983), pp. 9–10.
72. Cobbett, Old Hall, Friday, 16 November 1821, 'Journal from Gloucester . . . to Kensington', *Rural Rides*, 2 vols (1853; London: J.M. Dent and New York: E.P. Dutton, 1912), 1: 38.
73. Newby, *Green and Pleasant Land?*, p. 46.
74. See Bunce's chapters on 'A Place in the Country', pp. 77–110, 'The People's Playground', pp. 111–40, and 'The Countryside Movement', pp. 176–205.
75. Blome, *The Gentlemans Recreation. In Two parts. The First being an Encyclopedy Of The Arts and Sciences . . .; The Second Part, Treats of Horsmanship, Hawking, Hunting, Fowling, Fishing, and Agriculture* (London: Printed by S. Rotcroft, 1686), 2: 91.
76. See reports in *The Sporting Magazine* of carted stag-hunting with George III's hounds at Windsor, for example, for October 1796 in vol. 9 (1797), pp. 6–8. The Rothschilds still hunted carted stags at Waddesdon in Buckinghamshire in the early years of the twentieth century; Simon Blow, *Fields Elysian: A Portrait of Hunting Society* (London and Melbourne: J.M. Dent & Sons, 1983), p. 69.
77. Blome, *Gentlemans Recreation*, 2: 91.
78. *Ibid.*, 2: 87.
79. W.G. Hoskins, *The Making of the English Landscape* (1955; London and New York: Penguin, 1985), p. 196.
80. Around 1830, the third horn or 'leaping head' was introduced, attributed to a Frenchman, Jules Charles Pellier, or to an Englishman, Thomas Oldaker. For Pellier, see Stephanie Grant, 'Evolution of the Amazone in Life and Literature', *National Sporting Library Newsletter* 7 (December 1978): 1–2, 4–5; this passage p. 1. For Oldaker, see Charles Chenevix Trench, *A History of Horsemanship* (Garden City, NY: Doubleday, 1970), p. 277. On women's relegation to spectators rather than participants in field sports, beginning in the late eighteenth century, see Munsche, *Gentlemen and Poachers*, pp. 38–9, 200, n. 64. I have analyzed the significance of the side-saddle extensively in a forthcoming manuscript, *The Making of the English Hunting Seat*.

81. Itzkowitz, *Peculiar Privilege*, p. 3.
82. Woolf, 'Jack Mytton', *The Common Reader: Second Series*, ed. Andrew McNeillie (1932; London: Hogarth Press, 1986), pp. 126–7. Woolf was reviewing 'Nimrod's' *Memoirs of the Life of the Late John Mytton, Esq. of Halston, Shropshire, formerly M.P. for Shrewsbury, High Sheriff for the Counties of Salop and Merioneth, and Major of the North Shropshire Yeomanry Cavalry; with Notices of His Hunting, Shooting, Driving, Racing, Eccentric and Extravagant Exploits* (2nd edn, 1837) (London: Methuen, 1903, 1915).
83. Trollope, ed., *British Sports and Pastimes* (London: Virtue & Co., 1868), p. 74.
84. *Ibid.*, p. 75.
85. *Ibid.*, p. 73.
86. Itzkowitz, *Peculiar Privilege*, pp. 123–4.
87. Carr, *English Fox Hunting*, pp. 222, 229, n. 15.
88. Hobsbawm and Ranger, eds, *The Invention of Tradition* (Cambridge: Cambridge University Press, 1983; Canto edn, 1992), p. 1.
89. See C.E. Hare, *The Language of Sport* (London: Country Life, 1939), pp. 32–3, 39, 43, 117, n. 2.
90. Carr, *English Fox Hunting*, p. 46.
91. *Ibid.*, p. 62, n. 2.
92. See the portraits of William Evelyn of St. Clere (1770), Sir Frederick Evelyn (?) riding with a hound (1771), Thomas Smith, huntsman of the Brocklesby, and his father Thomas Smith (1776), and John Musters and the Rev. Philip Story riding out from the stableblock at Colwick Hall (1777) in Judy Egerton, *George Stubbs 1724–1806* (London: Tate Gallery Publishing, 1984; rpt 1996), pp. 152, 153, 156, 159.
93. *Ibid.*, pp. 52–3.
94. Carr, *English Fox Hunting*, p. 136.
95. *Ibid.*, p. 136.
96. Merrily Harpur, 'Brush with Death', *The Listener* 20/27 (1990), pp. 40–1; this passage p. 40.
97. David W. Macdonald, Fran H. Tattersall, Paul J. Johnson, Chris Carbone, Jonathan C. Reynolds, Jochen Langbein, Steve P. Rushton and Mark D.F. Shirley, *Managing British Mammals: Case Studies from the Hunting Debate* (Oxford: Wildlife Conservation Research Unit, 2000), pp. vi, x, 39, 46.
98. *Ibid.*, p. 34.
99. *Wordsworth's Guide to the Lakes*, 5th edn (1835), ed. Ernest de Sélincourt (Oxford: Oxford University Press, 1977), p. 92. The *Guide* had first appeared anonymously in 1810, accompanying sketches by the Rev. Joseph Wilkinson, and was first published under Wordsworth's name with some of his poems in 1820. It first appeared independently in a third edition of 1822. The 1835 edition was Wordsworth's final text.
100. Addison, *Spectator* No. 411 (Saturday, 21 June 1712) in *The Spectator*, ed. Donald F. Bond, 5 vols (Oxford: Clarendon, 1965), 3: 538.
101. All quotations from Wordsworth's poetry will be, unless otherwise stated, from Stephen Gill, ed., *Oxford Authors: William Wordsworth* (Oxford and New York: Oxford University Press, 1984).
102. William Gilpin, whose guidebook Wordsworth carried with him, had observed, 'The banks of the Wye consist, almost entirely either of wood, or of pasturage; which I mention as a circumstance of peculiar value in landscape'; *Observations on the River Wye, and several parts of South Wales, &c. relative chiefly to Picturesque*

Beauty; made In the Summer of the Year 1770 (London: Printed for R. Blamire, 1782), p. 29.
103. McKeon, 'The Pastoral Revolution', in Kevin Sharpe and Steven N. Zwicker, eds, *Refiguring Revolutions: Aesthetics and Politics from the English Revolution to the Romantic Revolution* (Berkeley, Los Angeles, London: University of California Press, 1998), pp. 267–89; this passage p. 289.
104. *Ibid.*, p. 289.
105. De Bruyn, 'From Virgilian Georgic to Agricultural Science: An Instance in the Transvaluation of Literature in Eighteenth-Century Britain', in Albert J. Rivero, ed., *Augustan Subjects: Essays in Honor of Martin C. Battestin* (Newark, DE and London: University of Delaware Press and Associated University Presses, 1997), pp. 47–67; this passage p. 53.
106. Laura Caroline Stevenson, *Praise and Paradox: Merchants and Craftsmen in Elizabethan Popular Literature* (Cambridge, 1984), pp. 16, 140–1.
107. Margaret Grainger, ed., *The Natural History Prose Writings of John Clare* (Oxford: Clarendon, 1983), Appendix Va, 'Volumes in Clare's Library relating to Natural and Garden History', p. 359.
108. Karen O'Brien, in 'Imperial Georgic, 1660–1789', in MacLean, Landry and Ward, eds, *Country and City Revisited*, pp. 160–79, finds georgic essentially dead by 1789.
109. '[T]oday the bulk of the population of most of our rural villages are able to live in the countryside and work in nearby towns and cities'; Newby, *Green and Pleasant Land?*, p. 22.
110. The ad for the NFU's Mutual Insurance 'Countryside' policy appeared in the weekly magazine *Horse and Hound* on 5 August 1993 and has been reprinted many times since.
111. James S. Ackerman, *The Villa: Form and Ideology of Country Houses*, A.W. Mellon Lectures in the Fine Arts, 1985; Bollingen Series 35: 34 (Princeton: Princeton University Press, 1990), p. 9.
112. Newby, *Green and Pleasant Land?*, pp. 13–24.
113. Malcolm Andrews, *The Search for the Picturesque: Landscape Aesthetics and Tourism in Britain, 1760–1800* (Aldershot: Scolar Press, 1989).
114. Tom Stephenson, *Forbidden Land* (Manchester: Manchester University Press, 1989), p. 57.
115. Quoted in Andrews, *Search for the Picturesque*, p. 82.
116. Kathleen Coburn, ed., *The Notebooks of Samuel Taylor Coleridge*, in 3 vols (London: Routledge, 1957–73), 1 (Text): entry 760 (June–July 1800).
117. On the sexual politics of the picturesque, see Vivien Jones, ' "The coquetry of nature": Politics and the Picturesque in Women's Fiction', in Stephen Copley and Peter Garside, eds, *The Politics of the Picturesque: Literature, Landscape, and Aesthetics since 1770* (Cambridge: Cambridge University Press, 1994), pp. 120–44.
118. Gilpin, *Three Essays: on Picturesque Beauty; on Picturesque Travel; and on Sketching Landscape: to which is added a poem, on Landscape Painting* (London: Printed for R. Blamire, 1792), p. 48.
119. Gilpin, *Three Essays*, p. 48.
120. Liu, *Wordsworth: The Sense of History* (Stanford: Stanford University Press, 1989), pp. 63–4.
121. [William Combe], *Doctor Syntax in Search of the Picturesque* (1809) (London: C. Daly, n.d.).

122. Massingham, *Remembrance: An Autobiography* (London: Batsford, 1941), pp. 20–1. On Massingham, see Patrick Wright, 'An Encroachment Too Far', in Anthony Barnett and Roger Scruton, eds, *Town and Country* (London: Jonathan Cape, 1998), pp. 18–33.
123. Massingham, *Remembrance*, pp. 21–2.
124. Newby, *Green and Pleasant Land?*, pp. 15–24.
125. John Barrell, *The Dark Side of the Landscape: The Rural Poor in English Painting 1730–1840* (Cambridge: Cambridge University Press, 1980), p. 89.
126. *Ibid.*, p. 95.
127. J. Hassell, *Memoirs of the late George Morland* (London, 1806), p. 131, quoted in *ibid.*, p. 95.
128. Robinson, ed., *Clare's Autobiographical Writings*, p. 44.
129. Gilpin, 'On Landscape Painting. A Poem', in *Three Essays*.
130. Fiennes, *Illustrated Journeys*, p. 32.
131. See the episode described in Grainger, ed., *Natural History Prose Writings*, p. 234, and the fine essay by John Goodridge and Kelsey Thornton, 'John Clare: The Trespasser', in Hugh Haughton, Adam Phillips and Geoffrey Summerfield, eds, *John Clare in Context* (Cambridge: Cambridge University Press, 1994), pp. 87–129.
132. Robinson, ed., *Clare's Autobiographical Writings*, p. 33.
133. *Ibid.*, p. 70.
134. Eric Robinson and David Powell, eds, *John Clare By Himself* (Ashington, Northumberland and Manchester: Mid-NAG/Carcanet Press, 1996), pp. 83–4.
135. Buchan, 'A Reputation' (1898), *The Watcher by the Threshold: Shorter Scottish Fiction*, Andrew Lownie, ed. (Edinburgh: Canongate Classics, 1997), pp. 107–17; this passage p. 109.
136. Buchan, 'Politics and the May-Fly' (1896), *The Watcher by the Threshold*, pp. 35–42; this passage p. 40.
137. Steven N. Zwicker, *Lines of Authority: Politics and English Literary Culture, 1649–1689* (Ithaca and London: Cornell University Press, 1993), pp. 72–3.
138. See Tom Williamson and Liz Bellamy, *Property and Landscape: A Social History of Land Ownership and the English Countryside* (London: George Philip, 1987), p. 205, and Williamson, *Polite Landscapes: Gardens and Society in Eighteenth-Century England* (Baltimore: The Johns Hopkins University Press, 1995), pp. 131–2.
139. Rackham, *The History of the Countryside* (London: Weidenfeld and Nicolson, 1995), pp. 26, 4–5. By contrast with the Planned, what Rackham calls the Ancient Countryside is the product of at least 1,000 years of continuity, and most of it has altered little since 1700. A country of hamlets and ancient isolated farms, ancient mixed irregular hedges, many footpaths, winding lanes and sunken drovers' roads, many small woodlands and ponds and patches of heath, the Ancient, or anciently enclosed, Countryside contrasts strikingly with the 'England of big villages, few, busy roads, thin hawthorn hedges, windswept brick farms, and ivied clumps of trees in corners of fields', a 'mass-produced, drawing-board landscape, hurriedly laid out parish by parish', pp. 4–5.
140. Rackham claims that the extent of new quick-set hedges planted thus equalled all those planted in the previous 500 years, *History of the Countryside*, p. 190.
141. Macdonald, *Running with the Fox* (New York and Oxford: Facts on File Publications, 1987), p. 173.
142. Macdonald and Johnson, 'Impact of Sport Hunting', in Taylor and Dunstone, eds, *Exploitation of Mammal Populations*, pp. 203–4.
143. Sandra E. Baker and David W. Macdonald, 'Foxes and Fox-hunting on Farms in Wiltshire: A Case Study', *Journal of Rural Studies* 16 (2000): 185–201.

144. D.W. Macdonald, J.C. Reynolds, C. Carbone, F. Mathews and P.J. Johnson, 'The Bio-economics of Fox Control', unpublished chapter. My thanks to the authors for letting me read it in manuscript.
145. Macdonald, *Running with the Fox*, p. 209.
146. Macdonald, Reynolds, Carbone, Mathews, and Johnson, 'Bio-economics of Fox Control', pp. 1, 16.
147. The late Poet Laureate Ted Hughes argued that 'the spell of the hunt' regarding the red deer of Exmoor had resulted in one of 'English conservation's most impressive success stories'. Hughes concluded that, 'Before we ban these hunts, perhaps we should make sure we have a new method of control (of the people) and preservation (of the animals) half as effective and simple as this strange system that our history has produced for us. That is, if we really do want the animals'; 'The Hart of the Mystery', *Guardian*, Saturday, 5 July 1997, p. 21. The chairman of the Joint Nature Conservation Committee, Lord Selborne, would agree: 'In fact, the best conservation happens where landowners have an interest in game', and 'Most people involved in field sports are in the business of creating habitats.' The British Ministry of Agriculture, Fisheries and Food (MAFF) has revealed that 18 per cent of the English and Welsh farmers it interviewed stated that country sports were either the primary or secondary reason for conserving wildlife on their farms. Even a former chief executive of the League Against Cruel Sports, now heading up the Wildlife Network, James Barrington, has proclaimed field sports the lesser of two evils in the battle for wildlife conservation: 'The alternative to hunting is to take away what is one form of fox killing, in effect quite a public one, as opposed to letting a number of other forms – which are far less accessible and accountable – take place to a greater degree. And it can't be denied that pro-hunting landowners want small coverts and hedgerows left. Who knows whether they would keep them like that if foxhunting were banned?'; Rebecca Austin, 'Why the Greens Need Us', *The Field*, November 1996, pp. 42–5.
148. *The True Levellers Standard Advanced* (April 1649), in *The Works of Gerrard Winstanley*, ed. George Sabine (Ithaca, NY: Cornell University Press, 1941), p. 251; quoted in Loewenstein, 'Digger Writing', in MacLean, Landry and Ward, eds, *Country and City Revisited*, p. 78.

Chapter 1

1. Peter Singer in *Animal Liberation*, new rev. edn (1975; New York: Avon, 1990), disavows pet-keeping and animal-loving in order that animals might be granted equal consideration of their interests, 'Preface to the 1975 edition', pp. i–iii. Keith Tester characterizes animal rights as a fetish, in that it is not about animals at all, but rather about human perfectability: 'The main questions in the invention of animal rights have been: what is it to be a human who lives an urban life? What should we do, how ought we to behave?' Urban separation from animals, unlike the sometimes bloody proximity to animal nature in the countryside, yields the answer 'to be human is to be humane, it is not to touch animals. We feel happy to enfranchise them morally'; *Animals and Society: The Humanity of Animal Rights* (London and New York: Routledge, 1991), pp. 192–3. For vegetarianism, see Chapter 5.
2. Alan Rabinowitz, in his study of jaguars in Belize – which was devoted explicitly to their conservation and to establishing a large enough rainforest preserve for them to survive in – did not extend his fieldwork beyond the initial time frame on the ethical grounds that his trapping and radio-collaring of jaguars had led to a

number of early deaths, and that since he had enough data, he should stop; *Jaguar: One Man's Struggle to Establish the World's First Jaguar Preserve* (1986; Washington, DC and Covelo, CA: Island Press/Shearwater Books, 2000), p. 339. See also Jeffrey Mousaieff Masson and Susan McCarthy, *When Elephants Weep: The Emotional Lives of Animals* (New York: Delta by Bantam Doubleday Bell, 1995); and Vicki Hearne, *Adam's Task: Calling Animals by Name* (New York: Alfred A. Knopf, 1986).

3. On relations between farmers and animals in traditional, or non-industrial, husbandry, see Ted Benton, *Natural Relations: Ecology, Animal Rights and Social Justice* (London and New York: Verso, 1993), who argues: 'In general, traditional practices of animal husbandry involve a practical acknowledgement that more-or-less autonomous animal social-processes are a precondition of the achievement of human purposes . . . That the animals incorporated into these human social practices are themselves living natural beings, with organic, social and ecological conditions of survival and thriving is therefore a practical recognition built into their intentional structure. It may also be, and generally is, experienced by the human agents involved as an affective disposition – something they choose to acknowledge, and gain satisfaction from – as well as a normative requirement', p. 153.

4. A 'bioregion' may be identified 'by its mountain ranges and rivers, its vegetation, weather patterns or soil types, or its patterns of animal habitats, whether birds, ground mammals, or humans'; Brian Tokar, *The Green Alternative: Creating an Ecological Future* (San Pedro, CA: R. & E. Miles, 1987), p. 27. Historical ecologists 'wish to stress that a false dichotomy between "natural" and anthropogenic causation glorifies a nonexistent "pristine" nature. No spot on the earth is unaffected by humans', as Carole L. Crumley puts it; 'Epilogue', in Crumley, ed., *Historical Ecology: Cultural Knowledge and Changing Landscapes* (Santa Fe, NM: School of American Research Press, 1994), pp. 239–41; this passage p. 239.

5. For Foreman, see Steve Chase, 'Introduction: Whither the Radical Ecology Movement?', in Murray Bookchin and Dave Foreman, *Defending the Earth: A Dialogue Between Murray Bookchin and Dave Foreman* (Boston: South End Press, 1991), pp. 7–24; this passage p. 21; and Foreman's own contributions, esp. pp. 37–46, 107–19. Benton, *Natural Relations*, pp. 194–221.

6. Andrew Dobson lucidly distinguishes 'ecologism' from the ' "managerial" ' mentality of environmentalism in *Green Political Thought: An Introduction* (London: Unwin Hyman, 1990), p. 13.

7. Jonathan Bate's *Romantic Ecology: Wordsworth and the Environmental Tradition* (London and New York: Routledge, 1991) and *The Song of the Earth* (London: Picador, 2000) are exemplary in this respect. See Peter J. Manning's review of *Romantic Ecology*, 'Reading and Writing Nature', *Review* 15 (1993), pp. 275–96, and Morton, *Shelley and the Revolution in Taste*, p. 229.

8. Barrell, *Idea of Landscape*, p. 187.

9. Richard Pickard has recently argued for what he calls 'Augustan ecology', a contradictory discourse derived on the one hand from eighteenth-century 'georgic ecology', a 'land ethic that seeks to accommodate society's need to consume natural resources with nature's need not to be consumed', and from contemporary protests against this very 'attitude of compromise' on the other; 'Augustan Ecology: Environmental Attitudes in Eighteenth-Century Poetry' (Dissertation; University of Alberta, 1998), pp. 16–17.

10. If commentators have repeatedly traced the history of English socialism back to William Morris, as Julia Swindells and Lisa Jardine have argued in *What's Left? Women in Culture and the Labour Movement* (London and New York: Routledge, 1990), pp. 47–68, the history of English naturalism and ecological awareness has been as invariably linked to Gilbert White. Raymond Williams recognized the importance of White's scientific writing in *The Country and the City*, but distinguished it from the new 'green language' of Clare and Wordsworth, pp. 118–19, 133–4.
11. Richard Mabey, 'Introduction' to Gilbert White, *The Natural History of Selborne* (1788–89; Harmondsworth: Penguin, 1977), pp. vii–xxii; this passage p. xii.
12. Walter Johnson, ed., 'Gilbert White and His Village', in *Gilbert White's Journals* (1931; rpt London: Routledge & Kegan Paul and Cambridge, MA: MIT Press, 1970), pp. xv–xxxvi; this passage p. xix.
13. *Ibid.*, p. xx.
14. Wyatt, *Wordsworth and the Geologists* (Cambridge: Cambridge University Press, 1995), p. 6.
15. *Ibid.*, p. 7.
16. Gilpin, *Remarks on Forest Scenery, and other Woodland Views, (Relative chiefly to Picturesque Beauty) Illustrated by The Scenes of New-Forest in Hampshire. In Three Books* (London: Printed for R. Blamire, 1791), p. 277.
17. [Pennant], *The British Zoology: Class I. Quadrupeds. II. Birds,* Published under the Inspection of the Cymmrodorion Society, Instituted For The Promoting Useful Charities, and the Knowledge of Nature, among the Descendants of the Ancient Britons (London: Printed by J. and J. March and Sold for the Benefit of the British Charity-School on Clerkenwell-Green, 1766), p. 4.
18. Aikin, *An Essay on the Application of Natural History to Poetry* (Warrington and London: Printed by W. Eyres for J. Johnson, 1777), p. v.
19. *Ibid.*, p. iv.
20. [Pennant], *British Zoology*, p. 6.
21. Ashworth, *The Economy of Nature: Rethinking the Connections Between Ecology and Economics* (Boston and New York: Houghton Mifflin, 1995), p. 237.
22. Gilpin, *Forest Scenery*, pp. 268–9.
23. White, *The Natural History and Antiquities of Selborne, in the County of Southampton* (London: Printed by T. Bensley for B. White and Son, 1789; rpt Menston: Scolar Press, 1970), pp. 21–2.
24. Jefferies, *The Gamekeeper at Home: Sketches of Natural History and Rural Life* (1878), in *The Gamekeeper at Home and The Amateur Poacher*, ed. Richard Fitter (Oxford and New York: Oxford University Press, 1978), p. 83.
25. Mabey, 'Introduction', *Food for Free* (London: HarperCollins, 1992), pp. 11–15; this passage p. 12.
26. Matt Cartmill has argued that the modern blurring of the distinction between people and beasts, or between human culture and the wild, has undercut the conceptual foundations of hunting in such a way that hunting in the modern world is hardly possible anymore. As Cartmill puts it, 'Giving up the distinction between these two worlds means discarding the whole system of symbolic meanings that have distinguished hunting from mere butchery and given it a special importance in the history of Western thought. If the edge of nature is a hallucination, then hunting is only animal-killing'; *A View to a Death in the Morning: Hunting and Nature Through History* (Cambridge, MA and London: Harvard University Press, 1993), pp. 243–4. This argument accounts only for hunting's

separating of human and animal, and not its symbolic merging, so important in early modern representations.

27. [Gascoigne], *The Noble Arte of Venerie or Hunting* (London: Henry Bynneman for Christopher Barker, 1575), [Bodleian Library shelfmark: Douce T 247(1)], reprinted as *Turbervile's Booke of Hunting, 1576* (Oxford: Clarendon Press, 1908). Formerly thought to be by George Turbervile, because it is usually bound with Turbervile's companion book, *The Book of Faulconrie, or Hauking* (1575), issued by the same printer, Henry Bynneman, for Christopher Barker, *The Noble Arte* is now generally attributed to George Gascoigne; see Jean Robertson, 'George Gascoigne and "The Noble Arte of Venerie and [sic] Hunting" ', *Modern Language Review* 37: 4 (October 1942): 484–5; Charles and Ruth Prouty, 'George Gascoigne, *The Noble Arte of Venerie*, and Queen Elizabeth at Kenilworth', in *Joseph Quincy Adams Memorial Studies*, eds James G. McManaway, Giles E. Dawson and Edwin E. Willoughby (Washington, DC: Folger Shakespeare Library, 1948), pp. 639–65; and Marcia Vale, *The Gentleman's Recreations: Accomplishments and Pastimes of the English Gentleman 1580–1630* (Cambridge: D.S. Brewer and Totowa, NJ: Rowman & Littlefield, 1977), pp. 31–3.
28. Rothstein, '*Discordia Non Concors*: The Motif of Hunting in Eighteenth-Century Verse', *Journal of English and Germanic Philology* (1984): 330–54; this passage p. 354.
29. Drayton, *The Poly-Olbion: A Chorographicall Description of Great Britain* (London: Printed for John Marriott, John Grismand, and Thomas Dewe, 1622; rpt New York: Burt Franklin, 1970), and [Gascoigne], *Noble Arte*, p. 246. All quotations from Gascoigne are taken from the (misattributed and misdated but textually accurate) Clarendon reprint because of its accessibility.
30. On medieval and later anxieties about killing animals, see Thomas, *Man and the Natural World*, pp. 143–91.
31. Dame Juliana Berners may have been the author of *The Boke of St. Albans* (St. Albans, 1486; 2nd edn, Westminster, 1496), or at least the compiler of the hunting section, but 'behind her stands the legendary originator of all hunting lore, Sir Tristram', popularly supposed to have been one of King Arthur's knights. When Gascoigne, or after him, Sir Thomas Cockaine, author of *A Short Treatise of Hunting: Compyled for the delight of Noblemen and Gentlemen* (1591), refer to Sir Tristram's book 'they mean the successive editions of *The Boke of St. Albans*'; Vale, *Gentlemen's Recreations*, pp. 30–1. See also Anne Rooney, *Hunting in Middle English Literature* (Cambridge: The Boydell Press, 1993), pp. 7–11.
32. Vale, *Gentleman's Recreations*, p. 31.
33. [Gascoigne], *Noble Arte*, p. 134.
34. 'The wofull wordes of the Hart to the Hunter' is based on a French original in *La Vénerie*, but the complaints of the hare, fox and otter are Gascoigne's own; C. and R. Prouty, 'George Gascoigne', pp. 646, 648–9.
35. Tennyson, *In Memoriam* (1833–50), section 56, l. 15.
36. Thompson, *Whigs and Hunters*, p. 30.
37. M[arkham], *Countrey Contentments*, p. 31.
38. [Nicholas Cox], *The Gentleman's Recreation, In Four Parts; (viz.) Hunting, Hawking, Fowling, Fishing. Collected From ancient and modern Authors Forrein and Domestick, and rectified by the Experience of the most Skilfull Artists of these times* (London: Printed by C. Flesher for Maurice Atkins and Nicholas Cox, 1674), p. 40.
39. John Boswell, *Christianity, Social Tolerance and Homosexuality* (Chicago: University of Chicago Press, 1980), p. 356; Salisbury, *Beast Within*, p. 44.
40. During the politically turbulent seventeenth century, after which the name disappeared, 'Wat' might well have recalled Wat Tyler, who during the Great Revolt of

1381 demanded that the poor as well as the rich be able to 'hunt the hare in the field'; *Chronicon Henrici Knighton*, 2: 137.

41. In 1735, William Somervile worried about what his audience would feel about hunting 'the poor hare', a 'puny, dastard animal', and 'So mean a prey', in *The CHACE. A POEM* (London: Printed for G. Hawkins and Sold by T. Cooper, 1735), 2: 295–9. From here on, I shall be quoting from the edition I happen to possess, *The Chase; A Poem*, by William Somervile, Esq. (London: Printed by W. Bulmer and Co., 1802).

42. John Smallman Gardiner, Gent., *The Art and the Pleasures of Hare-Hunting. In Six Letters to a Person of Quality* (London: Printed for R. Griffiths, 1750), p. 2.

43. Ibid., p. 4.

44. Budgell, *Spectator* No. 116 (Friday, 13 July 1711) in Bond, ed., *The Spectator*, 1: 475–9. It is hard to know what to make of Budgell's account of this imaginary chase, since it conflates hare -and stag-hunting practices. Sir Roger has reputedly replaced his fox 'beagles' with '*Stop-Hounds*', which Percy (*OED*) says are peculiar to stag-hunting. Beagles were more often used for hare-hunting, though a fast northern 'fox beagle' is sometimes referred to. Peter Beckford, in *Thoughts upon Hunting. In a series of Familiar Letters to a Friend* (1781), 3rd edn (Sarum: Printed by E. Easton, 1784), recommends hounds 'between the large slow hunting harrier, and the little fox beagle' for hare-hunting, p. 135. Certainly, the heavyweight George III used to have hounds stopped so he could stay in the running while stag-hunting at Windsor. Budgell also sets the hare hunt in July, during the royal stag-hunting season; hares were officially hunted only after Michaelmas, when harvesting was over.

45. M[arkham], *Countrey Contentments*, p. 33.

46. 'Nun's Priest's Tale', in John H. Fisher, ed., *The Complete Poetry and Prose of Geoffrey Chaucer* (New York: Holt, Rinehart and Winston, 1977), ll. 398–400.

47. Sir Thomas Cockaine, Knight, *A Short Treatise of Hunting: Compyled for the delight of Noblemen and Gentlemen* (London: Thomas Orwin for Thomas Woodcocke, 1591), Sig. B2v.

48. Charles and Ruth Prouty speculate that this illustration, which does not appear in the French original and is unlike any other woodcut in the volume, may have been hastily introduced into Gascoigne's text from some previous book, 'George Gascoigne', p. 663.

49. See, for example, [William Taplin], *The Sportsman's Cabinet; or, a correct delineation of the various Dogs Used In The Sports of The Field: including the canine race in general. Consisting of A Series of Engravings of every distinct breed, from original paintings, taken from life . . .*, by a Veteran Sportsman, 2 vols (London: Printed and Published for the Proprietors, by J. Cundee; Sold by T. Hurst, T. Ostell, and Chapple, 1803–4).

50. Moody, 'To a Gentleman Who Invited Me to Go A-Fishing', from *Poetic Trifles* (1798), in Roger Lonsdale, ed., *Eighteenth-Century Women Poets: An Oxford Anthology* (Oxford and New York: Oxford University Press, 1990), p. 404.

51. See, for example, 'The Petition of the Hard Parishes' from *Two-Penny Trash; Or, Politics for the Poor* (July 1832): 'We complain, that notwithstanding the misery and half starvation to which we are reduced, the law, under severe imprisonment and heavy fine, forbids us to take for our own use the wild birds and animals that inhabit the woods and fields, or the fish that swim in the water; those being kept not for the service, but for the sports of the rich'; quoted in Ian Dyck, *William Cobbett and Rural Popular Culture* (Cambridge: Cambridge University Press, 1992), Appendix II, pp. 222–7; this passage p. 225. Elizabeth K. Helsinger analyzes Cobbett's 'textual poaching' in *Rural Scenes and National Representation:*

Britain, 1815–1850 (Princeton, NJ: Princeton University Press, 1997), pp. 105, 112–19.
52. Dyck, *Cobbett and Rural Popular Culture*, pp. 46–7.
53. *Ibid.*, p. 47.
54. Nattrass, *William Cobbett: The Politics of Style* (Cambridge: Cambridge University Press, 1995), p. 216.
55. Cobbett, Canterbury, Thursday Afternoon, 4 September 1823, 'From Dover to the Wen', *Rural Rides*, 1: 248.
56. 'We should never lose sight of the fact that the project of human liberation has now become an ecological project, just as, conversely, the project of defending the Earth has also become a social project'; Murray Bookchin, 'Where I Stand Now', in Bookchin and Foreman, *Defending the Earth*, pp. 121–33; this passage p. 131.
57. Cobbett, Old Hall, Friday, 16 November 1821, 'Journal from Gloucester . . . to Kensington', *Rural Rides*, 1: 32.
58. Itzkowitz, *Peculiar Privilege*, p. 200, n. 22.
59. Barrell, *Dark Side of the Landscape*, pp. 173–4, n. 99.
60. Daniels and Watkins, 'The Picturesque Landscape', in Stephen Daniels and Charles Watkins, eds, *The Picturesque Landscape: Visions of Georgian Herefordshire* (Nottingham: Department of Geography, University of Nottingham, 1994), pp. 9–14; this passage p. 10.
61. There were limits to the democracy of Cobbett's vision in another respect as well: the question of political rights for women. See Catherine Hall, *White, Male and Middle-Class: Explorations in Feminism and History* (Cambridge: Polity Press, 1992), pp. 124–50.
62. Cobbett, *Rural Rides*, 1: 38.
63. Attributed to Smith by Hare, *Language of Sport*, p. ix.
64. Thomas Smith, Esq., Late Master of the Craven hounds, *Extracts from the Diary of A Huntsman* (London: Whittaker & Co., 1838), p. 35.
65. *The Life Of A Fox, written by himself*, with illustrations by Thomas Smith, Esq., author of 'Extracts from the Diary of a Huntsman' (London: Whittaker and Co., 1843), pp. 6–7, 107–9, 114–15.
66. Carr, *English Fox Hunting*, p. 100.
67. *Ibid.*, p. 100.
68. Macdonald and Johnson, 'Impact of Sport Hunting', p. 203.
69. See Masson and McCarthy, *When Elephants Weep*, for a critique of the behaviorist eschewing of anthropomorphism.
70. Macdonald, *Running with the Fox*, p. 132.
71. Other examples of this knowledge sharing include Macdonald's time on the farm of John and Pamela Gee, where he 'learnt more, and more quickly, than in any other period of my life', *ibid.*, p. 17; using a terrier to bolt a fox there, pp. 23–4, and in the Cumbrian Fells, pp. 155–6, 159, 174–5; Edwin Dargue, and the other shepherds of the Cumbrian Fells, pp. 152–75; learning from 'a gamekeeper friend' that, having shot a 'milky' vixen, he had observed a 'dry' vixen apparently bringing food to the cubs, p. 92.
72. Jefferies, *The Gamekeeper at Home*, p. 75.
73. Hearne, *Adam's Task*, pp. 48–9.
74. Macdonald et al., *Managing British Mammals*, p. 130. The authors discuss the bases for assessing welfare, suffering and cruelty in hunting foxes, deer, hares and mink with dogs, pp. 121–41.

75. On the pleasures of fox-hunting shared between horse and rider, see Budiansky, *The Nature of Horses: Exploring Equine Evolution, Intelligence, and Behavior* (New York and London: The Free Press, 1997), pp. 6, 170.
76. Ibid., pp. 68–70.
77. Goldschmidt, *Bridle Wise: A Key to Better Hunters, Better Ponies* (London and New York: Country Life, Ltd. and Charles Scribner's Sons, 1927), p. 61.
78. Budiansky, *Nature of Horses*, p. 105.
79. Murray Bookchin, *The Ecology of Freedom: The Emergence and Dissolution of Hierarchy*, rev. edn (Montreal and New York: Black Rose Books, 1991), p. 362.
80. Ibid., p. 362.
81. 'Even large carnivores that prey upon large herbivores have a vital function in selectively controlling large population swings by removing weakened or old animals for whom life would in fact become a form of "suffering" ', ibid., p. 362.
82. Taplin, *Observations on the Present State of the Game in England, in which The late Methods of Preservation are clearly refuted and condemned: The Real Cause, or Causes of the Deficiency Demonstrated; And Proposals Offered for its more sure and effectual Preservation* (London: Printed for T. Davies, 1772), p. 8.
83. Ibid., p. 37. In his eagerness to preserve 'the Game', Taplin would sacrifice many a poor man's dog: 'An annual Tax of five Shillings *per* head on every Dog in the Kingdom should immediately take place; and by that Means the many thousands of those useless Animals would be destroyed', pp. 38–9. By the time he came to publish, as a 'Veteran Sportsman', *The Sportsman's Cabinet* (1803–4) – a rapturous celebration of sporting dogs, including lurchers – his view that they were 'useless' would seem to have changed.

Chapter 2

1. Shared agrarian culture gave literate laborers an opportunity to become published poets. See Landry, *The Muses of Resistance: Laboring-Class Women's Poetry in Britain, 1739–1796* (Cambridge and New York: Cambridge University Press, 1990); and John Goodridge, *Rural Life in Eighteenth-Century English Poetry* (Cambridge and New York: Cambridge University Press, 1995).
2. Everett, *Tory View of Landscape*, p. 39.
3. Howard Newby calls eighteenth-century landscape architecture 'one of the indisputably great English contributions to Western art and aesthetics', in *Country Life*, p. 16.
4. Joan Thirsk, 'The Farming Regions of England', in Thirsk, ed., *The Agrarian History of England and Wales, Volume IV, 1500–1640* (Cambridge: Cambridge University Press, 1967), pp. 1–112; this passage p. 2. See also her discussion of 'Enclosing and Engrossing', pp. 200–55.
5. [Gascoigne], *Noble Arte*, p. 89. On ecology in the period, and the politics of forestry, see Robert Markley, ' "Gulfes, Deserts, Precipices, Stone": Marvell's "Upon Appleton House" and the Contradictions of "Nature" ', in MacLean, Landry, and Ward, eds, *Country and City Revisited*, pp. 89–105.
6. Cowper, *The Task, A Poem, in Six Books* (London: J. Johnson, 1785) in John D. Baird and Charles Ryskamp, eds, *The Poems of William Cowper*, 3 vols (Oxford: Clarendon, 1981–95), 2: 125.

7. 'By 1800 England was one of the least wooded of all north European nations. Despite this, or probably because of it, English enthusiasm for trees and woodland seems never to have been higher'; Stephen Daniels, 'The Political Iconography of Woodland in Later Georgian England', in Denis Cosgrove and Stephen Daniels, eds, *The Iconography of Landscape: Essays on the Symbolic Representation, Design and Use of Past Environments* (Cambridge: Cambridge University Press, 1988), pp. 43–82; this passage pp. 43–4. See also Tim Fulford, 'Cowper, Wordsworth, Clare: The Politics of Trees', *The John Clare Society Journal* 14 (July 1995): 47–59.
8. Mark Overton, *Agricultural Revolution in England: The Transformation of the Agrarian Economy 1500–1850* (Cambridge: Cambridge University Press, 1996), p. 8. Overton disputes Eric Kerridge's sixteenth- and seventeenth-century agricultural revolution, claiming that the most important growth in productivity occurred after 1750, when the most 'dramatic and unprecedented improvements in output, land productivity and labour productivity' allowed the population 'to exceed the barrier of 5.5 million people for the first time', pp. 198, 206.
9. Kerridge's argument, in *The Agricultural Revolution*, is that the fundamental technological innovations – the floating of water-meadows, the regular rotation of tillage and grass (replacing areas of permanent tillage and permanent grass, or simply shifting cultivation), the introduction of new fallow or fodder crops and selected grasses, marsh drainage, manuring, and stock-breeding – produced a great spurt in productivity between 1540 and 1700, that would remain unmatched in the period between 1750 and 1880, pp. 40, 336.
10. Allen, *Enclosure and the Yeoman*, p. 21.
11. Newby, *Country Life*, pp. 22, 23.
12. Turner, *Politics of Landscape*, p. 158.
13. Williams, *Culture and Society* (Harmondsworth: Penguin, 1961), p. 322. For Williams, the dominative mode, whether exercised upon human beings or upon the natural world, is equally suspect. Both attitudes of domination must be 'unlearned', as 'the price of survival', p. 322.
14. Jonson, 'Penshurst', in *Poems*, Ian Donaldson, ed. (London: Oxford University Press, 1975).
15. McRae, *God Speed the Plough*, p. 5.
16. Dryden, dedication of the *Georgics* 'To the Right Honourable Philip Earl of Chesterfield, &c.', *The Poems of John Dryden*, ed. James Kinsley, 4 vols (Oxford: Clarendon, 1958), 2: 913.
17. Letter from Hannah More to Elizabeth Robinson Montagu, 27 August 1784; Huntington Library manuscript MO 3986, p. 3. Yearsley owned a copy of Dryden's translation of the *Georgics*. See *Catalogue of the Books, Tracts, &c. contained in Ann Yearsley's Public Library, No. 4, Crescent, Hotwells* (Bristol: Printed for the Proprietor, 1793), p. 27; British Library shelfmark s.c. 726 (9.).
18. O'Brien, 'Imperial Georgic, 1660–1789', in MacLean, Landry and Ward, eds, *Country and City Revisited*, pp. 162–3.
19. This is the terminus established by O'Brien, *ibid.*, pp. 172–6.
20. Addison, 'An Essay on Virgil's *Georgics*' (1697), in Scott Elledge, ed., *Eighteenth-Century Critical Essays*, 2 vols (Ithaca, NY: Cornell University Press, 1961), 1: 1–7; this passage p. 6.
21. Dyer, Book 1, *The Fleece, a Poem in Four Books* (London, 1757); quoted in Goodridge, *Rural Life*, p. 163.
22. Goodridge, *Rural Life*, p. 164.

23. For some of the imperial dimensions of developments in botany, see David Philip Miller and Peter Hanns Reill, eds, *Visions of Empire: Voyages, Botany, and Representations of Nature* (Cambridge: Cambridge University Press, 1996).
24. Leonore Davidoff and Catherine Hall, *Family Fortunes: Men and Women of the English Middle Class, 1780–1850* (Chicago: University of Chicago Press, 1987), p. 157.
25. Douglas Gordon, *Dartmoor in All its Moods* (London: John Murray, 1931), p. 23.
26. Young had written, '[A]lways remember, that the raising dung in the winter is the grand pillar of your husbandry', in *The Farmer's Kalendar; or, a Monthly Directory for all sorts of Country Business* (London: Robinson and Roberts, 1771), Sig. a2v, p. 12. See Landry, 'Mud, Blood, Muck: Country Filth', in a special issue, 'The Culture of Filth', Richard A. Barney and Grant Holly, eds, *Genre* 27: 4 (Winter 1994): 315–32.
27. Goodridge, *Rural Life*, p. 155.
28. Tusser, *Five Hundred POINTS of Husbandry: Directing What Corn, Grass, &c. is proper to be sown; what Trees to be planted; how Land is to be improved: With whatever is fit to be done for the Benefit of the FARMER in every Month of the YEAR . . . To which are added, Notes and Observations explaining many obsolete TERMS used therein, and what is agreeable to the present Practice in several Counties of this Kingdom. A WORK very necessary and useful for Gentlemen, as well as Occupiers of LAND, whether Wood-Ground or Tillage and Pasture* (London: Printed for M. Cooper; Sold by John Duncan 1744), p. 151. I have chosen this edition to illustrate Tusser's continuing popularity.
29. Scott, 'Eclogue II. Rural Business; or, The Agriculturalists', *Amoebaean Eclogues*, in *The Poetical Works* (London: Printed for J. Buckland, 1782; rpt Farnborough, Hants: Gregg International, 1969).
30. This refrain originated with Kenneth Williams and Kenneth Horne in 'Round the Horne' on BBC radio.
31. This felicitous phrasing is Goodridge's; *Rural Life*, p. 164.
32. Sir Egerton Brydges, *The Retrospective Review* (1825); quoted in Carr, *English Fox Hunting*, p. 64, n. 21.
33. A. Henry Higginson, *Peter Beckford Esquire, Sportsman, Traveller, Man of Letters: A Biography* (London: Collins, 1937), pp. 16–24, 37–9.
34. *Ibid.*, p. 145.
35. Beckford, *Thoughts upon Hunting*, p. 130.
36. Macdonald, *Running with the Fox*, p. 125.
37. Macdonald, *ibid.*, p. 125, cites J. David Henry, *Red Fox: The Catlike Canine* (Washington, D. C. and London: Smithsonian Institution Press, 1986), for the 'bookkeeping' function; see Henry, p. 120. Macdonald discusses fox odors in more technical terms in his chapter on 'The carnivores: order Carnivora', in Richard E. Brown and David W. Macdonald, eds, *Social Odours in Mammals*, 2 vols (Oxford: Clarendon, 1985), 2: 619–722. The following summary draws on Macdonald's account in both.
38. Carr, *English Fox Hunting*, pp. 24–5.
39. Munsche, *Gentlemen and Poachers*, p. 32.
40. Dodsley, Canto III, *Agriculture* (London: R. and J. Dodsley, 1753). This poem was to be the first of three comprising *Public Virtue. A Poem in Three Books. 1–Agriculture; 2–Commerce; 3–Arts*, but was the only one to be published. For the best recent treatment of Dodsley, see Pickard's 1998 University of Alberta PhD Diss., *Augustan Ecology*, pp. 96–107.
41. Fiennes, *Illustrated Journeys*, p. 36.

264 *The Invention of the Countryside*

42. Rackham, *History of the Countryside*, p. 190.
43. Ibid., pp. 4–5, 26.
44. McRae, *God Speed the Plough*, p. 151.
45. Walpole, *Anecdotes of Painting in England; With some Account of the principal Artists; And incidental Notes on other Arts; Collected by the late Mr. George Vertue; And now digested and published from his original MSS . . . To which is added The History of The Modern Taste in Gardening*, 4 vols (Strawberry-Hill: Printed by Thomas Kirgate, 1771), 4: 149. See Richard E. Quaintance, 'Walpole's Whig Interpretation of Landscaping History', in Roseann Runte, ed., *Studies in Eighteenth-Century Culture* 9 (1979): 25–30.
46. Walpole, *Anecdotes and History*, 4: 137.
47. Barrell, *Idea of Landscape*, pp. 3–12.
48. Andrews, *Search for the Picturesque*, pp. 67–73.
49. Thompson, *Whigs and Hunters*, p. 184.
50. Clare, Saturday, 5 March 1825, *The Journal*, in Grainger, ed., *Natural History Prose Writings*, pp. 226–7.
51. Hoskins, *Making of English Landscape*, p. 196.
52. Patten, 'Fox Coverts for the Squirearchy: The Chase and the English Landscape – II', *Country Life* 40: 3876 (23 September 1971): 736–8; this passage p. 738. See also John Patten, 'How the Deer Parks Began: The Chase and the English Landscape – I', *Country Life* 40: 3875 (16 September 1971): 660–2.
53. Hoskins, *Making of English Landscape*, p. 197.
54. Ellis, *Leicestershire and the Quorn*, pp. 60–5.
55. Hoskins, *Making of English Landscape*, p. 197.
56. Williamson and Bellamy, *Property and Landscape*, p. 205.
57. Williamson, *Polite Landscapes*, pp. 131–2.
58. Harry Hopkins, *The Long Affray: The Poaching Wars 1760–1914* (London: Secker & Warburg, 1985), p. 70; Anthony Vandervell and Charles Coles, *Game & the English Landscape: The Influence of the Chase on Sporting Art and Scenery* (London: Debrett's Peerage Ltd., 1980), pp. 41–5.
59. The appearance of a ring-necked pheasant in a Roman mosaic from the fourth century AD at Woodchester in Gloucestershire suggests this species might also have been introduced much earlier; Vandervell and Coles, *Game & the English Landscape*, pp. 44–5. Pennant categorized pheasants as belonging to the same genus as quail, pointing out that they are not wild but cultivated, in *The British Zoology*, p. 87.
60. Williamson and Bellamy, *Property and Landscape*, pp. 136, 70–1, who conclude that, 'although deer parks were a luxury, they were also essentially functional'.
61. Barrell, *Idea of Landscape*, p. 48.
62. Walpole, *Anecdotes and History*, 4: 149. On Brown, see Williamson, *Polite Landscapes*.
63. Everett, *Tory View of Landscape*, p. 39.
64. Hoskins, *Making of English Landscape*, pp. 174–6.
65. Delabere P. Blaine, Esq., *An Encyclopaedia of Rural Sports or Complete Account (Historical, Practical, and Descriptive) of Hunting, Shooting, Fishing, Racing, &c., &c.*, 3rd edn, 2 vols (London: Longmans, Green, Reader, and Dyer, 1880), 2: 856.
66. Williamson, *Polite Landscapes*, p. 138. See Forsyth, 'Game Preserves and Fences', *Journal of the Horticultural Society* 1 (1846): p. 201.
67. Hopkins, *Long Affray*, p. 75.
68. Williamson, *Polite Landscapes*, pp. 139–40.

69. *The Woodland Trust's Woodland Management Principles* (Grantham, Lincs.: The Woodland Trust, n.d.), p. 4.
70. Rackham, *History of the Countryside*, p. 31.

Chapter 3

1. Teresa Michals, ' "That Sole and Despotic Dominion": Slaves, Wives, and Game in Blackstone's *Commentaries*', *Eighteenth-Century Studies* 27: 2 (Winter 1993–94), pp. 195–216; this passage p. 215, n. 34. Douglas Hay has uncovered in the parishes he studied as many as 15 times the number of 'prosecutions in those parishes at Quarter Sessions and assizes, by all prosecutors, for all other thefts in the same years' being brought by a single landowner for poaching; 'Poaching and the Game Laws on Cannock Chase', in Douglas Hay, Peter Linebaugh, John G. Rule, E. P. Thompson and Cal Winslow, *Albion's Fatal Tree: Crime and Society in Eighteenth-Century England* (New York: Pantheon, 1975), pp. 189–253; this passage p. 251.
2. *A DIALOGUE Between A Lawyer and A Country Gentleman, upon the subject of the GAME LAWS, relative to Hares, Partridges, and Pheasants. With a LETTER to JOHN GLYNN, Esq: Serjeant at Law, and Representative of the County of Middlesex, Upon the PENAL LAWS of this Country By A GENTLEMAN of LINCOLNS-INN, A Freeholder of Middlesex* (London: Printed for J. Wilkie; and P. Uriel, 1771), pp. xi–xii. Hoping to turn the tables in favor of poachers but against the monied but unqualified classes who continued to pay for illegal game, the Gentleman proposed that a fine of £50 be levied upon anyone who purchased an illegally killed hare or partridge, with the seller, the poacher or unqualified supplier, as the informant: 'destroy the market and you will ruin the poachers, for it is the receiving of stolen goods that encourageth theft', p. xiii; British Library shelfmark 518.i.15.(6.).
3. Blackstone, *Commentaries on the Laws of England* (1765–69), 4 vols (fac. rpt Chicago and London: University of Chicago Press, 1979), 4: 175.
4. Munsche, *Gentlemen and Poachers*, pp. 3–7; Hay, 'Poaching and the Game Laws on Cannock Chase', p. 189.
5. Munsche, *Gentlemen and Poachers*, pp. 156, 181.
6. Rabbit skins had long been an important English export in the Levant trade. In the late sixteenth century, the 'main exports to the Levant were kerseys, coloured cloths, tin and coney-skins, while the principal imports into England were spices, silks, drugs, cotton-wool and currants'; Sir Percival Griffiths, *A Licence to Trade: The History of English Chartered Companies* (London and Tonbridge: Ernest Benn, 1974), p. 46.
7. For all these distinctions, see Munsche, *Gentlemen and Poachers*, p. 5.
8. Michals, 'Slaves, Wives, and Game in Blackstone's *Commentaries*', p. 211.
9. Munsche, *Gentlemen and Poachers*, pp. 13–14.
10. *Ibid.*, p. 13.
11. *Ibid.*, pp. 11–12.
12. John Allen Stevenson, 'Black George and the Black Act', *Eighteenth-Century Fiction* 8: 3 (April 1996), pp. 355–82; this passage p. 360.
13. Thirsk, ed., *Agrarian History V.ii*, p. 371.
14. Brewer, *The Sinews of Power: War, Money and the English State, 1688–1783* (London: Unwin Hyman, 1989), pp. 206, 199–206.
15. *Ibid.*, p. 206.

16. Rosenthal, *British Landscape Painting* (Ithaca, NY and Oxford: Cornell University Press/Phaidon Press, 1982), p. 42.
17. Bermingham, *Landscape and Ideology: The English Rustic Tradition, 1740–1860* (Berkeley and Los Angeles: University of California Press, 1986), pp. 28–9.
18. Cain and Hopkins, *British Imperialism: Innovation and Expansion 1688–1914* (London and New York: Longman, 1993), p. 12.
19. *Ibid.*, p. 24.
20. *Ibid.*, p. 101.
21. Bermingham, *Landscape and Ideology*, p. 201, n. 45.
22. *Ibid.*, p. 31.
23. Hayes, *Gainsborough Paintings and Drawings* (London: Phaidon, 1975), p. 203, n. 12.
24. Bermingham, *Landscape and Ideology*, pp. 29, 202, n. 47.
25. Cormack, *The Paintings of Thomas Gainsborough* (Cambridge: Cambridge University Press, 1991), p. 46.
26. Deuchar, *Sporting Art*, p. 84.
27. Vandervell and Coles, *Game & the English Landscape*, p. 58.
28. Alan Everitt, 'Farm Labourers', in Thirsk, ed., *Agrarian History IV*, pp. 396–465; this passage p. 405.
29. Neeson, *Commoners*, p. 34.
30. *Ibid.*, p. 34.
31. Turner, *Politics of Landscape*, p. 158.
32. McRae, *God Speed the Plough*, p. 8.
33. This observation of Turner's, *Politics of Landscape*, p. 185, has been recently echoed by Goodridge, *Rural Life*, p. 2. Raymond Williams found the positing of a hypothetical harmonious Golden Age within English rural society equally difficult to locate, always just one floor further on the historical escalator of the past in *The Country and the City*, pp. 9–12.
34. Thirsk, ed., *Agrarian History V.ii.*, p. 367.
35. Manning, *Hunters and Poachers*, p. 12.
36. Scott, *Rob Roy* (1817) (London: J. M. Dent & Sons, 1976), p. 42.
37. Manning, *Hunters and Poachers*, p. 62.
38. See, for example, *Cony-Catchers and Bawdy Baskets: An Anthology of Elizabethan Low Life*, ed. Gamini Salgado (Harmondsworth: Penguin, 1972), esp. the works by Robert Greene.
39. Manning, *Hunters and Poachers*, p. 2.
40. *James Hawker's Journal: A Victorian Poacher*, ed. Garth Christian (London: Oxford University Press, 1961), p. 109.
41. See Stevenson on gamekeepers and poachers, 'Black George', pp. 358–82; Stevenson makes an interesting case for a doubling between George Seagrim, the disreputable gamekeeper in *Tom Jones* (1749), and Henry Fielding, the author.
42. Kingsley, *The Water Babies: A Fairy Tale for a Land Baby* in *The Works*, 28 vols (London: Macmillan, 1884–85), 9: 20.
43. Deuchar, *Sporting Art*, p. 119.
44. *Ibid.*, p. 119.
45. Manning, *Hunters and Poachers*, p. 3.
46. Hawker, *Victorian Poacher*, pp. 62–3.
47. Cobbett, Canterbury, 4 September 1823, 'From Dover to the Wen', *Rural Rides*, 1: 249.

48. Cobbett, Hurstbourn Tarrant, 11 October 1826, 'From Burghclere to Lyndhurst', *Rural Rides*, 2: 149. For more on Charles Smith, James Turner, and other poaching incidents, see Harry Hopkins, *Long Affray*, pp. 17–35, 96–108.
49. *Dialogue Between A Lawyer and A Country Gentleman*, pp. vi–vii.
50. Anne F. Janowitz investigates the literary figure of the Gypsy in ' "Wild outcasts of society": The Transit of the Gypsies in Romantic Period Poetry', in MacLean, Landry and Ward, eds, *Country and City Revisited*, pp. 213–30.
51. Robinson and Powell, eds, *John Clare By Himself*, p. 78.
52. *Ibid.*, pp. 83–4.
53. Robinson, ed., *John Clare's Autobiographical Writings*, pp. 69–70.
54. *Ibid.*, pp. 85–6.
55. Clare, 'The Journal', Saturday 16 April 1825, in Grainger, ed., *Natural History Prose Writings*, p. 234.
56. Buxton, *A Botanical Guide to the Flowering Plants, Ferns, Mosses, and Algae, found indigenous Within Sixteen Miles Of Manchester, with some information as to their Agricultural, Medicinal, And Other Uses. Together with A Sketch Of The Author's Life; And Remarks On The Geology Of The District* (London: Longman and Co. and Manchester: Abel Heywood, 1849); James Cash, *Where There's a Will There's a Way! or, Science in the Cottage: an account of the Labours Of Naturalists In Humble Life* (London: Robert Hardwicke, 1873), pp. 77–89, 94–107; Tom Stephenson, *Forbidden Land* (Manchester: Manchester University Press, 1989), pp. 62–3.
57. Cash, *Where There's a Will*, pp. 78–80, 106–7.
58. Ian Niall, *The Poacher's Handbook* (London and Melbourne: William Heinemann, 1950), Note.
59. Hawker, *Victorian Poacher*, pp. 23, 78.
60. Sir Francis Hill, in *Georgian Lincoln* (Cambridge: Cambridge University Press, 1966), p. 152, quotes Robert Bell as reporting that the oldest copy of this ballad he had seen was dated 'about 1776' and printed at York; Bell, ed., *Ancient Poems, Ballads and Songs of the Peasantry of England* (1862), p. 216. See also Munsche, *Gentlemen and Poachers*, p. 63.
61. Mitford, 'Tom Cordery', *Our Village: Sketches of Rural Character And Scenery* (London: G. and W.B. Whittaker, 1824), pp. 164–76; this passage pp. 165–6.
62. Macdonald, *Running with the Fox*, p. 132.
63. Landry, *Muses of Resistance*, pp. 273–80. For constraints on the production and reception of working-class autobiographies, see Regenia Gagnier, *Subjectivities: A History of Self-Representation in Britain, 1832–1920* (New York and Oxford: Oxford University Press, 1991), pp. 41–54, 138–70.
64. Cobbett, Thursley, Wednesday, 26 October 1825, 'From Chilworth, in Surrey, to Winchester', *Rural Rides*, 1: 280.
65. Clare, Letter to [? Richard Newcomb], [early 1819] and Letter to Isaiah Knowles Holland [October? 1819] in Mark Storey, ed., *The Letters of John Clare* (Oxford: Clarendon, 1985), pp. 4, 15.
66. John E. Archer, *By a Flash and a Scare: Incendiarism, Animal Maiming, and Poaching in East Anglia 1815–1870* (Oxford: Clarendon, 1990), p. 255.
67. In a letter to James Augustus Hessey (Tuesday, 4 July 1820), Clare reported that he had been given '3 vols calld "Percys Relics" ', and 'there is some sweet Poetry in them & I think it the most pleasing book I ever happend on', Storey, ed., *Letters*, p. 82. On Percy's influence, see Nick Groom, *The Making of Percy's Reliques* (Oxford: Clarendon, 1999).

268 *The Invention of the Countryside*

68. Clare, 'Sports of the Field', in Eric Robinson and David Powell, eds, Margaret Grainger, assoc. ed., *The Early Poems of John Clare 1804–1822*, gen. ed. Eric Robinson, 2 vols (Oxford: Clarendon, 1989), 1: 378–9.
69. Clare, 'The Milton Hunt' (1819–20, revised 1821) in Robinson and Powell and Grainger, eds, *Early Poems*, 2: 198–9.
70. Clare, Letter to John Taylor [c. Saturday, 17 February 1821] in Storey, ed., *Letters*, pp. 154–5.
71. Clare, 'To Day the Fox Must Dye: A Hunting Song' (1818) in Robinson and Powell and Grainger, eds, *Early Poems*, 1: 400–1.
72. Clare, 'Hunting Song', in Eric Robinson and David Powell, eds, and Margaret Grainger, assoc. ed., *The Later Poems of John Clare 1837–1864*, gen. ed. Eric Robinson, 2 vols (Oxford: Clarendon, 1984), 1: 434–5.
73. Clare, 'Careless Rambles', in Robinson and Powell, eds, *Oxford Authors: John Clare*, p. 103.
74. David Perkins recognizes this ambivalence in Clare's attitude toward blood sports in 'Sweet Helpston! John Clare on Badger Baiting', *Studies in Romanticism* 38: 3 (Fall 1999): 387–407.
75. Bloomfield, 'Hunting-Song', in *Rural Tales, Ballads, and Songs* (London: Printed for Vernor and Hood and Longman and Rees, 1802) in Lawson, ed., *Collected Poems*.
76. Bloomfield, *May Day with The Muses* (London: Printed for the Author and for Baldwin, Cradock, and Joy, 1822), in Lawson, ed., *Collected Poems*.
77. Salisbury, *Beast Within*, p. 131.
78. Smith, *Diary of A Huntsman*, p. 157.

Chapter 4

1. Anon., *The Institucion of a Gentleman* (London: Thomas Marshe, 1568), fo. 45r. In the British Library copy, shelfmark 519, there is an inscription on the flyleaf by Joseph Haslewood, dated 2 August 1814: 'I can scarcely refer to any volume in my possession of equal curiosity with this as it is an original work and the earliest I know in our language upon the character and amusements of an Englishman.'
2. 'Introduction', *Ystradffin, a Descriptive Poem, with an Appendix Containing Historical and Explanatory Notes*, by Mrs. Bowen (London: Longman, Orme, Brown, Green and Longmans; Llandovery: W. Rees, 1839), p. iii. Copy courtesy of the Special Collections Department of the University of Colorado at Boulder Libraries.
3. Tim Fulford, *Landscape, Liberty and Authority: Poetry, Criticism and Politics from Thomson to Wordsworth* (Cambridge: Cambridge University Press, 1996), p. 117.
4. [Apperley], 'Dorset and Devon', *Nimrod's Hunting Tours, interspersed with characteristic anecdotes, sayings, and doings of Sporting Men, including notices of the principal Crack Riders of England, with analytical contents and general index of names. To which are added Nimrod's Letters on Riding to Hounds* (London: M. A. Pittman, 1835), p. 161.
5. Rev. Richard Warner, of Bath, *A Walk Through Wales, In August 1797* (Bath: Printed by R. Cruttwell and Sold by C. Dilly, London, 1798), pp. 94–5.
6. Johnson, *A Dictionary of the English Language*, 2 vols (London, 1755); quoted in Malcolmson, *Popular Recreations*, p. 4.

7. Malcolmson, *Popular Recreations, passim*; and for a good regional study, Mark Stoyle, *Loyalty and Locality: Popular Allegiance in Devon During the English Civil War* (Exeter: University of Exeter Press, 1994).
8. Malcolmson, *Popular Recreations*, p. 15.
9. Milton, 'L'Allegro', *Complete Poems and Major Prose*, ed. Merritt Y. Hughes (New York: Macmillan, 1957). I thank Margaret Ferguson for suggesting that Milton was relevant to this inquiry.
10. Marcus, *The Politics of Mirth: Jonson, Herrick, Milton, Marvell, and the Defense of Old Holiday Pastimes* (Chicago and London: University of Chicago Press, 1986), pp. 1–23.
11. Malcolmson, *Popular Recreations*, pp. 56–71.
12. Robert Greene's cony-catching texts of the 1590s are written in much the same irreverent, deliberately 'low', spirit as his contemporary Thomas Nashe's *The Unfortunate Traveller and other works*, ed. J.B. Steane (Harmondsworth: Penguin, 1972).
13. Thompson, *Customs in Common*, p. 9.
14. On the most appropriate short title for this poem, commonly abbreviated as 'Tintern Abbey', see David Fairer, ' "Sweet native stream!": Wordsworth and the School of Warton', in Alvaro Ribeiro, SJ and James G. Basker, eds, *Tradition in Transition: Women Writers, Marginal Texts, and the Eighteenth-Century Canon* (Oxford: Clarendon, 1996), pp. 314–38, esp. pp. 316–17. Nicholas Roe recovers some aspects of the poem's contemporary politics in *Wordsworth and Coleridge: The Radical Years* (Oxford: Clarendon, 1988), pp. 268–75, and *The Politics of Nature: Wordsworth and Some Contemporaries* (Basingstoke and London: Macmillan, 1992), pp. 117–36.
15. See David Perkins, 'Wordsworth and the Polemic Against Hunting: "Hart-Leap Well" ', *Nineteenth-Century Literature* 52: 4 (March 1998): 421–45.
16. [Cox], *Gentleman's Recreation*, p. 2.
17. Blome, *Gentlemans Recreation*, 2: 67.
18. Manning, *Hunters and Poachers*, p. 28.
19. Smith, *Literature and Revolution*, pp. 327–36.
20. *Pteryplegia: Or, the ART of Shooting-Flying. A POEM*, by Mr. Markland, A B and formerly Fellow of St. John's College in Oxford (London: Printed and Sold by J. Roberts, 1717), p. 20.
21. Commander Mark Beaufoy, 'A Father's Advice to his Son', in Hare, *Language of Sport*, p. 69, ll. 3–4.
22. Markland, *Pteryplegia*, p. 31.
23. Munsche's *Gentlemen and Poachers* contains a useful summary of changes in field sports between 1660 and 1830, pp. 32–9. See also Charles P. Chenevix Trench, *A History of Marksmanship* (London, 1972).
24. David Elliston Allen, *The Naturalist in Britain: A Social History* (London: Allen Lane, 1976), p. 141.
25. *Ibid.*, pp. 141–2.
26. *Ibid.*, p. 142.
27. *Ibid.*, p. 142.
28. Cobbett, Thursley, 26 October 1825, 'From Chilworth, in Surrey, to Winchester', *Rural Rides*, 1: 281.
29. [Gascoigne], *Noble Arte*, p. 246.
30. Stephen Daniels, *Fields of Vision: Landscape Imagery and National Identity in England and the United States* (Princeton, NJ: Princeton University Press, 1993), p. 91.

270 *The Invention of the Countryside*

31. Repton, assisted by his son, J. Adey Repton, *Fragments on The Theory And Practice of Landscape Gardening* (London: Printed by T. Bensley and Son for J. Taylor, 1816), p. 204.
32. *Ibid.*, p. 206.
33. *Ibid.*, p. 206.
34. Deuchar notes that this picture seems to have inspired no imitators, *Sporting Art*, p. 81.
35. Montcrieff et al., *Farm Animal Portraits*, pp. 126–7.
36. Munsche comments regarding coursing, 'Certainly it seems likely that heavy betting did nothing to reduce attendance at these events', *Gentlemen and Poachers*, p. 35.
37. *Ibid.*, pp. 38–9.
38. *Ibid.*, pp. 38–9, 200, n. 64.
39. Fiennes, *Illustrated Journeys*, p. 169.
40. Morris, 'Introduction', *Illustrated Journeys*, pp. 10–31; this passage p. 23.
41. [Taplin], *Sportsman's Cabinet*, is 'Dedicated to the Ladies Patronesses, Vice Patronesses, and Members of the Swaffham, Ash-Down Park, Bradwell, and Flixton Coursing Societies; and to the Noblemen and Gentlemen of the various subscription hunts in every part of the United Kingdom'.
42. Proctor, 'Article IX. – Review of *Our Village: Sketches of Rural Character and Scenery*, by Mary Russell Mitford (1824)', *The Quarterly Review* 31 (April 1824), pp. 166–74; this passage pp. 167–8. My thanks to Clare Bainbridge for this reference.
43. Mitford, 'A Great Farm-House', *Our Village*, pp. 48–57; this passage pp. 51–2.
44. Dryden, 'To my Honour'd Kinsman, John Driden, Of Chesterton In The County of Huntingdon, Esquire', in *The Poems of John Dryden*, ed. James Kinsley, 4 vols (Oxford: Clarendon, 1958), 4: ll. 92–3.
45. *The Academick Sportsman, or A Winter's Day. A Poem*, By the Rev. Gerald Fitzgerald, F[ellow] [of] T[rinity] C[ollege] D[ublin], 2nd edn (Dublin: Printed for William Hallhead, 1780), p. 9.
46. My thanks to Marjorie Garson for this observation.
47. Nedham, *Mercurius Pragmaticus* (London, 1649); quoted in Smith, *Literature and Revolution*, p. 68.
48. [John Lilburne?], *The hunting of the foxes from New-market and Triploe-heaths to White-hall, by five small beagles, late of the armie, or The grandie-deceivers unmasked. Directed to all the free-people of England by R. Ward, T. Watson [and 3 others.]* (London, 1649). Bodleian Library shelfmark: G. Pamph. 111 7 (11). On this text, see David Loewenstein, *Representing Revolution in Milton and His Contemporaries: Religion, Politics, and Polemics in Radical Puritanism* (Cambridge: Cambridge University Press, 2001), pp. 39–42.
49. Marvell, 'The First Anniversary Of the Government under O.C.', in H.M. Margoliouth, ed., *The Poems and Letters of Andrew Marvell*, 3rd edn rev. by Pierre Legouis and E.E. Duncan-Jones, 2 vols (Oxford: Clarendon, 1971), 1: ll. 125–6.
50. Buchan, *A Prince of the Captivity* (1933; Edinburgh: B & W Publishing, 1996), p. 47.
51. See, for example, Smith, *Literature and Revolution*, pp. 321–5; Turner, *Politics of Landscape*, pp. 49–61; and Earl R. Wasserman, *The Subtler Language: Critical Readings of Neoclassic and Romantic Poems* (Baltimore: The Johns Hopkins University Press, 1959), pp. 45–88.
52. Denham, *Coopers Hill*, Draft I (1641), in O Hehir, *Expans'd Hieroglyphicks*.
53. Denham, *Coopers Hill*, 'B' Text (1655; 1668), in *ibid*.

54. Wallace, '*Coopers Hill*: The Manifesto of Parliamentary Royalism, 1641', *English Literary History* 41: 4 (Winter 1974): 494–540; this passage p. 532.
55. Smith, *Literature and Revolution*, pp. 322–3.
56. Underdown, *Revel, Riot, and Rebellion*, p. 160.
57. Evelyn Philip Shirley, *Some Account of English Deer Parks with notes on the management of deer* (London: John Murray, 1867), pp. 47–8.
58. Blome, *Gentlemans Recreation*, 2: 169.
59. [Taplin], *Sportsman's Cabinet*, 2: 173.
60. *Field-Sports. A POEM*, Humbly Address'd To His Royal Highness The PRINCE by William Somervile, Esq. (London: Printed for J. Stagg, 1742).
61. Blome, *Gentlemans Recreation*, 2: 170.
62. *An HIPPONOMIE or The Vineyard of Horsemanship: Devided into three Bookes*, By Michaell Baret, Practitioner and Professor of the same Art (London: Printed by George Eld, 1618), p. 101.
63. Beth Fowkes Tobin, 'The Politics of the Deferential Gaze: Portraits of English Gentlemen and Their Dogs', unpublished paper. My thanks to the author for allowing me to read it in manuscript.
64. Gray to Walpole, [Burnham, August 1736], Paget Toynbee and Leonard Whibley, eds, with corr. and add. by H.W. Starr, *The Correspondence of Thomas Gray*, 3 vols (Oxford: Clarendon, 1971), 1: 47. Jonathan Rogers, an attorney, had married Gray's aunt, Ann Antrobus.
65. David Brock, MFH, Sometime Master of the Thurles and Kilshane Foxhounds and of the East Sussex, *The Fox-Hunter's Week-End Book* (London: Seeley Service, n.d.), p. 326.
66. [Kendall], *Keeper's Travels in search of His Master* (London: Printed for E. Newbery, 1798), pp. iii–vi; this passage p. v.
67. *The Life and Adventures of Bampfylde-Moore Carew, the noted Devonshire Stroller and Dog-Stealer; As related by Himself, during his Passage to the Plantations in America. Containing, A great Variety of remarkable Transactions in a vagrant Course of Life, which he followed for the Space of Thirty Years and upwards* (Exeter: Printed by the Farleys, for Joseph Drew, 1745), p. 3. For the pleasures of this text, see John Barrell, 'Afterword: Moving Stories, Still Lives', in MacLean, Landry and Ward, eds, *Country and City Revisited*, pp. 231–50.
68. *Sportsman's Cabinet*, 2: 102.
69. Vesey-FitzGerald, *It's My Delight* (London: Eyre & Spottiswoode, 1947), pp. 130–4.
70. Hawker, *Victorian Poacher*, p. 52.

Chapter 5

1. J. Aikin, MD, 'A Critical Essay on Somerville's Poem of *The Chace*' in *The Chace, A Poem*, by William Somervile, Esq., A New Edition (London: Printed for T. Cadell, Jun. and W. Davies, 1796), pp. 1–25; this passage p. 23.
2. *Ibid*., pp. 22–3.
3. [Wetenhall Wilkes], *Hounslow-Heath, a POEM*, Inscribed to a NOBLEMAN (London: Printed for C. Corbett, 1747). British Library shelfmark 11630.b.3, 1–17 [item 2].
4. *Hounslow-Heath. A POEM*, by the Rev. Wetenhall Wilkes, MA, Minister of the Chapel at Hounslow, in the Patronage of Richard Bulstrode, Esq., 2nd edn (London: Printed for the Author and Sold by T. Gardner, 1748).

5. Print culture, readership and professional authorship have a vast bibliography. A very minimal list would include Roger Chartier, *The Order of Books: Readers, Authors, and Libraries in Europe Between the Fourteenth and Eighteenth Centuries*, trans. Lydia G. Cochrane (Stanford: Stanford University Press, 1994); James Raven, Helen Small and Naomi Tadmor, eds, *The Practice and Representation of Reading in England* (Cambridge: Cambridge University Press, 1996); Kevin Sharpe, *Reading Revolutions: The Politics of Reading in Early Modern England* (London: Yale University Press, 2000); Brean S. Hammond, *Professional Imaginative Writing in England, 1670–1740: 'Hackney for Bread'* (Oxford: Clarendon, 1997); and Clifford Siskin, *The Work of Writing: Literature and Social Change in Britain, 1700–1830* (Baltimore and London: The Johns Hopkins University Press, 1998).
6. Thomas, *Man and the Natural World*, p. 181.
7. Ibid., p. 182.
8. Ibid., pp. 182–3.
9. On the class dimensions of the movement against cruelty to animals, see Ritvo, *Animal Estate*, pp. 130–53, and Malcolmson, *Popular Recreations*, pp. 152–7 and 172–3. For critiques of Thomas's analysis, see Tester, *Animals and Society*, pp. 49–71, 92–3, and Benton, *Natural Relations*, pp. 73–5.
10. Hughes, 'A Vegetarian', *Wodwo* (New York and Evanston: Harper & Row, 1967), p. 28.
11. Vegetarianism is represented as the non-negotiable first step toward a non-speciesist relation to animals by Peter Singer in *Animal Liberation*, new rev. edn (1975; New York: Avon, 1990), pp. 159–83. For early modern English vegetarianism, see Nigel Smith's study of Thomas Tryon, 'Enthusiasm and Enlightenment: Of Food, Filth, and Slavery', in MacLean, Landry, and Ward, eds, *Country and City Revisited*, pp. 106–18; Morton, *Shelley and the Revolution in Taste*; and Carol J. Adams, *The Sexual Politics of Meat: A Feminist-Vegetarian Critical Theory* (New York: Continuum, 1994). Adams argues that feminism and vegetarianism are versions of the same radical social critique, since both expose the objectification necessary for women to be regarded as sexual objects (or 'meat') and for animals to be regarded as meat. These analogous forms of objectification both depend upon the metaphysical suppression of animal and female subjectivity, what Adams deconstructively calls 'the absent referent', pp. 39–82. Adams's is an extreme position as well as an academic one, but it reproduces, with scholarly apparatus, what many of the younger generation in both Britain and the United States appear to think. On animal rights as a legal desideratum, see Tom Regan, *The Case for Animal Rights* (Berkeley and Los Angeles: University of California Press, 1983), who argues for recognizing animals as subjects of their own lives: 'Those who satisfy the subject-of-a-life criterion themselves have a distinctive kind of value – inherent value – and are not to be viewed or treated as mere receptacles', p. 243.
12. James Sambrook, *James Thomson 1700–1748: A Life* (Oxford: Clarendon, 1991), p. 107.
13. Thomson, *The Seasons*, in *The Complete Poetical Works of James Thomson*, ed. J. Logie Robertson (London and New York: Oxford University Press, 1908). All quotations are from this edition, which includes the revisions of 1744.
14. The Rev. William Bingley, for example, quoted these passages and also one on the fox from Somervile's *The Chace* in *Animal Biography; or, Authentic Anecdotes of the Lives, Manners, and Economy, of the Animal Creation, arranged according to the system of Linnaeus*, 2nd edn, 3 vols (London: Printed for Richard Phillips, 1804), 1: 252–3, 479 and 2: 39–40.
15. Fulford, *Landscape, Liberty*, p. 26.

16. *The Correspondence of Alexander Pope*, ed. George Sherburn, 5 vols (Oxford: Clarendon, 1956), 1: 515.
17. Maynard Mack opines that in *Windsor-Forest* field sports are regarded from the point of view of the pathos of the animal victim, but nevertheless as preferable to war; *Alexander Pope: A Life* (New Haven: Yale University Press, 1985), pp. 73–5. This would put Pope in line with Somervile, who argued in *The Chace* that hunting offered the 'Image of war, without its guilt' (1: 15). But see also the more conventionally political readings of Wasserman, *The Subtler Language*, pp. 101–68; Howard Erskine-Hill, 'Literature and the Jacobite Cause: Was There a Rhetoric of Jacobitism?', in Eveline Cruickshanks, ed., *Ideology and Conspiracy: Aspects of Jacobitism, 1689–1759* (Edinburgh: John Donald, 1982), pp. 49–69; and Pat Rogers, ' "The Enamelled Ground": The Language of Heraldry and Natural Description in Windsor-Forest', *Studia Neophilologica* 45 (1973): 356–71.
18. *A Poem on the Cruelty of Shooting: with Some Tender Remarks on the 10th of MAY 1768, particularly on young Mr. ALLEN. Humbly dedicated to the SONS OF LIBERTY*, By John Aldington (London: Printed for the Author and Sold by C. Pyne; Mr. Hodges's Musick Shop, Southwark; and at the Pamphlet-Shops and News Carriers in Town and Country, 1769), p. 7.
19. *Ibid.*, p. 19.
20. On sensibility and anti-slavery, see Charlotte Sussman, *Consuming Anxieties: Consumer Protest, Gender, and British Slavery, 1713–1833* (Stanford: Stanford University Press, 2000).
21. Moira Ferguson, *Animal Advocacy*, pp. 1–6.
22. Ritvo, *Animal Estate*, pp. 1–2 and *passim*.
23. Lovibond, 'On Rural Sports', *Poems on Several Occasions* (London: J. Dodsley, 1785). My thanks to Jan Wellington for this reference.
24. Goldsmith, 'The Deserted Village', in Roger Lonsdale, ed., *The Poems of Thomas Gray, William Collins, Oliver Goldsmith* (London: Longmans, 1969).
25. Perkins, 'Cowper's Hares', *Eighteenth-Century Life* 20, n.s., 2 (May 1996): 57–69.
26. Appendix II, 'Cowper's Account of His Hares', *Gentleman's Magazine*, 28 May 1783, pp. 412–14, in Baird and Ryskamp, eds, *Poems* 2: 443–6.
27. 'Epitaph on a Hare', in H.S. Milford, ed., *The Poetical Works of William Cowper*, 3rd edn (London: Oxford University Press, 1926).
28. Tester, *Animals and Society*, p. 176.
29. Peter Singer, who is not a rights theorist *per se*, but a utilitarian philosopher and anti-speciesist influential in the animal liberation movement, finds the very term '"animal-lover" ' itself indicative of 'the absence of the slightest inkling that the moral standards that we apply among human beings might extend to other animals', 'Preface to the 1975 Edition', *Animal Liberation*, p. ii. Singer is committed to extending 'the basic moral principle of equal consideration of interests' to animals, and this extension is 'demanded by reason, not emotion', pp. ii, iii. Singer relates the story of the animal lover who, upon learning that the Singers had no pets, asked, while eating a ham sandwich, ' "But you *are* interested in animals, aren't you, Mr. Singer?" ' Singer comments, 'Neither of us had ever been inordinately fond of dogs, cats, or horses in the way that many people are. We didn't "love" animals. We simply wanted them treated as the independent sentient beings that they are, and not as a means to human ends – as the pig whose flesh was now in our hostess's sandwiches had been treated', p. ii. For a critique of Singer from an animal rights' perspective, see Regan, *Case for Animal Rights*, pp. 136–40, 206–31.
30. Tester, *Animals and Society*, p. 191.

31. Quoted in *ibid.*, p. 191, from Jon Wynne-Tyson, *The Extended Circle: A Dictionary of Humane Thought* (Fontwell: Centaur Press, 1985), p. 174.
32. [Kendall], *The Canary Bird*, pp. 70–1.
33. Hester Lynch [Thrale] Piozzi, *Anecdotes of the late Samuel Johnson, LL. D. during the last Twenty Years of His Life* (London: Printed for T. Cadell, 1786), p. 206.
34. *Ibid.*, p. 206.
35. *Ibid.*, pp. 206–7.
36. Johnson, 'Somervile', in George Birkbeck Hill, ed., *Lives of the English Poets*, 3 vols (Oxford: Clarendon, 1905), 2: 317–20; this passage p. 318.
37. Duckers and Davies, *A Place in the Country*, p. 155.
38. Tom Bowker, *Mountain Lakeland* (London: Robert Hale, 1984), pp. 18, 203–4.
39. Battles over public access to such uplands as Kinder Scout, 'affording in many ways the most exhilarating and picturesque scenery', have often meant establishing rights of way in the absence of any ancient tracks; P. A. Barnes, *Trespassers Will Be Prosecuted* (Sheffield, 1934), p. 5; quoted in Howard Hill, *Freedom to Roam: The Struggle for Access to Britain's Moors and Mountains*, with a preface by Alan Mattingly, the Ramblers' Association (Ashbourne, Derbyshire: Moorland Publishing, 1980), p. 35.
40. A.L. Beier, *Masterless Men: The Vagrancy Problem in England 1560–1640* (London: 1985), p. xxii.
41. Wycherley, *The Country Wife*, in *Restoration Plays*, ed. Brice Harris (New York: Modern Library, 1953), Act 2, pp. 73–4.
42. Congreve, *The Way of the World*, in Brice, ed., *Restoration Plays*, Act 4, p. 567. My thanks to James Grantham Turner for reminding me of these examples.
43. Rosamond Bayne-Powell, *Travellers in Eighteenth-Century England* (London: John Murray, 1951), pp. 180, 135.
44. Norman Nicholson, *The Lakers: The Adventures of the First Tourists* (London: Robert Hale, 1955), p. 112.
45. Langan, *Romantic Vagrancy: Wordsworth and the Simulation of Freedom* (Cambridge: Cambridge University Press, 1995), p. 17.
46. *Ibid.*, p. 17.
47. Ward, *Clarion Handbook* (1934–35); quoted in Hill, *Freedom to Roam*, p. 32.
48. Bagwell, *The Transport Revolution from 1770* (London: B.T. Batsford, 1974), pp. 41–2. See also Anne D. Wallace, *Walking, Literature, and English Culture: The Origins and Uses of Peripatetic in the Nineteenth Century* (Oxford: Clarendon, 1993), who draws on Bagwell to explain walking's new popularity.
49. Robin Jarvis, *Romantic Writing and Pedestrian Travel* (Basingstoke and London: Macmillan Press and New York: St. Martin's Press, 1997), p. 22.
50. *Ibid.*, p. 27.
51. *A Pedestrian Tour through North Wales, in a Series Of Letters*, by J. Hucks, BA (London: Printed for J. Debrett and J. Edwards, 1795), pp. 4–5.
52. *Ibid.*, p. 113.
53. Mary Moorman, ed., *Journals of Dorothy Wordsworth*, 2nd edn (Oxford and New York: Oxford University Press, 1971), p. 15.
54. *Ibid.*, p. 164.
55. Warner, *Walk Through Wales, August 1797*, p. 87.
56. [Plumptre], *The Lakers: A Comic Opera, in Three Acts* (London: Printed for W. Clarke, 1798), p. 59.
57. *Ibid.*, p. 59.
58. Anne Wallace discusses *The Lakers* in *Walking, Literature*, pp. 95–9. Plumptre himself, a fellow of Clare College, Cambridge, and a keen pedestrian, made three

tours of the Lakes between 1796 and 1799. In 1799, he made his way from Cambridge to the Highlands of Scotland and returned by way of the Lakes, covering 2,236 miles, of which 1,774¼ miles were on foot. See Peter Bicknell and Robert Woof, *The Discovery of the Lake District 1750–1810: A Context for Wordsworth* (Grasmere: Trustees of Dove Cottage, 1982), p. 36.
59. Johnson, *Lives of the Poets*, 1: 77.
60. Theodore Howard Banks, ed., 'Introduction', *The Poetical Works of Sir John Denham*, 2nd edn (Hamden, CT: Archon Books, 1969), pp. 1–57; this passage p. 55.
61. Quoted in *ibid.*, pp. 56–7.
62. [Crowe], *Lewesdon Hill. A Poem* (Oxford: Clarendon, 1788; fac. rpt Spelsbury, Oxford: Woodstock Books, 1989), p. 1.
63. This is 'a vision which must have increased the attractiveness of his poem to Coleridge and to Wordsworth, opposed as they were both to what they saw as Britain's unnecessary war with France and to the tyranny of the Jacobins in the revolution there'; Fulford, *Landscape, Liberty*, p. 226.
64. Gray to Walpole [Burnham, August 1736], Toynbee and Whibley, eds, *Correspondence*, 1: 47–8.
65. Armstrong, *The Art of Preserving Health* (1744), in George Gilfillan, ed., *The Poetical Works of Armstrong, Dyer, and Green* (Edinburgh: James Nichol, 1858).
66. Beattie, *The Minstrel; or, The Progress of Genius. A Poem. The First Book*, 3rd edn (London: Printed for Edward and Charles Dilly; and A. Kincaid and W. Creech, Edinburgh, 1772), p. 11.
67. See Morris Marples, *Shanks's Pony: A Study of Walking* (London: J.M. Dent & Sons, 1959) on seventeenth-century literary pedestrians such as Thomas Coryate, William Lithgow, John Taylor (the Water Poet), and Ben Jonson, pp. 1–19, and on late eighteenth-century 'heel and toe' or competitive walking exploits, pp. 20–9.
68. *Ibid.*, pp. 21–4.
69. *Ibid.*, pp. 24–9.
70. Letter to Robert Southey, Sunday, 13 July [1794], in Earl Leslie Griggs, ed., *Collected Letters of Samuel Taylor Coleridge*, 6 vols (Oxford: Clarendon, 1956–71), 1: 89.
71. *Sporting Magazine*, vol. 7 (October 1795), p. 22.
72. *Sporting Magazine*, vol. 7 (October 1795), p. 23.
73. *Sporting Magazine*, vol. 4 (April 1794), p. 7.
74. *Sporting Magazine*, vol. 4 (April 1794), p. 7.
75. *The Life of a Sportsman* (1832) by Nimrod [Charles James Apperley] (London: John Lehmann, 1948), p. 225.
76. *Ibid.*, pp. 225–35.
77. Perkins, 'Wordsworth and the Polemic Against Hunting: "Hart-Leap Well" '.
78. Jim Birkett recounts that he began climbing 'When I first started bird nesting – as soon as I was old enough to walk, practically', in Bill Birkett, 'Talking with Jim Birkett', *Climber and Rambler* (August 1982), quoted in David Craig, *Native Stones: A Book about Climbing* (London: Pimlico, 1996), p. 128.
79. Burns, 'On Seeing a Wounded Hare Limp By Me Which a Fellow Had Just Shot At', in *The Complete Poetical Works of Robert Burns*, eds W.E. Henley and T.F. Henderson (Boston and New York: Houghton Mifflin, 1897), p. 93, ll, 15–16. My thanks to Carol McGuirk for this reference. Burns wrote to his patron Mrs. Dunlop, '[W]hatever I have said of shooting hares I have not spoken one irreverent word against coursing them' (p. 93, head-note). Like Cowper, Wordsworth gives us an uncoursed hare as well as an unwounded one.

80. David Simpson comes closest to this reading in *Wordsworth's Historical Imagination: The Poetry of Displacement* (New York and London: Methuen, 1987), pp. 149–59. Previous critics have emphasized the relation between the poem's incompetent narrator and Wordsworth, and between Wordsworth and his readers, trying to account for tonal shifts, especially moments of awkward comedy, that make readers 'aware not only of Simon Lee but also of what it means to read "Simon Lee" ', as James H. Averill puts it; *Wordsworth and the Poetry of Human Suffering* (Ithaca, NY and London: Cornell University Press, 1980), p. 165; Andrew L. Griffin, 'Wordsworth and the Problem of Imaginative Story: The Case of "Simon Lee" ', *PMLA* 92: 3 (May 1977): 392–409; and Paul D. Sheats, *The Making of Wordsworth's Poetry, 1785–1798* (Cambridge, MA: Harvard University Press, 1973), pp. 188–92. All subsequent critics are indebted to John F. Danby's treatment of the poem's irony in *The Simple Wordsworth: Studies in the Poems 1797–1807* (London: Routledge & Kegan Paul, 1960), pp. 38–47.
81. Clare, 'To Wordsworth', in Robinson, Powell and Grainger, eds, *Later Poems*, 1: 25.
82. Alexander B. Grosart, ed., *The Prose Works of William Wordsworth*, 3 vols (London: Edward Moxon, Son, & Co., 1876; rpt New York: AMS Press, 1967), 3: 160.
83. Berta Lawrence, *Coleridge and Wordsworth in Somerset* (Newton Abbot: David & Charles, 1970), pp. 147, 165.
84. 'Wordsworth occupies the classic bourgeois site, an unstable and amorphous middle ground which disables him from validating *any* orthodox social role in a wholehearted manner'; Simpson, *Wordsworth's Historical Imagination*, p. 155.
85. Coleridge, Letter to Robert Southey, [17 July 1797], in Griggs, ed., *Collected Letters*, 1: 197; see also Simpson, *Wordsworth's Historical Imagination*, pp. 153–6.
86. Dorothy Wordsworth, Letter to Mary Hutchinson (?), 14 August 1797, in Ernest de Selincourt, ed., *The Early Letters of William and Dorothy Wordsworth (1787–1805)* (Oxford: Clarendon, 1935), pp. 170–1.
87. Glen, *Vision and Disenchantment: Blake's Songs and Wordsworth's Lyrical Ballads* (Cambridge, London, and New York: Cambridge University Press, 1983), p. 237. These are the first two lines of 'The Human Abstract' in Blake's *Songs of Experience*.
88. Simpson, *Wordsworth's Historical Imagination*, p. 155.
89. Paine, *Rights of Man, Part One* (1791) (Harmondsworth: Penguin, 1985), p. 58.
90. Ibid., p. 140.
91. Glen, *Vision and Disenchantment*, p. 342.
92. Deuchar, *Sporting Art*, pp. 156, 155.
93. On dissent at Cambridge and other radical political influences on Pantisocracy, see Roe, *Radical Years*, pp. 84–117.
94. Richard Holmes, *Coleridge: Early Visions* (London: Hodder & Stoughton, 1989), p. 61.
95. Ibid., p. 82.
96. Coleridge, 'To a Young Ass, Its Mother Being Tethered Near It', in E. H. Coleridge, ed., *Poetical Works*, p. 75, n.
97. According to Holmes, upon finding mice in the cottage belonging to his friend Thomas Poole, 'Coleridge, with his fraternal attitude to animals, found himself in a ludicrous quandary, which curiously bears on the symbolism he would later apply to the shooting of the albatross. To kill the mice would be to betray them. A domestic joke here contains the seed of metaphysical drama'; *Early Visions*, p. 138. David Perkins finds Coleridge's commitment to animal rights more self-serving and less sincere in 'Compassion for Animals and Radical Politics'.

98. Holmes remarks, 'In the symbolic killing of the albatross, he found what might be called a "green parable", the idea of man's destructive effect on the natural world, so that human moral blindness inadvertently introduces evil into the benign systems of nature, releasing uncontrollable forces that take terrible revenge'; *Early Visions*, p. 173.

Chapter 6

1. John Smyth, of Nibley, *The Berkeley Manuscripts*, ed. Sir John Maclean, 3 vols (Gloucester: John Bellows, 1883–85), 2: 363.
2. David Coombs, *Sport and the Countryside in English Paintings, Watercolours, and Prints* (Oxford: Phaidon, 1978), p. 90.
3. Amanda Vickery, *The Gentleman's Daughter: Women's Lives in Georgian England* (New Haven and London: Yale University Press, 1998), pp. 272–6.
4. Manning, *Hunters and Poachers*, p. 183. Manning notes, 'For more examples of female poachers, see PRO DL 1/203/H23; STAC 8/215/1', n. 37.
5. *The Sporting Magazine*, vol. 5 (January 1795), p. 215.
6. See Munsche, *Gentlemen and Poachers*, pp. 38–9, 200, n. 64.
7. Londa Schiebinger, *The Mind Has No Sex? Women in the Origins of Modern Science* (Cambridge, MA and London: Harvard University Press, 1989), pp. 37, 47.
8. Smyth, *Berkeley Manuscripts*, 2: 385.
9. Schiebinger, *The Mind Has No Sex?*, p. 59. Two developments, 'the privatization of the family and the professionalization of science', changed women's prospects for scientific study, Schiebinger claims. The professionalization of scientific fields occurred precisely through the exclusion of women as eternal amateurs, rationalized by 'the theory of sexual complementarity, a theory that articulated and justified the continued exclusion of women from science in terms acceptable to both liberal democratic theory and modern science', pp. 245, 244.
10. Allen, *Naturalist in Britain*, p. 28.
11. Lady Diana Shedden and Lady Apsley, *'To Whom the Goddess . . .'; Hunting and Riding for Women* (London: Hutchinson & Co., 1932) and Lady Apsley, *Bridleways Through History* (London: Hutchinson & Co., 1936).
12. Ann B. Shteir, *Cultivating Women, Cultivating Science: Flora's Daughters and Botany in England, 1760–1860* (Baltimore and London: The Johns Hopkins University Press, 1996), p. 47.
13. Allen, *Naturalist in Britain*, p. 29.
14. Shteir, *Flora's Daughters*, p. 47.
15. Henrietta Cavendish Holles was the only daughter and heiress of John, fourth Earl of Clare, created Duke of Newcastle, by Lady Margaret Cavendish, third daughter and co-heiress of Henry, second Duke of Newcastle (William Cavendish's younger son by his first wife; Margaret the poet had no children). Henrietta brought to the marriage with Edward Harley £500, 000 (*DNB*).
16. John Wootton, *Lady Henrietta Cavendish Holles, Countess Oxford hunting at Wimpole Park* (1716), private collection; and *Lady Henrietta Harley out Hunting with Harriers* (no date), collection of Lady Thompson. See Arline Meyer, *John Wootton 1682–1764: Landscapes and Sporting Art in Early Georgian England* (London: Greater London Council, 1984), pp. 37–8.

17. Lanyer, 'The Description of Cooke-ham', in *The Poems of Aemilia Lanyer: Salve Deus Rex Judaeorum*, ed. Susanne Woods (New York and Oxford: Oxford University Press, 1993), pp. 130–8.
18. Yet Lanyer hedges her bets slightly: Europe could not afford *much* more delight. Cookham was a crown manor leased to the Countess's brother William Russell of Thornhaugh and occupied by her periodically during her estrangement from her husband. The poet may be reminding her audience that the Countess of Cumberland is only periodically the mistress of her brother's estate: 'What was there then but gave you all content, / While you the time in meditation spent' (ll. 75–6). See *ibid.*, 'Introduction', pp. xxv and xxxv; and Barbara Lewalski, 'Imagining Female Community: Aemilia Lanyer's Poems', in *Writing Women in Jacobean England* (Cambridge, MA: Harvard University Press, 1993), pp. 212–41, and 'Rewriting Patriarchy and Patronage: Margaret Clifford, Anne Clifford, and Aemilia Lanyer', in *The Yearbook of English Studies* 21 (1991): 87–106.
19. *POEMS, and Fancies*, Written By the Right Honourable, the Lady Newcastle (London: Printed by T.R. for J. Martin and J. Allestrye, 1653; fac. rpt Menston, Yorks.: Scolar Press, 1972). 'The Hunting of the Hare' is included in Germaine Greer, Susan Hastings, Jeslyn Medoff and Melinda Sansone, eds, *Kissing the Rod: An Anthology of Seventeenth-Century Women's Verse* (New York: Noonday Press, 1989), pp. 168–72.
20. *The Life of the Thrice Noble, High and Puissant Prince William Cavendishe, Duke, Marquess, and Earl of Newcastle*, Written By the thrice Noble, Illustrious, and Excellent Princess, Margaret, Duchess of Newcastle, His Wife (London: Printed by A. Maxwell, 1667), pp. 92–3. 'Poots' were game birds, either blackcock or red grouse, probably the former; C. H. Firth, ed., *The Life of William Cavendish, Duke of Newcastle, to which is added The True Relation of my Birth Breeding and Life By Margaret, Duchess of Newcastle* (London: George Routledge and Sons; New York: E.P. Dutton, n.d.), p. 71, n. 2.
21. Firth, ed., *Life of William Cavendish and True Relation*, pp. 159, 175.
22. *The Philosophical and Physical Opinions*, Written by her Excellency, the Lady Marchioness of Newcastle (London: Printed for J. Martin and J. Allestrye, 1655), pp. 100–1.
23. *Orations of Divers Sorts, Accommodated to Divers Places*, Written by the thrice Noble, Illustrious and excellent Princess, the Lady Marchioness of Newcastle (London, 1662), p. 228.
24. *Ibid.*, p. 229.
25. *Philosophical Letters: or, Modest Reflections Upon some Opinions in Natural Philosophy, Maintained By several Famous and Learned Authors of this Age, Expressed by way of Letters*, By the Thrice Noble, Illustrious, and Excellent Princess, The Lady Marchioness of Newcastle (London, 1664), p. 18.
26. *Ibid.*, p. 43.
27. Rogers, *The Matter of Revolution: Science, Poetry, and Politics in the Age of Milton* (Ithaca, NY and London: Cornell University Press, 1996), p. 189.
28. *Ibid.*, p. 209. In *Margaret Cavendish and the Exiles of the Mind* (Lexington, KY: The University Press of Kentucky, 1998), Anna Battigelli discusses Cavendish's acquaintance with Hobbes in explaining his influence, pp. 63–73.
29. Rogers, *Matter of Revolution*, p. 209.
30. *Ibid.*, pp. 187–211, and Schiebinger, *The Mind Has No Sex?*, p. 58.
31. *Philosophical and Physical Opinions*, p. 100.
32. Smith, *Literature and Revolution*, p. 259.

33. Greer et al. in *Kissing the Rod* opine that poems against blood sports tend to be a female genre, p. 171. But the poems they mention, especially the single hare poem, Mary Jones's 'To Mrs. CLAYTON, With a Hare', in *Miscellanies in Prose and Verse* (Oxford: Printed and delivered by Mr. Dodsley, Mr. Clements and Mr. Frederick, 1750), pp. 50–2, could be read as belonging to hunting discourse in its self-reflexive and critical mode, as exemplified by Gascoigne's *The Noble Arte*.
34. On fashion and gender-crossing, see Erin Mackie, *Market à la Mode: Fashion, Commodity, and Gender in The Tatler and The Spectator* (Baltimore and London: The Johns Hopkins University Press, 1997), pp. 75–8, 116–18.
35. Bond, ed., *The Spectator*, 1: 241. Bond notes of *Andromache* that: 'Tatler 37 had described "Mrs. *Alse Copswood*, the *Yorkshire* Huntress, who is come to Town lately, and moves as if she were on her Nag, and going to take a Five-Bar Gate; and is as loud as if she were following her Dogs" ' (1: 241, n. 4).
36. Cohen, *The Unfolding of the Seasons* (London: Routledge & Kegan Paul, 1970), p. 178.
37. *Ibid.*, p. 192.
38. Rothstein calls Thomson 'fat and indolent' in *'Discordia Non Concors'*, p. 353. James Sambrook describes Thomson as 'awkward with women and serious about poetry', confirms his indolence and love of carousing, labels him 'a man's man', and speculates that the drunken cleric of tremendous paunch is a self-portrait in *James Thomson*, pp. 21, 207–11, 209, 231. *The Castle of Indolence: An Allegorical Poem* appeared in 1748, probably in May; Thomson died in August.
39. British Library Harley MSS. 7524, fol. 209r–v.
40. Letter to Robert Harley, Wimpole, 21 August 1716, British Library Additional MSS. 70239, BM Loan 29/45, fol. 2. Peggy had been born in February 1715.
41. Isobel Grundy, *Lady Mary Wortley Montagu* (Oxford: Oxford University Press, 1999), p. 21.
42. Robert Halsband, ed., *The Complete Letters of Lady Mary Wortley Montagu*, 3 vols (Oxford: Clarendon, 1965–67), 2: 54.
43. Chatsworth, Devonshire MSS 219.16. My thanks to Isobel Grundy for this reference. See Grundy, *Lady Mary Wortley Montagu*, p. 187, n. 41.
44. Somervile, *The Chace*, 2: 254.
45. John Smallman Gardiner, Gent., *The Art and the Pleasures of Hare-Hunting. In Six Letters to a Person of Quality* (London: Printed for R. Griffiths, 1750), p. 3.
46. Thus does Taplin describe hare-hunting with beagles, and then with the larger harriers, who were often a beagle and foxhound cross; [Taplin], *Sportsman's Cabinet*, 1: 133, 137.
47. Mrs. J. Stirling Clarke, *The Habit & The Horse; a treatise on Female Equitation* (London: Day & Son, 1860), pp. 211–12.
48. A. Aspinall, ed., *The Correspondence of George, Prince of Wales 1770–1812*, 8 vols (London: Cassell, 1963–71), 1: 75.
49. *The ROYAL CHASE; A POEM. Wherein Are Described Some humorous Incidents of a Hunt at Windsor* (London: Printed for G. Kearsly, 1782), p. 9.
50. *The Sporting Magazine*, vol. 7 (January 1796), p. 176. Lady Emily Mary Hill, Marchioness of Salisbury, was Master of Foxhounds at Hatfield from 1775 until 1819 and continued hunting into her seventies after that. See also David Cecil, *The Cecils of Hatfield House* (London: Constable, 1973), pp. 189–96; and Meriel Buxton, *Ladies of the Chase* (London: The Sportsman's Press, 1987), pp. 38–41.

51. See [Charles Pigott], *The Jockey Club, or a Sketch of the Manners of The Age*, Part I, 4th edn (London: H.D. Symonds, 1792), pp. 81–3.
52. Judy Egerton, *George Stubbs*, p. 177; Mark Evans, *The Royal Collection: Paintings from Windsor Castle* (Cardiff and London: National Museum of Wales and Lund Humphries, 1990), p. 108; John Robert Robinson, *The Last Earls of Barrymore 1769–1824* (London: Sampson Low, Marston & Co., 1894), pp. 60–2, 204, 250–1; Lewis Melville, *The Beaux of the Regency*, vol. 1 (London: Hutchinson, 1908), pp. 62–4; Buxton, *Ladies of the Chase*, p. 36.
53. Stella A. Walker, *Sporting Art: England 1700–1900* (London: Studio Vista, 1972), p. 68.
54. *The Sporting Magazine*, vol. 4 (June 1794), pp. 154–6.
55. *The Sporting Magazine*, vol. 9 (November 1797), p. 61.
56. Surtees, *Mr Sponge's Sporting Tour* (1853) (Gloucester: Alan Sutton, 1984), pp. 498, 579, 498.
57. *Ibid.*, p. 533.
58. Deleuze and Guattari's notion that a 'technical machine' is 'not a cause but merely an index of a general form of social production' is useful; Gilles Deleuze and Félix Guattari, *Anti-Oedipus: Capitalism and Schizophrenia*, trans. Robert Hurley, Mark Seem and Helen R. Lane (Minneapolis: University of Minnesota Press, 1983), p. 32. I have analyzed the history and significance of the side-saddle further in an unpublished manuscript, *The Making of the English Hunting Seat*.
59. Chenevix Trench, *History of Horsemanship*, p. 273.
60. Boudewijn F. Commandeur, 'The Historical Development of the Side-Seat on Horseback', *National Sporting Library Newsletter* 23 (December 1986): 1–7; this passage quoted p. 5.
61. Quoted in *ibid.*, p. 5.
62. *Ibid.*, pp. 1–2.
63. Lida Fleitmann Bloodgood, *The Saddle of Queens: The Story of the Side-Saddle* (London: J. A. Allen, 1959), pp. 16–17.
64. Commandeur, 'Historical Development', p. 3.
65. For Pellier, see Grant, 'Evolution of the Amazone', p. 1; for Oldaker, Chenevix Trench, *History of Horsemanship*, p. 277.
66. Buxton, *Ladies of the Chase*, pp. 34–5.
67. Mrs. Elizabeth Karr, *The American Horsewoman*, 3rd edn (1884; Boston and New York: Houghton Mifflin, 1890); and Theodore H. Mead, *Horsemanship for Women* (New York: Harper and Brothers, 1887).
68. Commandeur, 'Historical Development', p. 7: Fig. 9.
69. Montagu, Letter to Wortley, 7 August/27 July [1739], Halsband, ed., *Letters*, 2: 141–2.
70. Montagu, Letter to Wortley, 22 December [1741], *Letters*, 2: 263.
71. Montagu, Letter to Lady Bute, 5 January [1748], *Letters*, 2: 393.
72. By 1755, Montagu does not bother to remind her daughter of this custom; Montagu, Letter to Lady Bute, 24 July [1755], *Letters*, 3: 86.
73. The more famous Duchess of Cleveland was Anne Pulteney's mother-in-law, Barbara Villiers Palmer, mistress of Charles II. The more famous William Pulteney, who became Earl of Bath, Pope's Parliamentary Opposition friend, was the grandson of Anne Pulteney's father; see G. Steinman Steinman, *A Memoir of Barbara, Duchess of Cleveland* (London: Printed for private circulation, 1871), p. 203, and *DNB*.

74. Montagu, Letter to Lady Bute, 1 October [1749], *Letters*, 2: 444.
75. Michael McKeon, 'Historicizing Patriarchy: The Emergence of Gender Difference in England, 1660–1760', *Eighteenth-Century Studies* 28: 3 (Spring 1995): 295–322.
76. See Landry, 'Learning to Ride at Mansfield Park', in You-me Park and Rajeswari Sunder Rajan, eds, *The Postcolonial Jane Austen* (London and New York: Routledge, 2000), pp. 56–73.
77. Carr, *English Fox Hunting*, pp. 8, 175, 243, and Buxton, *Ladies of the Chase*, p. 170.
78. Germaine Greer has characterized a girl's pleasure in riding as love of 'an *other* which is responding to her control. What she feels is a potent love calling forth a response. The control required by riding is so strong and subtle that it hardly melts into the kind of diffuse eroticism that theorists . . . would have us believe in . . . George Eliot knew what she was doing when she described Dorothea Brooke's passion for wild gallops over the moors in *Middlemarch*. It is part and parcel of her desire to perform some great heroism, to be free and noble'; *The Female Eunuch* (1970; New York: McGraw-Hill, 1971), pp. 73–4. Melissa Holbrook Pierson comes to much the same conclusion in *Dark Horses and Black Beauties: Animals, Women, A Passion* (New York and London: W.W. Norton, 2000).
79. Scruton, *On Hunting* (London: Yellow Jersey Press, 1998), p. 115.

Chapter 7

1. W.H. Auden, *Collected Shorter Poems (1927–1957)* (London: Faber and Faber, 1969), p. 74.
2. *An Essay on Hunting*, by a Country Squire (London: Printed for J. Roberts, 1733), p. 5. All quotations from this page.
3. Colley, *Britons: Forging the Nation 1701–1837* (New Haven and London: Yale University Press, 1992), p. 172.
4. Raymond Williams usefully identified changes in cultural practice by the terms 'dominant', 'residual' and 'emergent' in *Marxism and Literature* (Oxford: Oxford University Press, 1977), pp. 121–7.
5. Jane Ridley, *Fox Hunting* (London: Collins, 1990), p. 165.
6. [Apperley], 'Dorset and Devon', *Nimrod's Hunting Tours*, p. 161.
7. Sir John Palmer, fifth Baronet, was MP for Leicester; Moncrieff et al., *Farm Animal Portraits*, p. 92.
8. William Leeke, *The History of Lord Seaton's Regiment at the Battle of Waterloo*, 2 vols (London, 1866), 1: 197; quoted in Colley, *Britons*, p. 172.
9. Daniel, *Rural Sports*, 2 vols (London: Bunny & Gold, 1801–2), 1: 236–9.
10. Deuchar, *Sporting Art*, pp. 5, 78; Itzkowitz, *Peculiar Privilege*, pp. 74–5; and Ellis, *Leicestershire and the Quorn*, pp. 10–11.
11. Much of the growth in high-status leisure in English provincial towns had happened between 1680 and 1760, 'injecting a new degree of refinement into their social life and raising their cultural image from the depressed level of the pre-Restoration years, so that by the reign of George III many were in the vanguard of fashion'; Borsay, *The English Urban Renaissance: Culture and Society in The Provincial Town 1660–1770* (Oxford: Clarendon, 1989), pp. 195–6.
12. Coombs, *Sport and the Countryside*, p. 71. Coombs reproduces this hand-colored etching on p. 70. It is not in the collection of the British Museum, as Coombs

claims, but in the collection of Birmingham Museums and Art Gallery, though it usually hangs in a private collection.
13. Coombs comments, *ibid.*, '[D]oubtless the ladies have ridden out in the middle of the day to meet the returning hunters', p. 71, but other pictures in this series include women in the field, galloping and jumping.
14. [Apperley,] 'Nimrod's Yorkshire Tour', *Nimrod's Hunting Tours*, p. 355.
15. Frederick Watson, *Robert Smith Surtees: A Critical Study* (London: George G. Harrup & Co., 1933), p. 147.
16. Coburn, ed., *Notebooks*, 1 (Text): entry 1262.
17. Carr, *English Fox Hunting*, p. 71.
18. *Ibid.*, p. 30.
19. *Ibid.*, p. 69.
20. *Ibid.*, p. 71; Ridley, *Fox Hunting*, p. 18.
21. Earl of March, *Records of the Old Charlton Hunt* (London: Elkin Mathews, 1910), pp. 61–5.
22. Anonymous verses of February 1737/38, *ibid.*, p. 18.
23. [Surtees], *Handley Cross; or, Mr. Jorrocks's Hunt* (1st pub. 1843; London: Bradbury and Evans, 1854; rpt Methuen, 1939), pp. 205–12.
24. Carr, *English Fox Hunting*, p. 73.
25. *Ibid.*, p. 76. Carr writes 'the damned hounds', but I have heard this anecdote many times with the wording 'these'.
26. Letter from Knightley quoted by William Scarth Dixon, *A History of the Bramham Moor Hunt* (Leeds: Richard Jackson, 1898), pp. 210–11.
27. Scruton, *On Hunting*, pp. 4–5.
28. Thirsk, ed., *Agrarian History V.ii.*, p. 578.
29. *Ibid.*, p. 578.
30. The National Pig Breeders Association was formed only in 1884; Moncrieff et al., *Farm Animal Portraits*, p. 229.
31. Lord Ribblesdale, Master of the Buckhounds from 1892 to 1895, *The Queen's Hounds and Stag-Hunting Recollections* (London: Longmans, Green, and Co., 1897), p. 301.
32. *Ibid.*, p. 301, n.
33. See Landry, 'Trafficking in Arabs, Barbs, and Turks: The English Trade in Eastern Blood Horses in the Seventeenth and Eighteenth Centuries', forthcoming in *Graeco-Arabica* 9/10.
34. Thirsk, ed., *Agrarian History V.ii.*, p. 578.
35. Arline Meyer, *John Wootton*, pp. 34, 35. Sir Gervas Clifton requested a 'leap' from the Bloody Shouldered Arabian on behalf of 'Mr. Willoughby of Aspaly' in letters to Edward Harley of 9 April 1726 and 7 May 1726, B.L. Add. MSS. 70374/Portland Papers, fol. 87r and 89r. James Bramston reported on 8 October 1729 that he had been promised one for the next spring, Add. MSS. 70373/Portland Papers, fol. 150v.
36. *Ibid.*, p. 34.
37. '[T]hat great authority, Mr. Meynell, only considered the produce of brothers and sisters as being bred "in-and-in," and not those produced from a union of a parent and offspring'; Robert R. Vyner, *Notitia Venatica: A Treatise on Fox-Hunting embracing the general management of hounds* (1841), 7th edn by William C. A. Blew, MA (London: John C. Nimmo, 1892), p. 51.
38. Chalker, *The English Georgic: A Study in the Development of a Form* (London: Routledge & Kegan Paul, 1969), p. 192.

39. On pit bulls in modern Britain, see Iain Sinclair, *Lights Out for the Territory: 9 Excursions in the Secret History of London* (London: Granta Books, 1997), pp. 55–87.
40. Carr, *English Fox Hunting*, p. 136.
41. Ibid., p. 136.
42. Ibid., pp. 136–7.
43. Sassoon, *Memoirs of a Fox-Hunting Man* (1929) in *The Memoirs of George Sherston* (New York: Literary Guild of America, 1937), pp. 165–6.
44. Beckford, *Thoughts upon Hunting*, p. 32.
45. [Apperley,] *Life of a Sportsman*, p. 37.
46. Ibid., p. 238.
47. [Surtees], *Handley Cross*, pp. 298–9.
48. Hare, *Language of Sport*, pp. 3, 51.
49. [Surtees,] 'The Yorkshireman and the Surrey', *Jorrocks' Jaunts and Jollities: The Hunting, Shooting, Racing, Driving, Sailing, Eccentric and Extravagant Exploits of that Renowned Sporting Citizen Mr. John Jorrocks* (1838), a new edn (London: George Routledge and Sons, n.d. [1880]), pp. 17–39; this passage pp. 28–9.
50. Johnson, 'Somervile', in Hill, ed., *Lives of the English Poets*, 2: 319.
51. *The Chace* appeared in 1800, 1802, 1804, 1807, 1817, 1830, 1873, 1886, 1892, 1896 and 1929.
52. 'Preface by The Author', in Bulmer's 1802 edition, pp. xvii–xxiii; this passage p. xviii.
53. See Christine Gerrard, *The Patriot Opposition to Walpole: Politics, Poetry, and National Myth, 1725–1742* (Oxford: Clarendon, 1994), pp. 210–23, and Chalker, *English Georgic*, pp. 180–95.
54. Chalker, *English Georgic*, p. 182, and Gerrard, *Patriot Opposition*, pp. 214–15, 218.
55. Gerrard, *Patriot Opposition*, p. 220. *Field-Sports. A POEM*, Humbly Address'd To His Royal Highness The PRINCE, by William Somervile, Esq. (London: Printed for J. Stagg, 1742) describes falconry, shooting, setting and angling; it was presented to Frederick posthumously.
56. Carr, *English Fox Hunting*, p. 71.
57. Blunt, 'The Old Squire', *The Poetical Works of Wilfrid Scawen Blunt*, 2 vols (London: Macmillan, 1914), 2: 11–13.
58. Watson, *Surtees*, p. 166.
59. Ibid., p. 170.
60. Aikin, 'Critical Essay', p. 13.
61. Brewer, *Sinews of Power*, pp. xvii–xxi.
62. Aikin, 'Critical Essay', pp. 15–16.
63. Richard Greville Verney, Lord Willoughby de Broke, *Hunting the Fox* (London: Constable, 1921), pp. 121–32; Greaves, *Yonder He Goes: A Calendar of Hunting Sketches* (London: Collins, 1935), pp. 13–20 and *passim*. For Sassoon and Kipling, see Chapter 8.
64. John Lawrence's many publications exemplify this trend; see, for example, *A Philosophical And Practical Treatise On Horses, and on the Moral Duties Of Man towards The Brute Creation*, 2 vols (London: Printed for T. Longman, 1796–98).
65. Vanbrugh, *A Journey to London* (1728), in A.E.H. Swain, ed., *Sir John Vanbrugh* (New York: A.A. Wyn, 1949), Act 1, scene 1, p. 446.
66. 'Prelude', *Hounslow-Heath. A POEM*, by the Rev. Mr Wetenhall Wilkes, MA, 2nd edn (London: Printed for the Author and Sold by T. Gardner, 1748), pp. ix–xvi; this passage p. x.
67. Aikin, 'Critical Essay', pp. 17–18.

68. Ibid., p. 18.
69. Vesey-FitzGerald, It's My Delight, p. 26.
70. ESSAYS on the GAME LAWS, Now existing in GREAT BRITAIN; and Remarks on their principal Defects: Also, Proposals for the better Preservation of the GAME in this Kingdom. With A PLAN for the Destruction of the VERMIN, By a SPORTSMAN (London: Printed for T. Becket, and P.A. DeHondt, 1770), p. 21.
71. Daniel, Rural Sports, 1: 522, 515.
72. [Apperley], Life of a Sportsman, p. 70.

Chapter 8

1. Blain, 'Editor's Introduction', Mr Sponge's Sporting Tour (St. Lucia and New York: University of Queensland Press, 1981), pp. ix–xvi; this passage p. xiii.
2. Gash, Robert Surtees and Early Victorian Society (Oxford: Clarendon, 1993), p. 9; Gash quotes G.M. Young, Portrait of an Age.
3. Watson, Surtees, pp. 68–9.
4. [Surtees], Handley Cross, p. 79.
5. Ibid., p. 69.
6. Kipling, ' "My Son's Wife" ', The English in England, ed. Randall Jarrell (Gloucester, MA: Peter Smith, 1972), pp. 103–35; this passage p. 120.
7. Ridley, Fox Hunting, p. 162.
8. Sassoon, Memoirs of a Fox-Hunting Man, p. 29.
9. Ridley, Fox Hunting, p. 162.
10. Ridley, Fox Hunting, p. 164.
11. [Surtees], 'The Swell and the Surrey', Jorrocks' Jaunts and Jollities, pp. 1–16; this passage p. 11.
12. 'Mr Jorrocks and the Surrey Staghounds', Jorrocks' Jaunts, pp. 60–75; this passage pp. 72–3.
13. 'The Swell and the Surrey', p. 3.
14. Beckford, Thoughts upon Hunting, p. 132.
15. Ibid., p. 211.
16. The Genevan Agasse studied in Paris and was brought to England by Lord Rivers, for whom he painted both sporting and natural historical subjects; Coombs, Sport and the Countryside, p. 100.
17. Cotton, 'The Meynell Hunt', in Peter Lewis, ed., A Fox-Hunter's Anthology (1934; Woodbridge, Suffolk and Dover, NH: The Boydell Press, 1985), pp. 287–90; p. 290, l. 89.
18. Peter Barnes and Rosemary Hooley, private communications to the author.

Chapter 9

1. Warner, Walk Through Wales, August 1797, p. 3.
2. Coburn, ed., Notebooks, 1 (Text): entry 1273 (November 1802).
3. Holmes, Early Visions, p. 342.
4. In Grainger, ed., Natural History Prose Writings, p. 59; also in Robinson and Powell, eds, Oxford Authors: Clare, p. 474.
5. Richard Mabey, 'Guest Editorial: Clare and Ecology', John Clare Society Journal 14 (July 1995), pp. 5–6; this passage p. 6. See also Douglas Chambers, ' "A love for

every simple weed": Clare, Botany and the Poetic Language of Lost Eden', in Haughton, Phillips, and Summerfield, eds, *Clare in Context*, pp. 238–58.
6. See Landry, 'Green Languages? Women Poets as Naturalists in 1653 and 1807', in Anne K. Mellor, Felicity Nussbaum and Jonathan Post, eds, *Huntington Library Quarterly* (forthcoming). On Dorothy Wordsworth, see Robert Mellin, ' "Some Other Ground": Dorothy Wordsworth, the Picturesque, Ecology', in Michael Branch, ed., *Critical Essays in Literature and Environment* (Boise: University of Idaho Press, 1998), pp. 81–97. I am grateful to Mellin for lending me this brilliant essay in typescript.
7. Allen, *Naturalist in Britain*, pp. 7–9.
8. Secord, 'Science in the Pub: Artisan Botanists in Early Nineteenth-Century Lancashire', *History of Science* 32: 3, No. 97 (September 1994): 269–315; this passage p. 276. See also Cash, *Where There's a Will*, pp. 1–2, 10.
9. Secord, 'Science in the Pub', p. 292.
10. Gaskell, *Mary Barton: A Tale of Manchester Life* (1848), ed. Stephen Gill (Harmondsworth: Penguin, 1970), p. 75.
11. Buxton, 'A Brief Memoir of the Author', *Botanical Guide*, pp. [iii]–xv; this passage p. vi.
12. Ibid., p. xiii.
13. *A Tour round North Wales, performed During the Summer of 1798: containing Not only the Description and local History of the Country, but also, a Sketch of the History of the Welsh Bards; An Essay on the Language; Observations on the Manners and Customs; and the Habitats of above 400 of the more rare Native Plants; intended as a Guide to future Tourists*, by the Rev. W. Bingley, BA FLS, of St. Peter's College, Cambridge, 2 vols (London: Printed by J. Smeeton [vol. 1] and T. Rikaby [vol. 2] and Sold by E. Williams and J. Deighton, Cambridge, 1800), 1: 218–19.
14. Jarvis, *Romantic Writing and Pedestrian Travel*, pp. 43–4.
15. Bingley found a sure-footed Welsh pony the next best thing to walking, as he remarked in his Preface: 'Next to being on foot, the tourist will find a horse the most useful; but in this case, if he intends to ramble much amongst the mountains, it will be necessary for him to take a Welsh poney, which, used to the stony paths, will carry him, without danger, over places where no English horse, accustomed to even roads and smooth turf, could stand with him'; *Tour round North Wales*, 1: [iii]–xvi; this passage pp. iv–v.
16. Mary Butts captures some of this sense of urban 'hikers' as trespassers – those who do not belong – but also classifies them as barbarians, 'the mass of our people, the unique barbarian we have been breeding for generations' in *Warning to Hikers* (London: Wishart, 1932), p. 17. See also Wright, *Living in an Old Country*, pp. 93–134.
17. Hucks, *Pedestrian Tour*, p. 7.
18. After Hucks and Coleridge had joined forces with Brookes and Berdmore, Hucks reported, 'One of my companions was a very skillful botanist, and his botanical furor induced him at all times to despise danger and difficulty, when in pursuit of a favourite plant'; ibid., p. 92.
19. Barrell has explored in definitive detail how Clare's sense of space derived from this unimproved landscape, which underwent enclosure during his lifetime; *Idea of Landscape*, pp. 95–109.
20. Ibid., pp. 121–2. Johanne Clare has usefully glossed Clare's use of dialect names and words by stressing that they were objected to by certain readers not because they were local per se, but because they were 'too obviously, too unabashedly

working-class'; *John Clare and the Bounds of Circumstance* (Kingston and Montreal: McGill-Queen's University Press, 1987), p. 127.
21. Bushaway, *By Rite*, p. 82.
22. Hucks, *Pedestrian Tour*, p. 6.
23. According to Jeffrey C. Robinson, no essay 'puts forward more restlessly' than Hazlitt's 'the Romantic walker's ambivalence toward entering the pleasurable idyll of the walk, first in nature and then on paper'; *The Walk: Notes on a Romantic Image* (Norman and London: University of Oklahoma Press, 1989), pp. 43–5.
24. Hazlitt, 'On Going a Journey', in P.P. Howe, ed., *The Complete Works of William Hazlitt*, 21 vols (London: J.M. Dent and Sons, 1930–34), 8: 181–9; this passage p. 187.
25. Wallace, *Walking, Literature*, p. 118.
26. White, 'An Invitation to Selborne', in *The Natural History and Antiquities of SELBORNE, in the County of Southampton*, ed. James Edmund Harting, 2nd edn (London: Bickers and Son, 1876), pp. 517–19.
27. Barrell, *Idea of Landscape*, p. 40.
28. See H. Pederson, *Linguistic Science in the Nineteenth Century: Methods and Results*, trans. J.W. Spargo (Cambridge, MA: Harvard University Press, 1931).
29. McDonagh, *De Quincey's Disciplines* (Oxford: Clarendon, 1994), pp. 86–7.
30. De Quincey, 'The Palimpsest of the Human Brain', *Suspiria De Profundis* (1845), in George Saintsbury, ed., *Confessions of an English Opium-Eater Together with their sequels* (London: Constable and Co., 1927), p. 245.
31. McDonagh, *De Quincey's Disciplines*, p. 87.
32. Kim Taplin, *The English Path* (Woodbridge, Suffolk: The Boydell Press, 1979), p. 28.
33. David McCracken, *Wordsworth and the Lake District: A Guide to the Poems and Their Places* (Oxford and New York: Oxford University Press, 1984), p. 3.
34. John Gaze, *Figures in a Landscape: A History of The National Trust* (London: Barrie & Jenkins, 1988), p. 10.
35. Adrian Phillips, 'Conservation', in Howard Newby, ed., *The National Trust: The Next Hundred Years* (London: The National Trust, 1995), pp. 32–52; this passage p. 32.
36. As reported in the *Manchester Guardian* of 7 October 1887; quoted in Hill, *Freedom to Roam*, p. 40. A better known report of the incident appears in Grosart, ed., *Prose Works*, 3: 425 and n.
37. Wallace, *Walking, Literature*, p. 117.
38. Jarvis connects walking and prosody, pedestrian poetry and blank verse, in *Romantic Writing and Pedestrian Travel*, pp. 87–8. He defends Wordsworth against Seamus Heaney's ironical charge of 'pedestrianism' on pp. 90–125.
39. To [James Augustus Hessey], [July 1823], in Storey, ed., *Letters*, p. 279.
40. Robinson, ed., *Clare's Autobiographical Writings*, p. 65.
41. See Ronald Blythe's inspiring essay, ' "*Solvitur ambulando*": John Clare and Footpath Walking', *John Clare Society Journal* 14 (July 1995): 17–27. Robin Jarvis briefly notes the relationship between perambulation and Clare's poetics in *Romantic Writing and Pedestrian Travel*, pp. 159–62, 176–91.
42. W. John Coletta defends Clare's ecologically grounded figuration as 'every bit as complex as the transcendental signifiers of a Shelley' in 'Ecological Aesthetics and the Natural History Poetry of John Clare', *John Clare Society Journal* 14 (July 1995): 29–46; this passage p. 38. See also James C. McKusick's ' "A Language that is ever green": The Ecological Vision of John Clare', *University of Toronto Quarterly* 61 (1991): 226–49.

43. Clare, 'Autumn', Northampton Ms 6, in Grainger, ed., *Natural History Prose Writings*, pp. 329–30.

Chapter 10

1. Hazlitt, 'My First Acquaintance with Poets', in Howe, ed., *Complete Works*, 17: 106–22; this passage p. 119.
2. Coburn, ed., *Notebooks*, 1 (Text): entry 1610 (24 October 1803).
3. Hazlitt, 'My First Acquaintance', p. 113.
4. *Notebooks*, 1 (Text): entry 1610.
5. *Ibid.*, entry 1610.
6. Modiano, *Coleridge and the Concept of Nature* (London and Basingstoke: Macmillan, 1985), pp. 37–8.
7. Craig, *Native Stones*, pp. 127, 130, 135.
8. Letter to Sara Hutchinson, Eskdale, Friday, 6 August 1802, in Griggs, ed., *Collected Letters*, 2: 841. See also *Notebooks*, 1 (Text): entries 1207–22.
9. Craig, *Native Stones*, p. 130.
10. Letter to Sara Hutchinson, 6 August 1802, in Griggs, ed., *Collected Letters*, 2: 841–3. The words in brackets, 'to my Navel', were inked out in the MS.
11. Holmes, *Early Visions*, p. 54.
12. Kelvin Everest, *Coleridge's Secret Ministry: The Context of the Conversation Poems 1795–1798* (Hassocks, Sussex: Harvester and New York: Barnes and Noble, 1979), p. 170. Robin Jarvis calls this poem Coleridge's 'finest piece of peripatetic verse', *Romantic Writing and Pedestrian Travel*, p. 148.
13. Fulford, *Landscape, Liberty*, p. 229.
14. Sara Fricker Coleridge, who went on the walk, was originally included as an addressee of the poem – 'my gentle-hearted Charles . . . My Sara and my Friends' – but was later dropped from the published text, as if, as Richard Holmes puts it, 'she had ceased to share in the experience'; *Early Visions*, p. 153.
15. Addison, *Spectator* No. 411 (Saturday, 21 June 1712), Bond, ed., 3: 538.
16. Holmes, *Early Visions*, p. 154.
17. Bond, ed., 3: 538.
18. Fulford, *Landscape, Liberty*, p. 243. Fulford describes as 'spiritual' what I call 'social', but I am in general agreement with his reading of Romanticism, and Coleridge in particular, as owing much to such eighteenth-century precedents as Thomson and Cowper.
19. Anne K. Mellor observes that the dell is picturesque, the hilltop beautiful and the sunset sublime, and that 'the ash, by the way, had been singled out by Gilpin as a particularly picturesque tree' in *Forest Scenery* (1791), pp. 32–3; 'Coleridge's "This Lime-Tree Bower My Prison" and the Categories of English Landscape', *Studies in Romanticism* 18:2 (Summer 1979): 253–70; this passage p. 257. I read the picturesque, in Gilpin but also in Uvedale Price and Richard Payne Knight, as a category designed to slide among and capture all three modalities of beauty and feeling.
20. By 27 March 1801, William Wordsworth and Coleridge had a copy of Withering and a botanical microscope each. See the letter from William to Messrs Longman and Rees in de Selincourt, ed., *Early Letters*, p. 265. Sometime after the books arrived, Coleridge allowed Sara Hutchinson to transcribe Withering's index of common English names into his notebook. Sara began the list with '*Adders Tongue*', although Withering's index included several plants ahead of it alphabetically; see *An*

Arrangement of British Plants; According to the latest Improvements of the Linnaean System. To which is prefixed, an easy Introduction to the Study of Botany, by William Withering, MD FRS, 3rd edn, 4 vols (Birmingham: Printed for the Author by M. Swinney; and Sold by G.G. and J. Robinson and B. and J. White, London, 1796), 3: 891. The dating of the entry is unclear – between 8 December 1800 and February 1801; see Coburn, ed., *Notebooks*, 1 (Text): entry 863.

21. Botany was associated with Coleridge's guilty passion for Sara Hutchinson, who was Wordsworth's sister-in-law, and with his vexed, often competitive friendship with Wordsworth. To Withering's 'Mouse-ear' in the transcribed index, Sara added '(= Forget me not)', an insertion 'charged with considerable emotional content', as Coburn notes; *Notebooks*, 1 (Notes): entry 863. Modiano connects Coleridge's relation to nature, his friendship with Wordsworth, and his love of Sara Hutchinson in *Coleridge and the Concept of Nature*, pp. 4, 29–50, 204–5.
22. E.H. Coleridge gives the date of 'Melancholy: A Fragment' as ?1794, presumably on internal evidence. Coleridge himself says, in a letter to William Sotheby, that he was 19 when he wrote these lines, which would mean the poem was written in 1791; Letter to Sotheby, Thursday 26 August 1802, in Griggs, ed., *Collected Letters*, 2: 856.
23. See Richard and Alastair Fitter, *The Wild Flowers of Britain and Northern Europe*, 4th edn (London and Glasgow: William Collins Sons and Co., 1985), p. 274.
24. Letter to Sotheby, Thursday, 26 August 1802, in Griggs, ed., *Collected Letters*, 2: 856.
25. E.H. Coleridge, ed., *Poetical Works*, p. 74, n. 1.
26. *The Annual Anthology, Volume II* (Bristol: Printed by Biggs and Co. for T.N. Longman and O. Rees, London, 1800), p. 141, n.
27. Judith Pascoe, 'Female Botanists and the Poetry of Charlotte Smith', in Carol Shiner Wilson and Joel Haefner, eds, *Revisioning Romanticism: British Women Writers, 1776–1837* (Philadelphia: University of Pennsylvania Press, 1994), pp. 193–209; this passage p. 201.
28. Mary Moorman, ed., *Journals of Dorothy Wordsworth*, 2nd edn (Oxford and New York: Oxford University Press, 1971), p. 128, and n. 4.
29. Everest, *Coleridge's Secret Ministry*, p. 171.
30. *Ibid.*, p. 171.
31. *Notebooks*, 1 (Text): entry 1610.
32. 'While Coleridge's poetry of these months brings the entire Quantocks landscape alive – the heath, the combes, the woodlands, the dells, the waterfalls, the hidden streams, Wordsworth's imagined countryside more often returns to the Lake District of his childhood. It was Coleridge alone who established the Quantocks as his poetic kingdom'; Holmes, *Early Visions*, p. 196.
33. Lamb wrote to Wordsworth on 30 January 1801, 'Separate from the pleasure of your company, I don't much care if I never see a mountain in my life. I have passed all my days in London, until I have formed as many and intense attachments as any of you mountaineers can have done with dead nature . . . So fading upon me from disuse, have been the "Beauties of Nature", as they have been confinedly called; so ever fresh and green and warm are all the inventions of men and assemblies of men in this great city'; Henry H. Harper, ed., *The Letters of Charles Lamb*, 5 vols (Boston, MA: The Bibliophile Society, 1906), 2: 298–9.
34. Etherege, *The Man of Mode; or, Sir Fopling Flutter*, in Brice, ed., *Restoration Plays*, Act 5.2, p. 242.
35. [Pennant], *British Zoology*, p. 76.
36. Gilpin, *Forest Scenery*, pp. 293, 295.

37. R.P. Knight, *The Landscape, A Didactic Poem. In Three Books*, Addressed to Uvedale Price, Esq. (London: Printed by W. Bulmer and Co. and Sold by G. Nicol, 1794), 1: 343, n.
38. Richard Jefferies, *Wild Life in a Southern County* (1879; London, Edinburgh, Dublin, and New York: Thomas Nelson & Sons, n.d.), pp. 271–2.
39. Mellor notes that 'the bridging flight of the rook spatially joins what Coleridge had already imaginatively united: himself and Charles Lamb; the lime-tree bower and the sublime sunset; the visible and the invisible . . .'; 'Coleridge's "Lime-Tree Bower" ', p. 269.

Chapter 11

1. DNP Authority, *Dartmoor National Park Management Plan*, p. 32.
2. *Ibid.*, p. 32.
3. Marion Shoard, 'The Lure of the Moors', in John R. Gold and Jacquelin Burgess, eds, *Valued Environments* (London: George Allen & Unwin, 1982), pp. 55–73; this passage p. 58.
4. Quoted in *ibid.*, p. 57.
5. *Ibid.*, p. 59.
6. Atkinson, *Dartmoor National Park Plan*, p. 116.
7. *Ibid.*, p. 112.
8. Shoard, 'Lure of the Moors', p. 60.
9. Quoted in *ibid.*, p. 60.
10. Helen Read, in *Veteran Trees: A Guide to Good Management* (Peterborough: English Nature, 2000), defines veterans as 'trees of interest biologically, aesthetically or culturally because of their age; trees in the ancient stage of their life; trees that are old relative to others of their species' (p. 13). The Woodland Trust's 'Position Statement 12. Ancient Woodland' (October 1999) specifies that in England and Wales 'ancient woods are those where there has been continuous woodland cover since at least AD 1600. Before this planting was uncommon, so a wood present in AD 1600 was likely to have developed naturally' (p. 1). In Scotland ancient woods are dated from AD 1750 because of a scarcity of accurate earlier maps.
11. Atkinson, *Dartmoor National Park Plan*, p. 36.
12. Fiennes, *Illustrated Journeys*, p. 213.
13. Defoe, *A Tour Through the Whole Island of Great Britain* (1724), eds P.N. Furbank and W.R. Owens (New Haven and London: Yale University Press, 1991), pp. 91–4.
14. *Ibid.*, p. 113.
15. Fuller, *The History of the Worthies Of England* (1662), ed. John Nichols, new edn, 2 vols (London: Printed for F.C. and J. Rivington et al., 1811), 1: 270.
16. Marshall, *The Rural Economy of the West of England: including Devonshire; and Parts of Somersetshire, Dorsetshire, and Cornwall. Together with Minutes In Practice*, 2 vols (London: Printed for G. Nicol, G.G. and J. Robinson and J. Debrett, 1796), 2: 19–20.
17. *Ibid.*, 2: 24.
18. J. Laskey, 'Three Days' Excursion on Dartmoor' (1795–96), in George Laurence Gomme, ed., *The Gentleman's Magazine Library: being a classified collection of the chief contents of the Gentleman's Magazine from 1731 to 1868*, English Topography, Part III (Derbyshire–Dorsetshire) (London: Elliot Stock, 1893), pp. 109–27; this passage p. 127.

19. Warner, *A Walk through some of the Western Counties of England* (Bath: Printed by R. Cruttwell; Sold by G.G. and J. Robinson, London, 1800), p. 160.
20. *Ibid.*, pp. 168, 169, 172.
21. In his *A Tour through Cornwall, in the Autumn Of 1808* (Bath: Printed by Richard Cruttwell; Sold by Wilkie and Robinson, London, 1809), Warner deviated from the pedestrian mode, hiring an 'Exeter hack' for part of the journey. The book closes with praise for equine virtues, including enhancing 'the pleasure of our sports', and accelerating 'the transactions of our business; but, what is more than all', the horse 'conveys us, with rapidity and safety, when separated from it by distance, into the bosom of that family, without whose participation no enjoyment can be complete', p. 361.
22. Fuller, *Worthies*, 1: 292.
23. Baring-Gould, *A Book of Dartmoor* (London: Methuen, 1900), p. 223.
24. Stephen H. Woods, *Dartmoor Stone* (Exeter: Devon Books, 1988), p. 83.
25. *Ibid.*, p. 83.
26. Baring-Gould, *Book of Dartmoor*, p. 223.
27. *Ibid.*, p. 223. Based on Dick Lloyd's findings, Ted Hughes argued that 'the spell of the hunt' has protected Exmoor's red deer herd over the centuries, and the same principle applies to areas where fox-hunting protects and promotes the fox population. In its absence, farmers and other country people 'revert to their common sense, and their market economy' where the only virtue of such animals 'is to be dead' and converted to cash; 'The Hart of the Mystery', *The Guardian*, Saturday, 5 July 1997, p. 21.
28. Coburn, ed., *Notebooks*, 1 (Notes): entry 454.
29. Coleridge, Letter to Thomas Poole, 16 September 1799, in Griggs, ed., *Collected Letters*, 1: 527–8.
30. Coleridge congratulated himself on having stuck to his food rather than engaging in quarrels with his relations: 'What occasion for it? – Hunger & Thirst – Roast Fowls, mealy Potatoes, Pies, & Clouted Cream – bless the Inventors thereof! An honest Philosoph may find therewith preoccupation for his *mouth*'; *ibid.*, p. 528.
31. See Underdown, *Revel, Riot, and Rebellion*, p. 73.
32. Stoyle, *Loyalty and Locality*, p. 17.
33. *Ibid.*, pp. 16–17.
34. The late fifteenth and early sixteenth centuries 'were truly the heyday of the Dartmoor tinning industry, which reached its peak production of 252 tons in the year 1524'; Helen Harris, *The Industrial Archaeology of Dartmoor* (1968; Newton Abbot: David & Charles, 1972), pp. 19, 37–66.
35. Ministry of Defence (MOD) pamphlet, *Dartmoor Armed Services Day 1996*, p. 9. My thanks to Colonel Tony Clark for this reference.
36. Harris, *Industrial Archaeology*, p. 20.
37. As with mountains, so also with moorland; lingering traces of fear remained part of the attraction of landscapes of 'delightful horror'; Simon Schama, *Landscape and Memory* (New York: Vintage, 1996), pp. 447–513.
38. Browne, *Britannia's Pastorals* (1613–16) (Menston: Scolar Press, 1969), pp. 1–2.
39. Drayton, 'The first Song', *Poly-Olbion*, pp. 12–13.
40. Cristall, *Poetical Sketches* (London: J. Johnson, 1795).
41. McGann, 'Conclusion: Starting from Death: The Poetry of Ann Batten Cristall', *The Poetics of Sensibility: A Revolution in Literary Style* (Oxford: Clarendon, 1996), pp. 195–206; this passage p. 203.

42. *Ibid.*, pp. 204–5. McGann states that the only known copy of Cristall's *Poetical Sketches* is the British Library copy (p. 195). The text is now available online from the Electronic Text Center of the University of Virginia Library.
43. Felicia Hemans, *Dartmoor*, in *Poems of Felicia Hemans* (Edinburgh: William Blackwood, 1852), and Joseph Cottle, *Dartmoor, and Other Poems* (London: Printed for T. Cadell by T.J. Manchee, Bristol, 1823). Nanora Sweet has analyzed Hemans's prize-winning poem as an 'attempt to play laureate to Britain's post-Napoleonic settlements', in 'History, Imperialism, and the Aesthetics of the Beautiful: Hemans and the Post-Napoleonic Moment', in Mary A. Favret and Nicola J. Watson, eds, *At the Limits of Romanticism: Essays in Cultural, Feminist, and Materialist Criticism* (Bloomington and Indianapolis: Indiana University Press, 1994), pp. 170–84; this passage p. 182.
44. Carrington, *Dartmoor: A Descriptive Poem* (London: Hatchard and Son and Devonport: R. Williams, 1826), p. cviii.
45. 'Although since the publication of the Perambulation many guides to the "wild and wondrous region" have appeared, none have been found so useful, none so full of information as this book'; J. Brooking Rowe, ed., 'Editor's Preface', *A Perambulation of the Antient and Royal Forest of Dartmoor and the Venville Precincts or a Topographical Survey of their Antiquities and Scenery* (1848), by the late Samuel Rowe, MA (Exeter: James G. Commin and London: Gibbings and Co., 1896), pp. ix–xii; this passage p. ix.
46. Baring-Gould, 'Preface', *Book of Dartmoor*, pp. ix–xii; this passage p. xi.
47. Dixon, *Castalian Hours. Poems* (London: Longman, Rees, Orme, Brown, and Green, 1829).
48. Holmes has made a persuasive case for Culbone Combe seen from Ash Farm as the origin of the 'erotic, magical geography' of 'Kubla Khan': 'Between the smooth curved flanks of the coastal hills, a thickly wooded gulley runs down to the sea (the "romantic chasm") enclosing a hidden stream which gushes beneath the tiny medieval chapel of Culbone, a plague-church and "sacred site" since Anglo-Saxon and possibly pre-Christian times'; *Early Visions*, p. 164, n.
49. My thanks to Kevin Sharpe for this formulation.
50. *A Journal of Ten Days Excursion on the Western And Northern Borders of Dartmoor*, By Miss Dixon, Author of 'Castalian Hours', &c, &c. (Plymouth: Printed by J. Williams, Old-Town, 1830) and *A Journal of Eighteen Days Excursion on the Eastern And Southern Borders of Dartmoor, and on the Western Vicinity of Exmoor; including Ilfracombe, Lynton*, &c., By Miss Dixon, Author of 'Castalian Hours', &c, &c. (Plymouth: Printed by T.J. Bond, 10, Whimple-Street, 1830).
51. Signed 'a. w.', this note appears before the title-page in Bodleian shelfmark G.A. Devon 8@ 584. Stamped on the inside front cover is 'Mr. Alfred Wallis, Regent's Park, Exeter', which appears again on the facing page, over a brown ink inscription dated 1835.
52. Dixon, *Western and Northern Borders*, entry for Friday, 21 May 1830, describing arrival in 'Oakhampton', p. 21.
53. Dixon, *Eastern and Southern Borders*, p. 49.
54. Newby, *Country Life*, pp. 203, 236.
55. *Ibid.*, p. 236.
56. Interviews conducted during December 1991 and January 1992, between May and August 1992, and between November 1999 and April 2000, with some Graziers of the Dartmoor Commoners Association, not all of whom approve of fox-hunting.
57. Without the spell of the hunt to provide symbolic compensation, farmers revert to economic self-interest, just like everybody else, and either shoot species regarded

as agricultural pests or fail to protect them. Deer must be tolerated and fed, hares must be saved from combine harvesters and hay-cutting machines. Because hares have their young above ground, 'Mechanization on the farm struck the hare much more severely than the rabbit which had its young beneath ground'; John Sheail, *Rabbits and Their History* (Newton Abbot: David & Charles, 1971), pp. 29–30.

58. In *The Nature of Dartmoor: A Biodiversity Profile* (Bovey Tracey: English Nature and Dartmoor National Park Authority, 1997), some 'key species for conservation attention' are the otter, dormouse, greater horseshoe bat, red grouse, golden plover, skylark, ring ouzel, salmon, blind freshwater shrimp, high brown fritillary, heather, wild daffodil and string-of-sausages lichen. Species were included on the grounds of being 'endemic to the UK', or 'threatened' or 'declining' in Great Britain or 'on a global or European scale', or 'highly characteristic of Dartmoor' and 'popular with the general public', p. 12.

Index

Abergavenny, Lord   196
Acteon   38
Adams, Carol J.   272 n. 11
Addison, Joseph   15, 17, 57, 153, 192, 211, 223
Aesop   8, 41
Agasse, Jacques-Laurent   284 n. 16
  Sleeping Fox   201
agrarian and agricultural improvement   5–6, 54–61, 169, 173–6, 209, 217–18, 244, 282 n. 37, 285 n. 19
agrarian complaint   56, 79, 249 n. 36
Agricultural Revolution   xv, 55–6, 176, 228, 262 n. 8, n. 9
  first or yeomen's   9–10, 56
  second or landlords'   9–10, 56, 57, 119–20
Aikin, John   31–2, 113–14, 185, 189, 192, 193
Aldington, John   118
Aleppo   174
Alfoxden   137–8, 223, 224
Allen, Robert C.   9, 56
Alvanley, Lord   173
amenity value of landscape   10, 11, 54–5
Ancient Countryside   47, 65, 254 n. 139
Andrews, Frances   77–8
Andrews, Robert   77–8
Andrews, Robert (senior)   77
Andromache   153, 279 n. 35
animal fables   xv, 8, 103
animal rights   xv, 8–9, 29, 119–25, 133–5, 152–3, 250 n. 49, 255 n. 1, 272 n. 11, 273 n. 29, 276 n. 97
animals   50–2, 255 n. 2, 261 n. 75, 272 n. 11
  changing human attitudes towards   7–9, 113–25, 133–5, 249 n. 46, 255 n. 1, 276 n. 97, 277 n. 98
Anne, Queen   175
Annual Anthology   222
anthropomorphism   260 n. 69
anti-cruelty   xiii, xv, 9, 51–2, 118, 120–4, 131–2, 133–6, 141–2, 190, 249 n. 36, n. 48, 250 n. 51, 260 n. 74, 272 n. 9, 279 n. 33, 283 n. 64

anti-slavery   xiii, 8, 118, 141, 273 n. 20
Antrobus, Ann   271 n. 64
Apperley, Charles James (Nimrod)   13, 93, 169, 171, 176–8, 230
  The Life of a Sportsman   133, 177–8, 191–2, 194
Armstrong, John, MD   130–1, 182
Ash Farm   291 n. 48
Ashbrook, Kate   231
Ashworth, William   32
Atherstone Hunt   46, 198
Aubrey, John   235
Auburn   119, 120
Auden, W. H.   168
Aurengzebe   187, 193
Austen, Jane   123
Austin, Rebecca   255 n. 147
Averill, James H.   276 n. 80

Badminton   146
Bagwell, Philip S.   126
Bakewell, Robert   9
Balliol College, Oxford   141
Band of Mercy   124
Barclay, Captain   132, 133
Baret, Michaell   108
Baring-Gould, Rev. Sabine   234, 237
Barker, Thomas   209
Barnstaple   145, 146
Barrell, John   1, 4–5, 47, 209, 212, 271 n. 67, 285 n. 19
Barrington, Daines   33
Barrington, James   255 n. 147
Bate, Jonathan   256 n. 7
Bath   209
Bathurst, Allen, first Earl   117
Battigelli, Anna   278 n. 28
Bayliss, Tom   100
Beattie, James   131–2, 238
beagles   xiv, 15, 104, 151, 159, 194, 259 n. 44, 279 n. 46
beating of the bounds   5, 217–18
Beaufort, Henry, third Duke of   65
Beaufort, Mary Somerset, Duchess of   146

293

Beaufoy, Commander Mark   269 n. 21
Beckford, Peter   13, 185, 189, 194, 200
  Thoughts upon Hunting   63–4, 177, 178–9, 183–4, 185–6, 189–91, 198, 201, 259 n. 44
Beckford, William
  Vathek   63
Bell, Robert   267 n. 60
Bellamy, Liz   69, 264 n. 60
Belvoir Hunt   46
Benton, Ted   29–30, 256 n. 3, 272 n. 9
Mr. Berdmore   132, 285 n. 18
Berger, John   7, 115
Berkeley, Henry, Lord   145
Berkeley, Lady (Katharine Howard)   145–6
Berkshire   47, 66
Bermingham, Ann   77, 78
Berners, Dame Juliana   258 n. 31
Billesdon Hunt   46
biodiversity   xvii, xviii, 142, 292 n. 58
Biodiversity Action Plan   15
bioregion   256 n. 4
Billesdon Coplow run of 1800   172
Bingley, Rev. William   208, 219, 272 n. 14, 285 n. 15
Birkett, Jim   133, 275 n. 78
Birmingham   179
Black Act of 1723   6
Blackator   241
Blackstone, William   2, 74
Blain, Virginia   195
Blair, Tony   xiii
Blake, William   139, 276 n. 87
Blencathra Hunt   49
Blenheim   70
Blome, Richard   12, 96, 107, 108
Bloodgood, Lida Fleitmann   165
bloodhounds   62–3, 241
Bloody Shoulder'd Arabian   174, 282 n. 35
Bloomfield, Robert   21, 86, 88–91, 173
  The Farmer's Boy   2–3, 17, 47–8, 61, 63
Blount, Martha   117
Blount, Teresa   117
Blunt, Wilfrid Scawen   184
Blythe, Ronald   286 n. 41
Bodleian Library   240, 291 n. 51
The Boke of St. Albans   36, 258 n. 31
Bolingbroke, Lord (Henry St. John)   5
Bolton, James   146
Book of Sports   94, 95

Bookchin, Murray   45, 53, 260 n. 56
Borrodale   220–1, 226
Borsay, Peter   170, 199, 247 n. 11, 281 n. 11
botany   146–7, 206–9, 217–18, 224–6, 263 n. 23, 285 n. 18, 287 n. 20, 288 n. 21, n. 27
Bowen, Melesina
  Ystradffin   92–4, 135, 140
Bradlaugh, Charles   84
Bramton Castle   157
Brantôme   165
Brewer, John   76, 188
Bridgeman, Charles   67, 68, 70
Brighthelmstone Downs   125
Bristol   57, 141
British Library   291 n. 42
Britishness   xv, 1, 247 n. 2
  rivalry with France   13, 15, 21–2, 247 n. 2
Broad Street, St. Giles   161
Mr. Brookes   132, 285 n. 18
Brotton, Jerry   250 n. 60
Brown, Launcelot 'Capability'   70
Browne, William   236
  Britannia's Pastorals   236
Browning, Elizabeth Barrett   212
Buchan, John   23, 104, 208
Buckinghamshire   108, 130, 147
Mr. Budd   47
Budgell, Eustace   40, 259 n. 44
Budiansky, Stephen   52
Budworth, Captain Joseph   125
Bulstrode   147
bulldogs   176, 241, 283 n. 39
Bunce, Michael   10
Burchardt, Jeremy   245 n. 2, 247 n. 1
Burdett, Sir Francis   69
Burghley   11, 70
Burnham   108, 130
Burnham Beeches   130
Burns, Lord, Inquiry into Hunting with Dogs   24
Burns, Robert
  'On Seeing a Wounded Hare Limp By Me Which a Fellow Had Just Shot At'   135, 275 n. 79
Burt, W., Esq.   237
Bushaway, Bob   5
Butts, Mary   285 n. 16
Buxton, Richard   83, 207
Byerley Turk   174
Byron, Lord (George Gordon)   238

Caer Idris 127
Cain, P.J. 77
Calvert, Raisley 135
Cambridge 108, 127, 276 n. 93
Cambridgeshire 49
Camden, William 232
Cardiganshire 137, 205
Carew, Bampfylde-Moore 109–10
Carlyle, Thomas 212
Carr, Raymond 14, 176
Carrington, T. 127
Carson, Rachel 245
Cartmill, Matt 257 n. 26
Chagford 233, 234
Chalker, John 176, 181
Chambers, Douglas 284 n. 5
Charles I xiv, 94, 104–6
Charles II 12, 41, 106
Charlie Fox 91
Charlton Hunt 14, 172–3
Chartier, Roger 272 n. 5
Chatterton, Thomas 87
Chaucer, Geoffrey
  'The Nun's Priest's Tale' 40
Chelsea 146
Chelsea Physic Garden 207
Chenevix Trench, Charles P. 251 n. 80, 269 n. 23
Chester 132
Cheyne, George 116
Child of Plimstock 234
Christ Church, Oxford 183
Cirencester 117
Civil War 3, 6, 41, 56, 76, 105–6
Clare, Johanne 285 n. 20
Clare, John 10, 11, 17, 21, 23, 68, 83, 86–8, 111, 136, 205, 206, 209–10, 216–18, 225, 257 n. 10, 267 n. 67, 285 n. 19, 286 n. 41
Clare College, Cambridge 274 n. 58
Clarion Ramblers of Sheffield 126
Clarke, Mrs. J. Stirling 159–60
Claude Glass 67, 231
Cleveland, Anne Pulteney Fitzroy, Duchess of 166, 280 n. 73
climbing 133, 221–2, 275 n. 78
Clipston-Park 148–9
Clogwyn Du'r Arddu 208
Cobbett, William 11, 44–8, 53, 82, 86, 88, 98, 99, 170, 177, 259 n. 51, 260 n. 61
  Rural Rides 45–7

Cockaine, Sir Thomas
  A Short Treatise of Hunting 41, 258 n. 31
Cohen, Ralph 154, 156
Coleman, Thomas 109
Coleridge, Hartley 235
Coleridge, John Taylor 213
Coleridge, Samuel Taylor 8, 31, 127, 132, 134–5, 140–1, 171, 205, 206, 209, 212, 220–9, 234–5, 275 n. 63, 276 n. 97, 277 n. 98, 285 n. 18, 287 n. 18, n. 20, 288 n. 21, n. 32, 289 n. 39, 290 n. 30
  'Kubla Khan' 238–9, 291 n. 48
  'Melancholy: A Fragment' 224–5, 288 n. 22
  The Rime of the Ancient Mariner 141–2, 276 n. 97, 277 n. 98
  Sibylline Leaves 225
  'To a Young Ass' 8, 141, 250 n. 50
  'This Lime-Tree Bower My Prison' 140–1, 222–9
Coleridge, Sara Fricker 223, 287 n. 14
Coles, Charles 78
Coletta, W. John 286 n. 42
Colley, Linda 169, 247 n. 2
Combe, William
  Doctor Syntax in Search of the Picturesque 20
Commandeur, Boudewijn F. 280 n. 60
common rights 65–6, 78–80
Commonwealth 56
Congreve, William
  The Way of the World 125–6
conservation 30, 231–2, 243–4, 292 n. 58
  and hunting xiv, xvi-xvii, xviii, 24, 53, 54, 68–9, 243–4, 255 n. 147
  and shooting 24, 53, 54, 69–72, 255 n. 147
Cookham 147–8, 278 n. 18
Coombs, David 277 n. 2, 281 n. 12, 282 n. 13
Cormack, Malcolm 78
Cornwall 232, 236
Cotehele 241–2
Cotswold hound 194
Cottesmore Hunt 46
Cottle, Joseph 237
Cotton, Frederick
  'The Meynell Hunt' 201
'the country' 1–2
'the countryside' 1–2, 4, 17, 247 n. 1

Index

Countryside and Rights of Way Act 2000 xiii, 245 n. 1
Countryside Commission 125
Countryside March xviii
Countryside Rally xvi
coursing (with greyhounds) xiv, 11, 37, 43–4, 75, 98–102, 178, 185, 270 n. 41, 275 n. 79
Coursing Club 100
Cowper, William 120–4, 142, 275 n. 79, 287 n. 18
  The Task 55–6, 58–60, 63, 120, 122–4, 131, 135
Cox, Nicholas 39, 96
Craig, David 221–2
Cranmere Pool 233
Cristall, Ann Batten 236–7, 239, 291 n. 42
Crockernwell 232
Cromwell, Oliver xiv, 56, 104, 105
crossbow xiv, 145, 229
Crowe, William
  Lewesdon Hill 129–30, 275 n. 63
Crowther, James 83
cucumbers 58–9
Culbone 235, 291 n. 48
Culbone Combe 291 n. 48
Cumberland, Margaret Clifford, Countess Dowager of 147–8, 278 n. 18
Cumberland, Duke of 160
Cumbrian Fells 14, 48, 49, 133, 221–2, 260 n. 71

Danby, James F. 276 n. 80
Daniel, Rev. William B. 169–70, 194
Daniels, Stephen 47, 262 n. 7
Dargue, Edwin 260 n. 71
Darley Arabian 174
Dart river 236, 237
Dartmoor 226, 230–44, 292 n. 58
Dartmoor Commoners Association 291 n. 56
Dartmoor National Park Authority 230, 232
Dartmoor Prison 10, 236, 237
Dartmouth 234, 235
Darwin, Charles 38
Davidoff, Lenore 263 n. 24
Davis, W. H.
  Colonel Newport Charlett's favourite Greyhounds at Exercise 100
De Bruyn, Frans 17
de Coverley, Sir Roger 2, 40

deer 6, 12, 36, 38, 41, 55, 64, 70, 74, 81, 106, 120, 149, 234, 244, 264 n. 60, 288 n. 27, 291 n. 57
de Foix, Gaston (Gaston Phoebus)
  Livre de chasse 36
  Maître de la Chasse 36
deforestation 5, 55–6, 123, 262 n. 7
Deleuze, Gilles and Félix Guattari 280 n. 58
de Medici, Catherine 165
Denham, Sir John
  Coopers Hill 35, 36, 104–6, 117–18, 128–9, 151, 176, 209
Derbyshire 41
De Quincey, Thomas 213
Deuchar, Stephen 78, 80, 99, 139, 270 n. 34
Devon 86, 109–10, 230, 232, 235
Devonshire, Duke of 172–3
Diana 38
Diggers 6, 25, 76, 248 n. 30
Dishley Grange Farm 9
Dixon, Sophie 237–43
Dobson, Andrew 256 n. 6
Dodsley, Robert
  Agriculture 64, 263 n. 40
dogs 92, 106–12, 140, 175–6, 241–2, 261 n. 83
Dorset, Anne Clifford, Countess of 148
Dorsetshire 66
Dove Cottage, Grasmere 135, 137
Downton 47
Drayton, Michael 169
  Poly-Olbion 36, 42–4, 236
dressage 52
Driden, John 103
Dryden, John 57, 90, 102, 103–4
Dublin 179
du Fouilloux, Jacques
  La Vénerie 36, 98
dung 17, 55, 59–61, 263 n. 26
Dunlop, Mrs. 275 n. 79
Dürer, Albrecht 165
Durham 171
Dyck, Ian 44–5
Dyer, John
  The Fleece 58–9, 63
  Grongar Hill 129

Earl of Abergavenny 216
East India Company 216, 223
ecocentrism 246 n. 19

ecology   xvii, 53, 91–2, 118, 124, 142, 207, 217, 229, 231, 232, 261 n. 81, 261 n. 5, 286 n. 42
  and hunting   xiv, 29–53, 54, 68–9, 207, 217–18, 243
  as distinct from environmentalism   30, 256 n. 6
  ecological theories   29–30, 256 n. 4, 276 n. 98
  deep ecology   52–3
  social ecology   45, 53
Edward II   36
Edwards, Lionel   169
Edwards, Peter   250 n. 59
Ehret, G. D.   147
Elizabeth I   17, 36, 145
'Enclosure-Act Myth'   24
English Nature   243, 289 n. 10
Englishness   xv, 1, 13, 70, 268 n. 1
An Essay on Hunting   168–9
Eridge Hunt   196
Erskine-Hill, Howard   273 n. 17
Escott, John   109
Etherege, Sir George
  The Man of Mode   228
Eton College   183
Euston Hall   47
Everest, Kelvin   225–6
Everett, Nigel   55, 70
Exeter   232, 234, 290 n. 21
Exe river   232
Exmoor   230, 232, 234, 238, 290 n. 27

Fairer, David   269 n. 14
Farnham, Hampshire   3
Fenwick, Isabella   137
Ferguson, Moira   249 n. 48
Ferneley, John   169
Fernie Hunt   46
ferrets   110
field sports   97–104
Fielding, Henry
  Joseph Andrews   7
  Tom Jones   7, 266 n. 41
Fiennes, Celia   22, 46, 65, 66, 69, 100–1, 169, 232
Fillis, James   165
Finch, Lady Isabella   158, 160, 161
fishing   23, 44, 97
Fitzgerald, Rev. Gerald   102–3, 107
Fonthill   63
Foreman, Dave   29, 52–3
Forsyth, Alexander   71, 97

Foucault, Michel   249 n. 46
Fox, Charles James   91
foxes   49–50, 53, 64, 72, 84, 88, 103–4, 185, 201, 244, 260 n. 71, 263 n. 37
fox-hunting   41, 52, 68–9, 100, 103–4, 108–9, 124–5, 154–7, 161, 163, 187–92, 201, 217–18, 234, 240–1, 246 n. 19, 291 n. 56
  and class antagonism   xv, 46
  as 'invented' tradition   14, 179
  as vermin control   xiv, 40, 74, 185
  coterie language of   14, 176–9
  foot-followers   47–8, 173
  modern mounted   xiv, xv, 12–15, 120, 169, 172–9, 185, 199–201, 218–19, 240–1
  subscription packs   14, 47, 170, 270 n. 41
  symbolic power of   xv, 13, 35–42, 91, 201
  uniform dress for   14, 176, 217–18
  with footpacks   14–5, 40, 48, 49, 133
Foxley   47
Franck, Richard
  Northern Memoirs   97
Frederick, Prince of Wales   77, 180–1, 283 n. 55
Freeport, Sir Andrew   2
French Revolution   3
Fudge, Erica   8
Fulford, Tim   116, 222, 223, 262 n. 7, 275 n. 63, 287 n. 18
Fuller, Thomas   232, 234

Gagnier, Regenia   267 n. 63
Gainsborough, Thomas
  Mr and Mrs Robert Andrews   77–8, 108
Galen   180
game   6, 53, 54, 74, 77–8, 81, 99–100, 261 n. 83, 278 n. 20
gamekeepers   49, 80, 209, 260 n. 71, 266 n. 41
Game Act of 1671   1, 4, 5, 73–6, 80–4, 265 n. 1
  repeal in 1831   1, 6, 74
Gascoigne, George   53, 55, 258 n. 31
  The Noble Arte of Venerie or Hunting   35–42, 98, 153, 258 n. 27
Gash, Norman   195
Gaskell, Elizabeth   207
Gee, John and Pamela   260 n. 71
Gentleman's Magazine   121, 129

298 *Index*

gentry   4–6, 10, 12, 46, 64–72, 75–8, 248 n. 30
George II   180, 193
  as Prince of Wales   158
George III   12, 251 n. 76, 259 n. 44, 281 n. 11
George IV
  as Prince of Wales   160–1, 163
georgic poetry   16, 47, 57–8, 179–94, 198, 214
  and empire   57, 58–9, 175–6, 184, 188
  georgic ethos   30
Gerrard, Christine   181
Gidleigh   234
Gidleigh Park   233
Gilpin, William   19–20, 47, 72, 139, 211, 228–9, 240, 252 n. 102, 287 n. 19
  Three Essays: on Picturesque Beauty; on Picturesque Travel; and on Sketching Landscape   21, 213, 240
Glasgow   179
Glastonbury   234
Glen, Heather   138–9
Glenridding Walks   213
Glisson, Francis   149
Godolphin Arabian   174
Goldschmidt, S. G.   52
Goldsmith, Oliver
  'The Deserted Village'   119–20
Goodridge, John   58, 254 n. 131, 266 n. 33
Goodwood   14
Gottolengo, south of Brescia   166
Grafton, Duke of   47
Granger, Rev. James   9, 249 n. 48
Grasmere   135, 225
Gratius   180
Gray, Thomas   108, 130
  'Elegy Written in a Country Churchyard'   130, 186
Great Kneeset   241
Great Revolt of 1381   3, 258 n. 40
Greaves, Ralph   189
Green, John   169
Greene, Robert   266 n. 38, 269 n. 12
greenhouses   58–9, 120
Greer, Germaine   281 n. 78
Greer, Germaine and Susan Hastings, Jeslyn Medoff, and Melinda Sansone   278 n. 19, 279 n. 33
greyhound racing   99
Groom, Nick   267 n. 67
Ground Game Act of 1880   6, 74

gypsies   21–2, 23, 82–3, 110, 139, 161, 267 n. 50

Hall, Catherine   260 n. 61, 263 n. 24
Mr. Hamilton   125
Hammond, Brean   272 n. 5
Hampshire   65
Hampton Court   147
Handforth, Howe, Glanville and Richardson coach   126
Hanley Court, Worcestershire   100
Hanway, Mary Anne   19
Hare, C. E.   29, 178
hares   53, 64, 74, 75, 81, 84, 99–100, 104, 119–22, 135–6, 150, 178, 185, 186, 244, 275 n. 79, 291 n. 57
hare-hunting   xiv, 12, 15, 37, 38–40, 45–6, 95, 99, 103–4, 133, 150–3, 158–60, 168–9, 177–8, 185–7, 193, 234, 259 n. 44, 279 n. 46
Harley, Nathaniel   174
Harpur, Merrily   176
Harrison's Stickle, Great Langdale   125
Harvey, Graham   246 n. 5
Harvey, William   149
Haslewood, Joseph   268 n. 1
Hatfield   161
Hawker, James   80, 82, 84–5, 110
hawking   xiv
Hawley, Brigadier   172
Hay, Douglas   265 n. 1
Hayes, John   78
Hazlitt, William   210–11, 214, 220–1, 226, 286 n. 23
Heaney, Seamus   286 n. 38
Hearne, Vicki   50–1
Helpston, Northamptonshire   11, 206, 209
Helsinger, Elizabeth K.   259 n. 51
Hemans, Felicia   237, 291 n. 43
Henry II of France   165
Henry, J. David   263 n. 37
Herbert, A. P.
  Tantivy Towers   54, 73
Herefordshire   45–7, 157
Hessy, James Augustus   267 n. 67
Hill, Sir Francis   267 n. 60
Hoare, Captain   132
Hobbes, Thomas   149–50, 278 n. 28
Hobsbawm, Eric   14, 179
Holford   223
Holford Combe   223, 224
Holmes, Richard   141, 206, 223, 276 n. 97, 287 n. 14, 288 n. 32, 291 n. 48

Homer 221
Hoole, John 133
Hopkins, A. G. 77
Horne, Kenneth 263 n. 30
horses 52, 190, 250 n. 59 n. 60, 281 n. 78, 283 n. 64, 285 n. 15, 290 n. 21
Hoskins, W. G. 68–9
Houghton, Norfolk 68
hounds 52, 53, 62–4, 96, 108–9, 112, 113, 145, 175–6, 189, 193, 241–2, 282 n. 25
Hucks, Joseph 127, 132, 141, 205, 208, 210, 285 n. 18
Hughes, Ted 113, 115, 234, 255 n. 147, 290 n. 27
hunting 257 n. 26, 273 n. 17
  and knowledge of natural history 33–4, 36–7, 48–50, 53, 54, 150–3, 218, 243
  as democratizing 11, 46–7, 98, 170–1, 177
  versus walking xiii, 120, 122–4, 125–8, 131–2, 133–42, 208, 218, 211, 218–19, 222, 234, 242
Hunting Bill xiii, 244, 245 n. 1
'hunting the borough' 5, 47, 209–10
husbandry xiv, 60, 66, 256 n. 3
  and poetry 55–61
Hutchinson, Sara 221, 287 n. 20, 288 n. 21
Hyde Park xvi
Hyde Park Corner 133
The Institucion of a Gentleman 92, 268 n. 1

intellectuals 114–15, 117–19
Interregnum 76
Ireland 107
Island of Thanet, Kent 45
Isle of Wight 98
Itzkowitz, David 46–7
Ives, Billy 172

Jacobins 6, 275 n. 63
Jacobite Rebellion 6
James VI and I 75, 94, 148
James II 12, 41, 172
Janowitz, Anne F. 1, 267 n. 50
Jardine, Lisa 250 n. 60, 257 n. 10
Jarvis, Robin 126–7, 208, 286 n. 38, n. 41, 287 n. 12
Jefferies, Richard 34, 50
Jesus College, Cambridge 141

Jesus Piece, Cambridge 141
John, King 3
Johnson, Paul J. xviii, 51, 246 n. 19, 252 n. 97
Johnson, Samuel 94, 124–5, 129, 179
Jones, Mary 279 n. 33
Jones, Vivien 253 n. 117
Jonson, Ben 275 n. 67
  'Penshurst' 56–7, 115
  Volpone 91
Jorrocks, John (Surtees' hero) 171, 173, 177–8, 184, 189, 199–201, 211, 241

Kendall, Edward Augustus 9, 109, 124
Kent 66
Kent, William 67–8, 70
Kerridge, Eric 56, 262 n. 8, n. 9
Kestor Rock 233
Kinder Scout 274 n. 39
Kingsley, Charles 80
Kipling, Rudyard 189, 196
Knight, Richard Payne 47, 228–9, 287 n. 19
Knightley, Sir Charles 173
Koehler, William 51

laborers 10, 11, 13, 23, 47–8, 57, 78–91, 111, 126, 207, 229, 261 n. 1, 267 n. 63, 285 n. 20
Labour Party xiii, xvi
Lade, Sir John 161
Lade, Laetitia Darby, Lady 161–3
Lake District 15, 96, 100, 125, 127–8, 220–2, 288 n. 32
Lake of Keswick 221
Lamb, Charles 223, 226, 227, 228, 288 n. 33, 289 n. 39
Lancashire 207
Langan, Celeste 126, 231
Langland, William
  Piers Plowman 79
Lanyer, Aemilia
  'The Description of Cooke-ham' 147–8, 278 n. 18
larches 70
Lasky, J. 233
Launceston 232
Lawrence, John 283 n. 64
Lee, Ronnie 124
Leicestershire 24, 46, 68–9, 170, 172, 191, 198, 199, 218
Levant trade 173–4, 250 n. 59, n. 60, 265 n. 6

Index

Levellers xiv, 8, 104
Lewalski, Barbara 278 n. 18
Lewes, G. H. 213
Lewin, William 147
Lightfoot, Rev. John 147
Lilburne, John 104
Lynton 235
'The Lincolnshire Poacher' 84–5, 267 n. 60
Linnean Society 30
Liu, Alan 20
LLandovery 92
Lloyd, Dick 290 n. 27
Loder, Norman 197
Loewenstein, David 248 n. 30, 270 n. 48
London xvi, 47, 66, 126, 132, 170, 223, 226, 288 n. 33
  City of 77, 198–9
longbow xiv, 145
Lorrain, Claude 67, 209, 212
Lovibond, Edward 118–19
Lowther 213
Luddism 98
lurchers 110–11, 261 n. 83
Lydford 233
Lyrical Ballads 136

Mabey, Richard 34–5, 284 n. 5
Macdonald, David W. xviii, 24, 49–50, 51, 53, 63–4, 246 n. 19, 252 n. 97, 263 n. 37
  Running with the Fox 49–50, 260 n. 71
Macdonald, Jenny 49
MacEwen, Malcolm 232
Machiavelli, Niccolò 8, 91, 103
Mack, Maynard 273 n. 17
Mackie, Erin 279 n. 34
MacLean, Gerald 247 n. 11
Malcolmson, Robert 94, 250 n. 51
Mallalieu, Ann, QC (Baroness Studdridge) xvi-xvii, 88
Manchester 83, 126, 207
Manchester Botanical Society 83
Manning, Roger B. 80, 277 n. 4
manuals of hunting 36
manuals of husbandry 17
Mar, Lady 145, 158
Marcus, Leah S. 95
Markham, Gervase 79–80
  Countrey Contentments 2, 39, 54, 79, 97, 218

Markland, George 97
Markley, Robert 261 n. 5
Marples, Morris 275 n. 67
Marshall, William 209, 232–3
Martin, John 109
Marvell, Andrew 104
Mary, Queen 145
Mary-Tavy 236
Massingham, H. J. 20
McDonagh, Josephine 212–13
McGann, Jerome 236, 290 n. 41, 291 n. 42
McKeon, Michael 16, 281 n. 75
McKusick, James C. 286 n. 42
McRae, Andrew 66, 79
Mellin, Robert 284 n. 6
Mellor, Anne K. 287 n. 19, 288 n. 39
Melton Mowbray 172, 173, 183, 198, 200
Meyer, Arline 277 n. 16, 282 n. 35
Meynell, Hugo 9, 170, 172, 191, 282 n. 37
Michals, Teresa 265 n. 1
Mid Devon foxhounds xviii, xix
middle classes 17–20, 72, 92–4, 115, 126, 127
Midland shires xviii, 12, 14, 15, 24, 46–7, 68–9, 169
Milton, John 94–5, 123, 221
  Paradise Lost 102–3, 115
Milton Hunt 87
Milton, Lord (Charles Fitzwilliam) 87
Mitford, Mary Russell, 101–2
  Our Village 85–6, 101
Modiano, Raimonda 221, 288 n. 21
Montcrieff, Elspeth 100
Montagu, Lady Mary Wortley 145, 158, 160, 161, 166
Moody, Elizabeth
  'To a Gentleman Who Invited Me to Go A-Fishing' 44
Moorman, Mary 225
Morland, George 21
  The Benevolent Sportsman 21–2, 139–40
  Tavern Interior with a Sportsman Refreshing 80–1
The Morning Chronicle 141
The Morning Post 224
Morris, Christopher 101
Morris, William 257 n. 10
Morton, Timothy 250 n. 49, 272 n. 11
Mulberry Garden 125

Munsche, P. B.   6, 75, 100, 269 n. 23, 270 n. 36
Mytton, Jack   1, 13, 252 n. 82

Napoleonic Wars   2, 6, 97, 98, 235–6, 275 n. 63
Nashe, Thomas   269 n. 12
National Farmers' Union   17, 18
National Parks   xvi, 15, 213, 230
National Trust   213
Nattrass, Leonora   45
Nedham, Marchamont   104
Neeson, J. M.   79
Nemesianus   180
Nether Stowey   141, 223, 226, 234
Neuadd   92
Newbury   47
Newby, Howard   11, 20–1, 243, 261 n. 3
Newcastle, Margaret Lucas Cavendish, Duchess of   146, 148–53, 277 n. 15, 278 n. 28
  'The Hunting of the Hare'   39, 148, 150–3
  'The Hunting of the Stag'   148, 151
Newcastle, William Cavendish, Duke of   147, 148–9, 277 n. 15
New College, Oxford   63, 180
New Exchange   125
New Forest   9, 201
New Sporting Magazine   195
Newman, Edward   98
Newman, Gerald   247 n. 2
Newmarket   132
Newton Toney, Wiltshire   65
Niff, a fox   49
Nimrod (*see* Apperley, Charles James)
Norfolk   98–9
Norfolk, Duke of   145
North Devon   235
Northampton Asylum   88, 136
Northamptonshire   46, 68–9

Oadby, Leicestershire   84
O'Brien, Karen   57, 253 n. 108
Okehampton   232, 233
Oldaker, Thomas   165, 251 n. 80
Oppian   180, 182
Oriel College, Oxford   30
Orientalism   9
otter-hunting   41–2, 74, 187, 193–4, 234
Ottery St. Mary   235
Overton, Mark   262 n. 8

Overton, Richard   8
Ovid   20, 180
Oxford   176
Oxford, Covered Market in   39
Oxford, Edward Harley, second Earl of   147, 157, 174, 277 n. 15, 282 n. 35
Oxford, Lady Henrietta Cavendish Holles Harley, Countess of   147, 157–60, 277 n. 15
Oxford, Robert Harley, first Earl of   147, 157
Oxford Street   176

Paignton   234
Paine, Thomas   3–4, 139
  Rights of Man 3–4
palimpsest   212–13, 216, 219, 222, 226
Palmer, Sir John   169, 281 n. 7
Palmer, Philip   45
Palmer, Walter   45
Pantisocracy   8, 134–5, 141, 229, 250 n. 50, 276 n. 93
Parliament   xiii, 3, 65, 68, 79, 103, 105–6, 209, 235, 245 n. 1
partridges   74, 75, 78
Pascoe, Judith   288 n. 27
pastoral poetry   16, 30, 57
pathocentrism   246 n. 19
Patriot Opposition   180–1, 188
Patten, John   69
Patterson, Annabel   8
Payne, C. J. (Snaffles)   169
Peasants' Revolt of 1381   75
Peasants' War of 1524–5   3
pedestrianism (see walking)
Pellier, Jules Charles   165, 251 n. 80
Pellier, Jules-Théodore   165
Pennant, Thomas
  British Zoology   31–2, 70, 228–9, 264 n. 59
Penrith   171
Penshurst House   56–7
perambulation   5, 47, 209–10, 217–18, 222, 227–8, 242–3, 286 n. 41, 291 n. 45
Percy, Thomas
  Reliques of Ancient English Poetry   87, 267 n. 67
periphrasis   58–9
Perkins, David   121, 133, 250 n. 50, 268 n. 74, 276 n. 97
pets   7, 50–1, 112, 115, 121–2

pheasants 53, 69–71, 74, 78, 84, 86, 118, 264 n. 59
Pickard, Richard 256 n. 9, 263 n. 40
picturesque xiii, 16, 19–21, 47, 66–8, 70, 71, 72, 93, 128, 169, 199, 211, 213, 217–18, 223, 227, 229, 240, 253 n. 117, 274 n. 39, 287 n. 19
Pierson, Melissa Holbrook 281 n. 78
pigs 9, 79, 174, 282 n. 30
Planned Countryside 24, 47, 254 n. 139
Plimstock 234
Pliny 180
Plumptre, James 274 n. 58
  The Lakers 113, 127–8, 205, 235
Plymouth 232
Plymouth Chamber of Commerce 237
poaching xv, 34, 71, 72, 73–111, 94, 95, 98–9, 110–11, 146, 208, 265 n. 1, 265 n. 2, 266 n. 41, 267 n. 48
  as naturalistic fieldwork xiv, xv, 23, 84–91, 208, 209
  gangs 6, 73, 249 n. 32
Poole, Thomas 223, 234, 276 n. 97
Pope, Alexander 19, 71, 117–18, 123, 147, 249 n. 48
  Essay on Man 4–5, 60–1, 102, 116–17
  Windsor-Forest 117–18, 139, 175–6, 209, 273 n. 17
Porlock 235
Port Meadow, Oxford 132
Portland, Earl of 147
Portland, Margaret Harley Bentinck, Duchess of 146–7, 157
Portland, William Bentinck, Duke of 147
Powell, Foster 132
Price, Uvedale 47, 287 n. 19
Princetown 236, 237, 238
Proctor, George 101
'progress' 4
'prospect' 4
Protectorate 56
proto-industrialization 10, 251 n. 62
Prouty, Charles and Ruth 258 n. 27, n. 34, 259 n. 48
provincial, as opposed to shire, packs of hounds 46–8, 170–2, 198–201
Pughe, David 94
Pulteney, Richard 147
Pulteney, Sir William, of Misterton, Leicestershire 166, 280 n. 73
Pytchley Hunt 46, 169, 173
Pythagoras 116

Quaintance, Richard E. 264 n. 45
Quantock staghounds 137
Quantock Hills 141, 222, 226, 235, 288 n. 32
The Quarterly Review 101
Quorn Hunt 9, 46, 69, 170, 172

rabbits 74, 79, 81, 84–5, 95, 265 n. 6, 291 n. 57
Rabinowitz, Alan 255 n. 2
Rackham, Oliver 24, 47, 72, 254 n. 139, 254 n. 140
Radstock, Lord 21
Ramblers' Association 205, 231, 274 n. 39
Ranger, Terence 14, 179
Rann, John 161
Raven, James and Helen Small and Naomi Tadmor 272 n. 5
Reform Bill 6
Regan, Tom 272 n. 11, 273 n. 29
Repton, Humphry 10, 98–9
Restoration of 1660 41, 75, 106, 228, 281 n. 11
Reynard 88, 91
Reynolds, George W. M. 163
rhododendrons 70–1
Ribbesdale, Lord (Thomas Lister) 174
Richard II 3
Richmond Park 68, 158
Richmond, Duke of 14, 172
Ridley, Jane 197
Right to Roam xiii, 213–14, 231–2, 245 n. 2
Ritvo, Harriet 7, 118
Rivers, Lord 284 n. 16
Robinson, Jeffrey C. 286 n. 23
Roe, Nicholas 269 n. 14, 276 n. 93
Rogers, John 149–50
Rogers, Jonathan 108, 271 n. 64
Rogers, Pat 273 n. 17
Romanticism xiii, 16
rooks 228–9, 289 n. 39
Rosa, Salvator 67, 85
Rosenthal, Michael 77
Rothschild family 251 n. 76
Rothstein, Eric 35–6, 279 n. 38
Rowe, Samuel 237, 291 n. 45
Rowlandson, Thomas 170–1
Royal Society of Literature 237
Royalists xiv, 76, 97, 104–6, 148, 235
Runnymeade 3
Rutland 46

St. Albyn family   137
St. George's Hill   76
St. James's Park   125
St. John, Charles   13
Salisbury, Lady Mary Emily Hill,
    Marchioness of   161, 279 n. 50
Salisbury Plain   65, 69
Sambrook, James   279 n. 38
Sassoon, Siegfried   176, 189, 196–8
Sca Fell   206, 221–2
scent   62–4, 180
Schama, Simon   290 n. 37
Schiebinger, Londa   277 n. 7, n. 9
Scotland   243, 289 n. 10
Scott, John   60–1
Scott, Walter
    Rob Roy   80
Scottish borders   62
Scottish Highlands   19, 275 n. 58
Scruton, Roger   167, 173, 246 n. 10
Seaton, Lord   169
Secord, Anne   207
Selborne   30, 211–12
Selborne, Lord   255 n. 147
Severn valley   9
Seymour, James
    A Coursing Party   99–100
Shakespeare
    As You Like It   7
    A Midsummer-Night's Dream   113
    Venus and Adonis   39
    The Winter's Tale   80
Sharpe, Kevin   272 n. 11
Sherringham, Norfolk   98–9
Sherwood Forest   106
Shoard, Marion   231–2
    The Theft of the Countryside   245 n. 5
shooting   xiv, xv, 33–4, 53, 69–71, 97–8,
    100, 102–3, 118, 218, 227–9, 275 n.
    79
Shteir, Ann   147
side-saddle   12, 163–6, 251 n. 80
Sidney familly   57
Siege of Gibraltar   125
Simpson, David   137, 139, 276 n. 80,
    n. 84
Sinclair, Iain   283 n. 39
Singer, Peter   255 n. 1, 272 n. 11, 273 n.
    29
Siskin, Clifford   272 n. 5
Slut, a pig   9
Smith, Adam   55
Smith, Charles   82, 267 n. 48

Smith, Charlotte   225
Smith, Nigel   106, 152, 270 n. 51, 272 n.
    11
Smith, Thomas   29, 48, 53
    The Life of a Fox   48
Smithfield   3
Snelgrove, Barbara   145, 146
Snowdon   140, 214
Society of Apothecaries   207
Soke of Peterborough   209
Solander, Daniel   147
Somerset   xviii, 235
Somerset family   146
Somervile, William   125, 193–4, 259 n.
    41
    The Chace   14, 62–4, 91, 99, 113–14,
        125, 175–6, 176–7, 179–94, 201,
        259 n. 41, 273 n. 17, 283 n. 51
    Field-Sports   95, 107–8, 181, 283 n.
        55
    Hobbinol   95
Sotheby, William   288 n. 22
South Devon   232, 235
South Hams   232
Southern hound   194
Southey, Robert   141, 220, 222, 226,
    234–5
spaniels   107–8
Spectator   2, 15–16, 40, 72, 153, 223,
    279 n. 34, n. 35
The Sporting Magazine   xiii, 132–3, 145,
    161, 245 n. 3
sports   94–6
Mr. Spring   49
stag-hunting   xiv, 12, 31, 55, 104–6,
    118, 133–4, 151, 158, 160–1, 169–70,
    177–8, 181, 187, 192–3, 233–4, 251
    n. 76, 259 n. 44
Steele, Richard
    Tatler   19, 279 n. 34
Sterne, Laurence   8
Stevenson, John Allen   265 n. 12, 266 n.
    41
stewardship   11, 30, 61, 243
stoats   84
Stone, Lawrence and Jeanne C. Fawtier
    248 n. 30
Strafford, Earl of   104
Stubbs, George   14, 161, 176
    Laetitia, Lady Lade   161
Suffolk   47
Surrey foxhounds   178, 198–9
Surrey staghounds   198

Surtees, Robert Smith   13, 48, 163, 171, 173, 176–8, 184, 195–201, 217
   Handley Cross   177–8, 184, 189, 195, 199–201
Susquehanna river, Pennsylvania   141
Sussman, Charlotte   273 n. 20
Sweet, Nanora   291 n. 43
Swindells, Julia   257 n. 10
Swordy Well, Helpston   11

Tamar river   236
Taplin, William   53
   The Sportsman's Cabinet   101, 107, 110, 168, 169–70, 261 n. 83, 270 n. 41, 279 n. 46
Tarporley Hunt Club   14
Tattersall, Fran H.   51, 252 n. 97
Tavistock   234, 236
Taw river   236
Taylor, John   88
Teign river   236
Tennyson, Alfred, Lord   38
terriers   110, 111–12, 260 n. 71
Tester, Keith   124, 255 n. 1, 272 n. 9
Thames valley   148
Thirsk, Joan   6, 173
Thomas, Keith   7, 115, 272 n. 9
Thompson, E. P.   5, 95, 217
Thomson, James   116, 123, 157, 180, 216, 279 n. 38, 287 n. 18
   The Seasons 11, 35, 116, 153–7, 179, 188–9, 191, 192, 193, 209, 212, 216
Thoresby, Nottinghamshire   158
thoroughbred horses   9, 161, 167, 172–4, 250 n. 60
Thornton, Kelsey   254 n. 131
Thrale, Henry   124
Thrale, Hester Lynch   124–5
Throckmorton, John Courtney   55
tin miners   235, 290 n. 34
Tiney, a hare   121–2, 135
Tinners' Parliament   235
Tiverton School   109
Tobin, Beth Fowkes   271 n. 63
Tory party   xvi, 2, 14, 213, 248 n. 30
Totnes   234, 235
tourism and tourists   19–20, 92–4, 125–8, 205–6, 211, 221–2, 233–6, 237
Trickey, Christopher   137
Trinity College, Cambridge   183
Trollope, Anthony   13
Trouncer, a foxhound   47
Turberville, George   258 n. 27

Turner, James Grantham   56, 79, 267 n. 33, 270 n. 51
Turner, James   82, 266 n. 48
turnips   58
Tusser, Thomas   61, 64, 65–6, 79–80, 93
   Five hundreth points of good husbandry   17, 60
Twici, William
   The Art of Venery   36
Twickenham   19, 71, 133
Two Bridges   233
Tyler, Wat   3, 258 n. 40

Ullswater   100
Underdown, David   5, 81
Upcher, Abbot   99
Up-street near Canterbury   82
urbanization   xv, xvii, 7, 10, 55, 113–15, 117–19, 157, 183, 243, 253 n. 109
Utrecht, peace treaty of   139

vagrancy   125
Vanbrugh, Sir John   191
Vandervell, Anthony   78
vegetarianism   29, 34–5, 113, 115–18, 123, 127, 179, 272 n. 11, 273 n. 29
Vesey-FitzGerald, Brian   110
Virgil   17, 57–8, 180
   Aeneid   180
   Georgics   16, 57, 180
Vyner, Robert R.   282 n. 37

Waddesdon, Buckinghamshire   251 n. 76
Wales   14–15, 19, 40, 48, 92–4, 127, 141, 205, 206, 209, 210, 243, 285 n. 15, 289 n. 10
Walker, Stella   161
walking   275 n. 67
   and botany   206–9, 217–19
   as recreation   113, 125–42, 182, 205–19, 233–43, 286 n. 23
   as sporting wager   132–3, 275 n. 67
   gear needed for   205–6
   political connotations of   126–7, 213–14, 216–18, 229
Wallace, Anne D.   211, 214, 274 n. 48, n. 58
Wallace, John M.   105–6
Wallis, Alfred   291 n. 51
Walpole, Horace   66–8, 70–2, 108
Walpole, Sir Robert   6, 67–8, 180–1, 249 n. 31

Waltham Forest 3, 106
Walton, Isaak
  The Compleat Angler 23, 44, 97
Ward, G. H. B. 126
Ward, Joseph P. 247 n. 11
Warner, Rev. Richard 94, 127, 205–6, 233–4, 290 n. 21
Warwickshire Hunt 46
Wasserman, Earl R. 270 n. 51, 273 n. 17
Wat 39, 150–2, 258 n. 40
Watendlath 220–1, 226
Waterloo 169
Watkins, Charles 47
Watson, Frederick 196
Wedgwood, Tom 206
Welbeck Abbey 158
Wells, John xix
West, Thomas
  A Guide to the Lakes 19
West Country 42–3, 137, 169
West Okement valley 241
West Sussex 172
Weston Underwood 55
Whig party 2, 6, 14, 70, 91, 180–1, 188, 213, 248 n. 30
White, Gilbert xiv, 30–4, 49, 53, 124, 151, 206, 211–12, 257 n. 10
  The Natural History and Antiquities of Selborne 30–4
wilderness 231, 243
Wilkes, Rev. Wetenhall 114, 191
Wilkinson, Rev. Joseph 252 n. 99
William the Conqueror 182
William III 147
Williams, Kenneth 263 n. 30
Williams, Raymond 56, 247 n. 3, 257 n. 10, 262 n. 13, 266 n. 33, 281 n. 4
Williamson, Tom 69, 70–1, 264 n. 60
Willoughby de Broke, Lord (Richard Greville Verney) 189
Wiltshire 5, 169, 235
Wily, a New Forest fox 48
Wincanton, Somerset xviii, xix
Winchester 82
Winchester College 180
Windsor Castle 129
Windsor Forest 104, 106, 117–18, 158, 160–1, 175–6, 188, 192, 251 n. 76, 259 n. 44
Winstanley, Gerrard 3, 248 n. 30
wire fencing 13
Withering, William 224, 225, 287 n. 20
Wolmer-Forest 33

women
  and hunting and field sports 12, 60, 100–2, 145–67, 170, 188, 192, 251 n. 80, 279 n. 33, 282 n. 13
  and science 60, 146–7, 207, 277 n. 9, 288 n. 27
  and walking 125–6
  as poachers 145, 146, 277 n. 4
Woodchester, Gloucestershire 264 n. 59
woodland, woodcraft, and forestry 55–6, 70–2, 76, 89–91, 209, 262 n. 5, 262 n. 7, 289 n. 10
Woodland Trust 243, 289 n. 10
Woods, Susanne 278 n. 17, n. 18
Woolf, Virginia 1, 13, 240–1, 252 n. 82
Wootton, John 65, 147, 158, 159
  The Bloody Shoulder'd Arabian 174
  Lady Henrietta Cavendish Holles, Countess of Oxford hunting at Wimpole Park 147, 158, 159, 277 n. 16
  Lady Henrietta Harley out Hunting with Harriers 277 n. 16
Wordie, J. R. 10
Wordsworth, Dorothy 127, 133–4, 137–8, 172, 223, 225, 228
Wordsworth, John 214–16, 228
Wordsworth, William 15, 85, 111, 113, 127, 133–40, 213–16, 217, 220–2, 223, 228, 238, 257 n. 10, 275 n. 63, n. 79, 276 n. 84, 286 n. 38, 287 n. 20, 288 n. 21, 288 n. 32, n. 33
  Guide to the Lakes 252 n. 99
  'Hart-Leap well' 133, 228
  'Home at Grasmere' 96, 133–5
  'Lines written a few miles above Tintern Abbey' 16, 47, 83, 96, 269 n. 14
  The Prelude 96, 111, 133, 140, 214, 221
  'Resolution and Independence' 135–6, 214
  'Simon Lee, the Old Huntsman' 136–40, 276 n. 80
  'When first I journeyed hither' 214–16, 228
Wright, Patrick xvi, 254 n. 122, 285 n. 16
Wyatt, John 31
Wycherley, William
  The Country Wife 125
Wye valley 16, 19, 96, 252 n. 102

Xenophon 180

Yearsley, Ann 57
Yeats, Thomas 147
Yes Tor 241
York 132, 267 n. 60

York, Edward Plantagenet, Duke of *The Master of Game* 36
Yorkshire 66, 171, 178, 279 n. 35
Young, Arthur 60, 209, 263 n. 26

Zwicker, Steven 23